ENVIRONMENTAL RESEARCH ADVANCES

REGIONAL EIA AND RISK ASSESSMENT IN A FAST DEVELOPING COUNTRY

ENVIRONMENTAL RESEARCH ADVANCES

Additional books in this series can be found on Nova's website
under the Series tab.

Additional E-books in this series can be found on Nova's website
under the E-book tab.

ENVIRONMENTAL REMEDIATION TECHNOLOGIES, REGULATIONS AND SAFETY

Additional books in this series can be found on Nova's website
under the Series tab.

Additional E-books in this series can be found on Nova's website
under the E-book tab.

ENVIRONMENTAL RESEARCH ADVANCES

REGIONAL EIA AND RISK ASSESSMENT IN A FAST DEVELOPING COUNTRY

TASNEEM ABBASI
E. V. RAMASAMY
F. I. KHAN
AND
S. A. ABBASI

nova publishers
New York

For permission to use material from this book please contact us:
Telephone 631-231-7269; Fax 631-231-8175
Web Site: http://www.novapublishers.com

Library of Congress Cataloging-in-Publication Data

Abbasi, S. A. (Shahid A.)
Regional EIA and risk assessment in a fast developing country / S.A. Abbasi, E.V. Ramasamy, F.I. Khan.
p. cm.
Includes index.
ISBN 978-1-61942-234-6 (hardcover)
1. Environmental impact analysis--India. 2. Industrial accidents--Risk assessment--India. 3. Chemical industry--Environmental aspects--India. I. Ramasami, E. II. Khan, Faisal I. III. Title.
TD194.68.I4A33 2011
333.71'40954--dc23

2011045791

Published by Nova Science Publishers, Inc. † New York

The memory of father M.H. Ibrahim and mother Asma
-SAA

Neelu Auntie, Sophia Auntie, Rubi Auntie
-TA

Appa Thiru E. Venkatesaperumal; Amma Thirumathi V. Jayalakshmi
-EVR

My late father Irshad Ali Khan and mother Mahmouda Begum
-FIK

CONTENTS

PREFACE

India, arguably, is among the most technologically advanced nations. According to some forecasts (e.g Danial Lak, *India Express*, Penguin/Viking, 2008; pp 499) India is on course to become *numero uno*!

India is also among the first set of nations which gave primacy to environmental concerns in their State policy. Indeed protection of environment is one of the Indian Constitution's 'article of faith'!

But how things are on the ground? Is industrial growth in India able to reconcile with the preservation of the country's environmental health? Are the industries now any less likely to meet with major accidents than they were in 1984 when the Bhopal gas tragedy — World's worst-ever process industry accident — had occurred? In order to seek answers to these questions the authors have carried out a very extensive study of the environment in and around a cluster of industries situated in the suburbs of Chennai. The industries include a large refinery, several downstream petrochemical industries, and a few other industries. A large team of scientists and engineers, led by the authors, conducted field work, analytical work, and computations – including mathematical modeling and digital simulation – for several years to assess the situation as a function of time and space. Besides environmental impact assessment, a very elaborate risk assessment was also carried out for that region. The gist of these studies which typifies the Indian situation *vis-a-vis* the continued impact of industrial growth on the environment is presented in the book.

The book not only provides an authentic and elaborate picture of the environmental impact and risk of accidents of a typical industrial complex in the developing world, it also reveals the wide gap that still exists between the State policy and its implementation in even the technologically very advanced countries like India. For the students of environmental science, environmental engineering, and chemical engineering the book would serve as a mine of information as well as a do-how manual pertaining to EIA and risk assessment (RA). To the faculty the book would be a valuable teaching aid. To the professionals working in the petroleum, petrochemical, and other process industries, and to the consultancy firms, this volume would serve as a 'do-how manual' pinpointing the challenges and opportunities associated with EIA, RA, and environmental management.

Tasneem Abbasi
E.V. Ramasamy
F.I. Khan
S.A. Abbasi

FOREWORD

India is one among very few countries in the world whose Constitution carries environmental protection as one of the articles of faith. India was also among the firsts to institute a full-fledged Ministry of Environment. Even otherwise, concern for the environment and emphasis on sustainable use of resources are embedded in the teachings of the two most widely followed religions of India: Hinduism and Islam. Yet, due to several factors, primarily driven by a runaway growth in population, India is not able to achieve economic progress without straining its environment. This has necessitated a great deal of introspection on what has gone wrong and how, and to learn lessons for a better future.

In this context the present book gains tremendous contemporary relevance. It presents in graphic detail what is arguably the first-ever regional environmental impact assessment and risk assessment done in India, perhaps the first-ever of its kind in the third world. More importantly the book is archetypal; its utility value and message goes well beyond its immediate context because the story of the Manali Industrial Complex that it narrates can very easily be a story of any other industrial cluster in the third World.

I congratulate Nova Science Publishers for bringing this book out and wish it great success.

Prof.
J.A.K. TAREEN
Vice-Chancellor

Chapter 1

INTRODUCTION

India's is one of the most rapidly advancing economies in the world. India is also consolidating and enhancing its thrust in science and technology, which, even otherwise, has been quite strong.

The upward zoom of Indian economy began during the 1990s and by mid-1990 it was going full throttle. It was also the period during which environmental consciousness grew very rapidly. But did the increasing concern for protecting the environment translate into better environmental management of the rapidly developing India?

To seek an answer to this question the authors carried out during the late 1990s what has become one of the first 'regional' environmental impact assessment studies conducted in India. The studies covered the area of Manali, a Northern suburb of the Chennai metropolis (Figure 1.1). The area houses Manali Industrial Complex which incorporates a large petroleum refinery – the Madras Refineries Limited (MRL), now known as Chennai Petroleum Corporation Limited (CPCL) – and several major downstream petrochemical industries (Table 1.1). The industries exist cheek by jowl with residential colonies, shops, hospitals, schools and various public amenities and have several surface water-bodies all around which are freely used by people for cattle, fishing, bathing, and laundry (Plates 1.1-1.4). Solid waste can be seen anywhere and everywhere — on open land, near water bodies, by the roadside (Plates 1.5-1.8) just as one can see plumes of gaseous pollutants rising on the horizon (Plates 1.9 and 1.10). Besides suburban sprawl, villages and slums exist right next to the industrial complete (Plates 1.11 and 1.12). In many ways Manali Industrial Complex typifies the way Industrial growth has occurred in India – industries have either come up in very close proximity with densely populated neighborhoods or townships have grown around industrial establishments to take advantage of the opportunities of employment and commerce that arise due to the industrial activity. By either mechanism, distinction between designated industrial areas and residential–commercial localities often gets quite blurred.

In the Manali Industrial Complex, each of the industries has been established on the basis of individual environmental impact assessment (EIA), including risk assessment (RA), and each is supposedly meeting with the stipulated standards of air, water, and solid waste discharges arising from its operations. But the collective impact of the industries is far from benign. Indeed Manali has been regarded as one of India's most polluted areas since several years and is currently among the top 20 'critically polluted areas' in India as per a list released by India's Central Pollution Control Board (www.samachaar.in; last accessed

September 01, 2011). These authors took up a study to quantify how good or bad things are. The result is a body of information which provides a valuable historic record of the status of the Manali environment just before the beginning of the present millennium. It sets a frame of reference with which the impact of changing times and environmental management strategies on the area would be gauged. After the very detailed frame-of-reference EIA and RA presented in this book, we have been periodically reassessing the status of the environment in the Manali region by indicative sampling and analysis at randomly picked locations. These follow-up studies reveal that the situation hasn't changed for the better; indeed the 20th position of the region in the latest grading of 'critically polluted areas' indicates that it has changed for worse.

RISK OF ACCIDENTS

India has had the unfortunate experience of facing the worst-ever process industry disaster — the Bhopal gas tragedy of 1984. Even 27 years after it happened, the ill-effects of the toxic leak are continuing to jeopardize the life of humans and other life-forms. The vast areas that were toxified by the accident will take long to be restored, if at all. The HPCL refinery disaster (Serial explosions and fire) of 1997 at Vishakhapatnam, India, was the world's worst process industry disaster of the 1990s. Several process industry accidents of serious nature keep occurring in one or other part of India, often taking a heavy toll of life and property. In many ways Indian situation is illustrative of the entire third world[1,2] but the current pattern of runaway economic and population growth in India makes the situation particularly volatile. Hence we have done an elaborate risk assessment along with the EIA; perhaps the first-ever regional risk assessment conducted in India. It addresses individual industries as well as explores the collective risk, with special attention to 'domino effect' viz one accident triggering another, and yet another to lead to a 'cascading' effect.

Part 1 of the book covers the EIA and Part 2 the risk assessment.

Table 1.1. Madras Refineries Limited (now CPCL) and major downstream industries in Manali Industrial Complex, Chennai, India

Sl. No	Name of the Industry	Products manufactured tonnes/month	Manufacturing process in brief
1.	Madras refineries Limited (now Chennai Petroleum-Corporation)	Products 1. LPG 2. Petrol 3. Diesel 4. Kerosene 5. Asphalt 6. Match wax 7. Par. wax 8. HSD	Manufacturing process involves following stages : Crude tank – atmospheric Distillation unit – amine Treating unit – merox Treating unit – vacuum Distillation unit – fluid catalytic Cracking unit – final products.

[1] Abbasi, T and Abbasi S.A. (2005). The expertise and the practice of loss prevention in the Indian process industry some pointers for the third world, Process Safety and Environmental Protection. 83, (5B) 413-420.

[2] Abdulhamidzadeh, B., Rashtchian, Abbasi, T. and Abbasi, S.A. (2011). Domino effect in process-industry accidents – An inventory of past events and identification of some patterns. Journal of Loss Prevention in Process Industries (Elsevier). 24 (5). 575-593.

Sl. No	Name of the Industry	Products manufactured tonnes/month	Manufacturing process in brief
		9. LDO etc. 10. Naphtha 11. Industrial fuel 12. Lube base stocks1	
2.	Madras Fertilizers Limited	Products Urea 38,500 NPK 7,000 Intermediate products Ammonia 28,875 By products CO_2 24,100 Nm^3/hr	Ammonia is produced through naphtha steam reformation process, urea through total recycle process, NPK through DORR-oliver NPK granulation process
3.	Indian Additives Limited	Products Finished lubricating oil additives, 15,000 Intermediate products Succinimides: 6,300 Phenates: 3,000 Pibsa: 4,200	The process units planned for additives manufacture are 1. Succinimide plant 2. Phenate plant 3. Pibsa plant 4. Blending plant
4.	Manali Petrochemicals Limited	Products Propylene oxide: 1,000 Propylene glycol: 520 Polyols: 500 By products di-chloropropane: 120 di-propylene glycol: 40	Propylene oxide Propylene and chloride are reacted in water to form propylene chlorohydrin. The propylene chlorohydrin is saponified with lime to form propylene oxide. Propylene glycol Propylene oxide is reacted with water to form mixture of mono, di and tri-propylene glycols. The mixture is concentrated by triple effect evaporators and vacuum distilled to produce pure pharmaceuticals grade propylene glycol and di-propylene glycol
5.	Madras Petro Chemicals Limited	Products 1. Transformer oil: 900 2. Petroleum specialities: 200 3. Oleum-sulphuric acid: 1200 4. Petroleum jellies: 40	Process involves following stages Raw oil storage - Retreat settler - Neutraliser – Oil still – Alcohol recovery – Percolation – Spent brine – Batch still – Drain – Desalter – Sulfonate still – Sulfarate – Miley retort – So_2 purification – oleum Plant – oleum Sulphuric acid
6.	Celex Petro Chemicals Limited	Products Methyl ethyl ketone 4000 T/Annum By Products Secondary butyl ether: 49 Heavy fraction: 66 Sulphuric acid: 1097	Feed preparation, secondary butyl alcohol synthesis and secondary butyl ether recovery and methyl ethyl ketone synthesis.

Table 1.1. (Continued)

Sl. No	Name of the Industry	Products manufactured tonnes/month	Manufacturing process in brief
7.	Sri Ram Fibers Ltd.	Products Nylone moulding powder: 2000 T/A Nylone – 6/Nylone 6, 6 Polymer compounded and natural: 2000 T/A	The manufacturing process consists of mainly melting, prepolymerisation granulation, extractions, drying and weaving
8.	Tamil Nadu Petro Products Limited	Products Linear alkyl benzene: 6667 Heavy normal paraffin: 1667 Heavy alkylate: 600 Intermediate products Normal paraffin: 5417	The manufacturing unit consists of prefractionation, hydrotreator, molex, pacol and detergent alkylation.
9.	Indian Organic Chemicals Ltd.	Products Polyester staple fibre: 2500 T/M. Ployester filament yarn: 1250 T/M. By products Methanol: 80 T/M.	The manufacturing process involves melting, reaction, fractionation, polycondensation, extrusion, & spinning.
10.	Spic (heavy Chemicals Division) Limited	Products Caustic soda - 5500 T/m Liquid ammonia -3333 T/M Hydrochloric acid – 412 T/M Ammonium chloride – 1788 T/M By Products Hydrogen – 137.5 T/M	Brine preparation, cell operation, chlorine preparation, caustic flakes preparation, hydrochloric acid preparation, ammonium chloride preparation.
11.	Kothari Sugar and Chemicals Ltd.	Products Poly butene & light: 425 T/M Polmers: 120 T/M	Feed preparation, polymerisation of isobutylene, fractionation and storage.
12.	U.B. Petro products Ltd.	Products Polyols: 1000 T/M Propylene glycol: 659 T/M Tripropylene glycol: 10 T/M Propylene oxide: 1000 T/M By Products Propylene dichloride: 175 T/M Dipropylene glycol: 80 T/M	Hypochlorination, neutralisation, dehy- drochlorination, purification, reaction concentration and distillation, reaction neutralisation and filteration.
13.	South India Carbonic Gas Industries Ltd.	Products Carbon-di-oxide liquid: 600 T/M Solid: 300 T/M	The manufacturing process consists of absorption, compression, cooling, separation drying, liquefaction and storing.
14.	Balmer Lawrie & Co. Ltd.	Drum/Barrel – 700 T/M Lubricating grease specialities – 417 T/M Bituminous compounds – 83 T/M Di Tertiary butyl para cresol – 12.5 T/M Para tertiary butyl phenol – 25 T/M Leather auxidemies 180 T/M HCL – 120 T/M	Deceiting, shearing, levelling cutting grinding, welding, heating, agitation saponification cooling direction heating distillation, filtration, washing, drying & packing primary reaction, secondary reaction tertiary reaction, & blending operation.

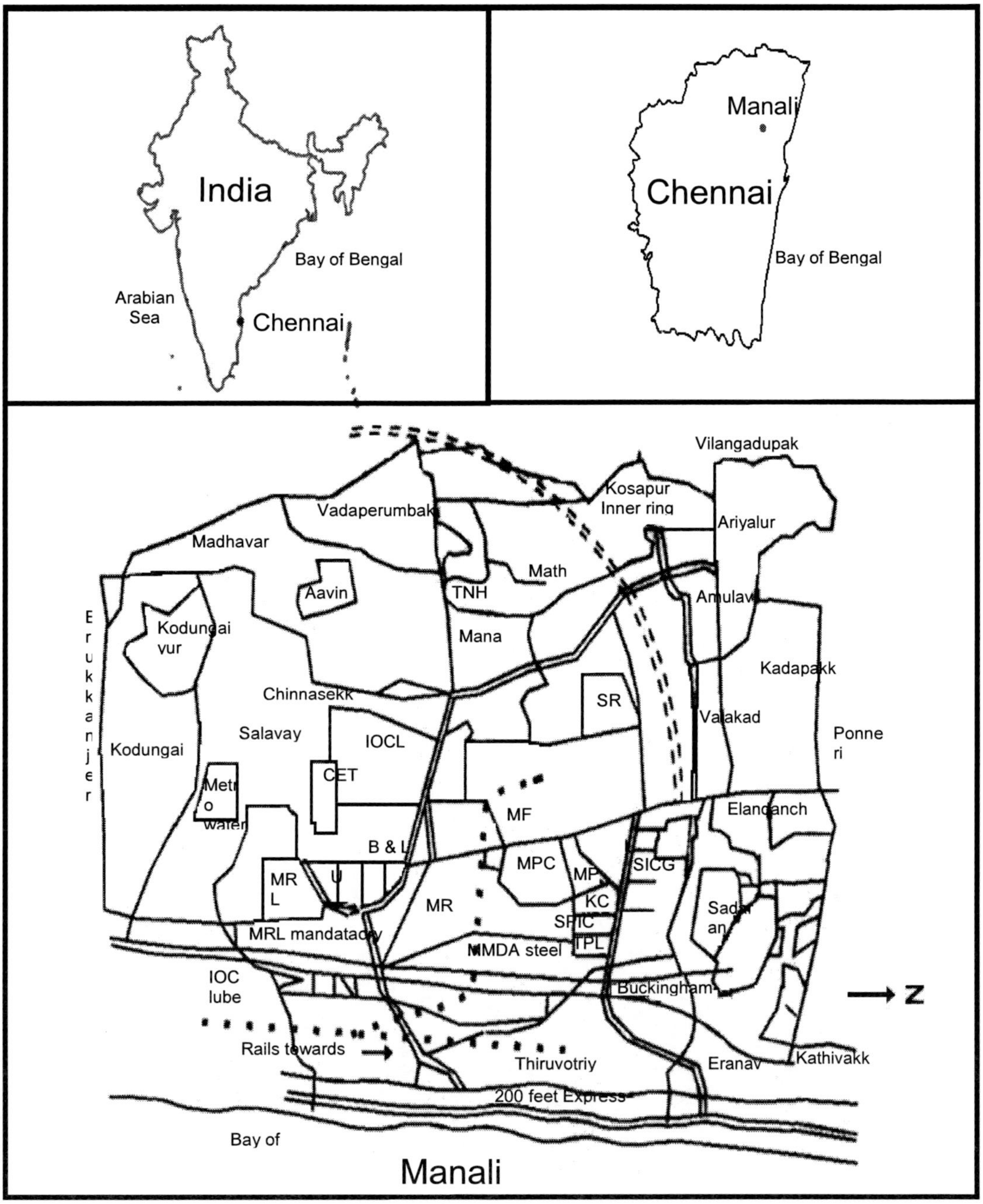

Figure 1.1. The study area.

Plates 1.1 and 1.2. A residential neighbourhood co-existing with dumps of industrial solid waste in the study area (above), and water-bodies adjacent to the industries (below).

Plates 1.3 and 1.4. A drying pond in the study area being used for bathing and laundry (above). Another pond used for cattle (below).

Plates 1.5 and 1.6. Industrial solid waste dumped indiscriminately on land in the study area (above and below).

Plates 1.7 and 1.8. A pond being overrun by industrial solid waste (above). Solid waste lying by the roadside (below).

Plates 1.9 and 1.10. Industrial chimneys can be seen belching gaseous pollutants everywhere (above and below).

Plates 1.11 and 1.12. A village (above) and a slum (below) right next to the industries.

PART 1

Chapter 2

WATER BODIES IN THE STUDY AREA AND THEIR IMPORTANCE

2.1. WATER–BODIES IN THE STUDY AREA

This chapter presents an overview of the water resources in the study area in terms of their availability and usage.

2.1.1. Ponds

Considering the high density of industries and of people living in and around Manali Industrial Complex, the study area still has a surprisingly large number of ponds. These ponds fill up during the monsoons and the water-spreads gradually shrink in size during the post-monsoon months. Due to this, the water-spread areas of the ponds can not be precisely defined, but a fair indication is presented in Figure 2.1.

The largest of these ponds (Madhavaram Tank) has an approximate water-spread of 10000 m^2 and a maximum depth of 2 m when full. Most other ponds have water-spreads of 2000-3000 m^2, and maximum depths close to 1m.

Usage

The ponds are used extensively for bathing, washing clothes, and by cattle. Eventhough none of the ponds is used to draw drinking water for humans there is a clear link with humans of the water quality of the ponds via the cattle, fishing, and the other forms of contact usage -- please see plates 2.1 and 2.2.

2.1.2. Dug-Wells and Borewells

The dug-wells and borewells form the principal source of water for most villages around the Manali Industrial Complex.

Indeed, for most people living in the study area these wells provide the *only* source of water – either directly or through municipal water taps.

2.1.3. Rivers and Canals

A creek passing through Amulavoil near the Manali Industrial Complex (MIC), and a stretch of the historical Buckingham canal are the two major lotic water bodies in the study area (Figure 2.2). Of these, the latter one is by far the most prominent and important.

The Buckingham Canal is a coastal channel running through the backwaters, depressions, and low-lying areas for 419 km between the towns of Peddagangam and Marakkanam situated at the east coast of the Indian peninsula. The canal runs close to the coast and is not more than one to two km away from the sea at any point along its length.

The canal was completed in 1897 and was the main means of communication between the rich deltaic areas of Coramandal coast and erstwhile Madras all through the later part of the 19th century and the early 20th century, when rail and road transport facilities were not well developed. Subsequently as the other modes of transport came up, the traffic in the Buckingham canal gradually declined.

Now, due to improper maintenance, effect of cyclonic storms, and high level of siltation hastened by pollution, the canal is in disuse at several places. It now exists in bits and pieces serving small stretches, notably in the city of Chennai (Madras).

One of the still useful portions of the Buckingham canal is running through the Manali industrial area, at Chennai (Madras), for a length of ~10 km (latitude 80.33°, longitude 13.13°). In this region the canal carries run-off water which is occasionally used for irrigation, laundry, and other domestic purposes by economically weak population living near its banks. They also occasionally fish in the canal waters.

2.2. The Importance of the Water Resources in the Study Area

From the foregoing it is evident that in spite of being located in a semi-arid region, Manali Industrial Complex and its surroundings has a large number of important water bodies. These water bodies continue to be used by people for both contact and non-contact purposes. It was thus considered very important to assess the status of the water quality of these resources.

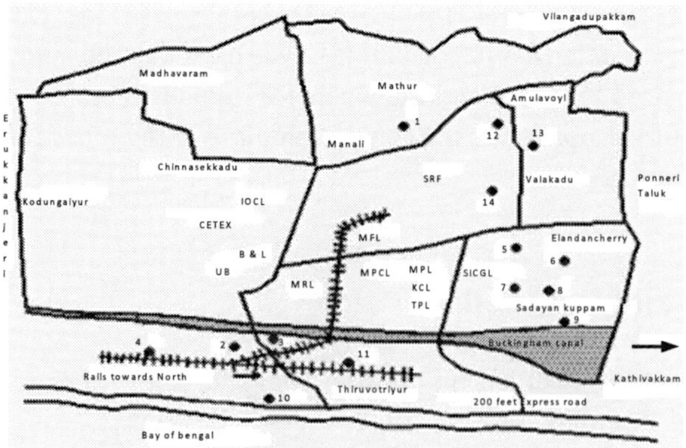

Figure 2.1. Location of pond-sampling sites [•].

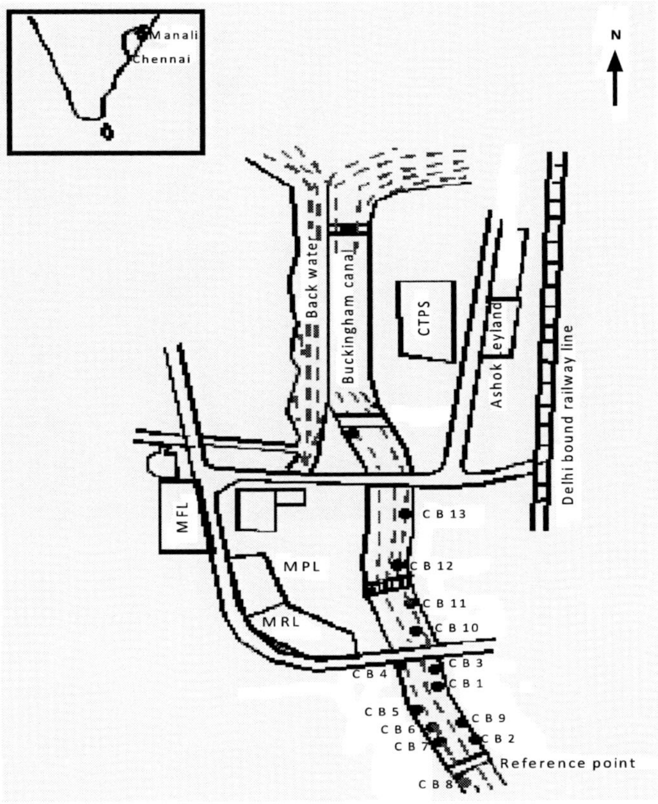

Figure 2.2. Sampling points in the Buckingham canal. All samples were collected from the middle of the canal, the sampling process strictly confirmed with the Standard methods.

Plate 2.1. Cattle wallowing in and drinking water from one of the ponds in the study area.

Plate 2.2. Fishing in a pond close to several industries.

Chapter 3

SAMPLING SITES AND THE METHODS OF SAMPLING AND ANALYSIS

3.1. GENERAL

Water samples were collected from surface water bodies (Figures 3.1 and 3.2), water taps and groundwaters (Figure 3.3) during three different seasons: pre-monsoon (April – July), monsoon (August – November) and post-monsoon (December – March). The surface water samples were taken from ponds and canals and groundwater samples were obtained from dug wells and bore wells.

The collection and analysis of samples was done in strict confirmity with the standard methods[*#]. Such of the parameters, which must be done *in situ,* were analyzed there and then with field equipment set up at our field station. Other parameters were analyzed with arrangements at Tamil Nadu Water and Drainage (TWAD) lab, and at the laboratories of Centre for Pollution Control & Energy Technology at Pondicherry. Certain other analyses were done in collaboration with IIT Chennai.

The physical parameters for each sample that were analyzed include colour, odour, temperature and turbidity. The physico-chemical and chemical parameters include pH, electrical conductivity (EC), total dissolved solids (TDS), alkalinity, acidity, total hardness, calcium hardness, total iron, ammonia (NH_3), nitrite (NO_2^-), nitrate (NO_3^-), chloride (Cl^-), fluoride (F^-), sulphate (SO_4^{2-}), phosphate (PO_4^{3-}), total kjeldahl nitrogen (TKN), chemical oxygen demand (COD), and biochemical oxygen demand (BOD). In addition some of the samples were analysed for trace metals and organics. The results have been discussed in the context of the following water quality standards.

1) Domestic water (For use as raw water for public water supply and bathing ghats) specification: IS 2296 (1974)
2) Drinking water specification: IS 10500 (1991)
3) Fresh water for fish culture: IS 13891 (1994)

---

[*] APHA(1995). Standard Methods for the Examination of Water and Wastewater

[#] Abbasi,S.A. (1998). Water Quality – Sampling and Analysis. Discovery Publishing House, NewDelhi.

The focus of the study was to ascertain:

1) the status of the quality of water resources in the context of the typical features of the study area.
2) the extent of resource degradation vis-à-vis different types of water use.

3.2. SURFACE WATER QUALITY SAMPLING STATIONS

3.2.1. Ponds

The location of ponds sampled in and around Manali Industrial Complex is presented in Figure 3.1. These ponds are used by humans for washing and bathing, and by cattle.

The sampling sites were chosen on the basis of their proximity to solid waste disposal sites and industrial or municipal wastewater streams (Plates 3.1 -3.4). Representative sites were chosen within and away from residential houses. Brief description of sampling locations is presented in Table 3.1.

The sampling sites P1-P4 are in the vicinity of solid waste dumpings. Effluents of some of the industries are received by P6 and P7 sampling sites. The site P11 belongs to a pond near a graveyard in Thiruvottriyur. It receives municipal and slaughterhouse wastes.

3.2.2. Canals

One of the stretches of Buckingham canal which has appreciable flow atleast during some months every year passes through the Manali Industrial Complex (MIC), running a length of 10 Km. This canal was monitored extensively and a model was developed to assess and forecast the self-purifying capacity of the flow and the impact of increase or decrease of pollutant inputs on this capacity. A creek passing near MIC through Amulavoil was also monitored. Eighteen sampling stations were set up (Figure 3.2, Table 3.2). The sites were chosen on the basis of use of the canal water and different wastewater inflows coming into the canal. The samples at sites CA1, CA2 and CA3 were collected from the Amulavoil canal. The site CA3 is located below Sadayankuppam bridge. Solid waste dumps are located near this site, and the canal water is used for drinking and bathing by the cattle. The site CA4 represents the location where an effluent stream enters the Buckingham canal. The samples CB13 and CB14 were collected from upstream and downstream of the effluent stream mixing with the canal. Other sites in the canal represented the gradual change in the canal water quality as it flows towards the Ennore creek. All samples were collected from the middle of the canal, in conformity with Standard Methods.

3.3. GROUNDWATER QUALITY SAMPLING STATIONS

The majority of people at Manali draw water for their essential needs either from dug wells or bore wells. The water supply aspect has been dealt in more detail in Chapter 12.

Twenty nine sampling points were set for these sources; the sites were chosen, once again, to cover all representative situations vis-a-vis use-pattern, and proximity to industrial sources of pollution (Figure 3.3, Table 3.3).

Of the bore wells, which provide most of the irrigation and domestic water needed in the area, seven were sampled (Figure 3.3, Table 3.4.).

Two of the water taps provided by the municipal corporation were also sampled.

All-in-all, 50 locations were sampled covering all types of water sources presented in Figures 3.1–3.3.

Table 3.1. Ponds sampled (cf Figure 3.1)

Sl. No.	Sample No.	Region	Water use
1	P1	Manali	Washing, bathing
2	P2	"	Cattle use
3	P3	"	"
4	P4	"	Washing
5	P5	Sadayankuppam	Cattle use
6	P6	"	"
7	P7	"	"
8	P8	"	"
9	P9	"	Washing, bathing
10	P10	Thiruvottriyur	No use
11	P11	"	Cattle use
12	P12	Vaikkadu	Washing, bathing
13	P13	"	"
14	P14	"	Cattle use

Table 3.2. Canals sampled; the location of sampling stations in Buckingham canal, which was modelled (Chapter 5), is given in Figure 3.2

Sl. No.	Sample No.	Region	Water use
1	CAM1	Amulavoil canal	Washing, bathing
2	CAM2	"	"
3	CAM3	Manali (wastewater effluent of MRL)	No use
4	CS4	Sadayankuppam canal	Cattle use
5	CB1	Buckingham canal, Manali	Transport
6	CB2	"	"
7	CB3	"	"
8	CB4	"	"
9	CB5	"	"
10	CB6	"	"
11	CB7	"	"
12	CB8	"	"
13	CB9	"	"
14	CB10	"	"
15	CB11	"	"
16	CB12	"	"
17	CB13	"	"
18	CB14	"	"

Table 3.3. Dug wells sampled (cf Figure 3.3)

Sl. No.	Sample No.	Region	Water use
1	DW1	Amulavoil	Washing, bathing and drinking
2	DW2	Chinnamathur	Washing, bathing
3	DW3	"	Washing, bathing
4	DW4	Madhavaram	Washing, bathing and drinking
5	DW5	"	Washing, bathing
6	DW6	Manali	Washing, bathing and drinking
7	DW7	"	"
8	DW8	"	"
9	DW9	"	"
10	DW10	"	"
11	DW11	"	"
12	DW12	"	"
13	DW13	Sadayankuppam	Washing, bathing
14	DW14	"	"
15	DW15	"	"
16	DW16	"	Washing, bathing and drinking
17	DW17	"	Washing
18	DW18	Tiruvottiyur	No use
19	DW19	"	Washing, bathing and drinking
20	DW20	"	Irrigation
21	DW21	Vaikkadu	Washing, bathing and drinking

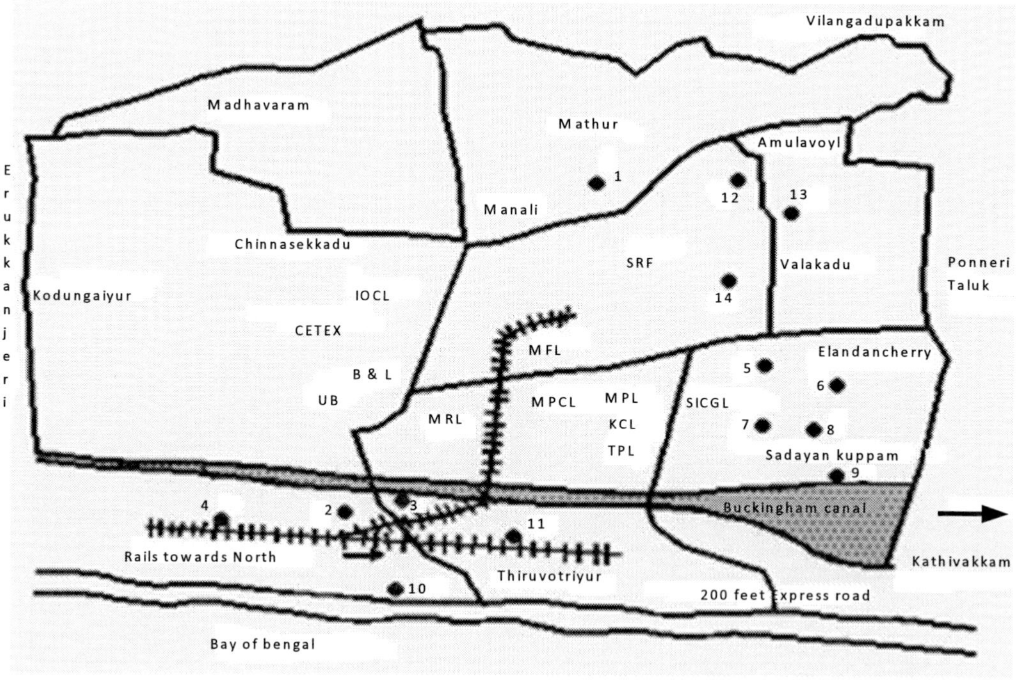

Figure 3.1. Location of pond – sampling sites [O].

Table 3.4. Bore wells and municipal taps sampled (cf Figure 3.3)

Sl. No.	Sample No.	Region	Water use
1	BW1	Chinnamathur	Washing, bathing and drinking
2	BW2	Madhavaram	"
3	BW3	Thiruvottriyur	Washing, bathing
4	BW4	"	"
5	BW5	"	Washing, bathing and drinking
6	BW6	"	Washing, bathing
7	BW7	"	Washing, bathing and drinking
8	TW1	Madhavaram	"
9	TW2	Vaikkadu	"

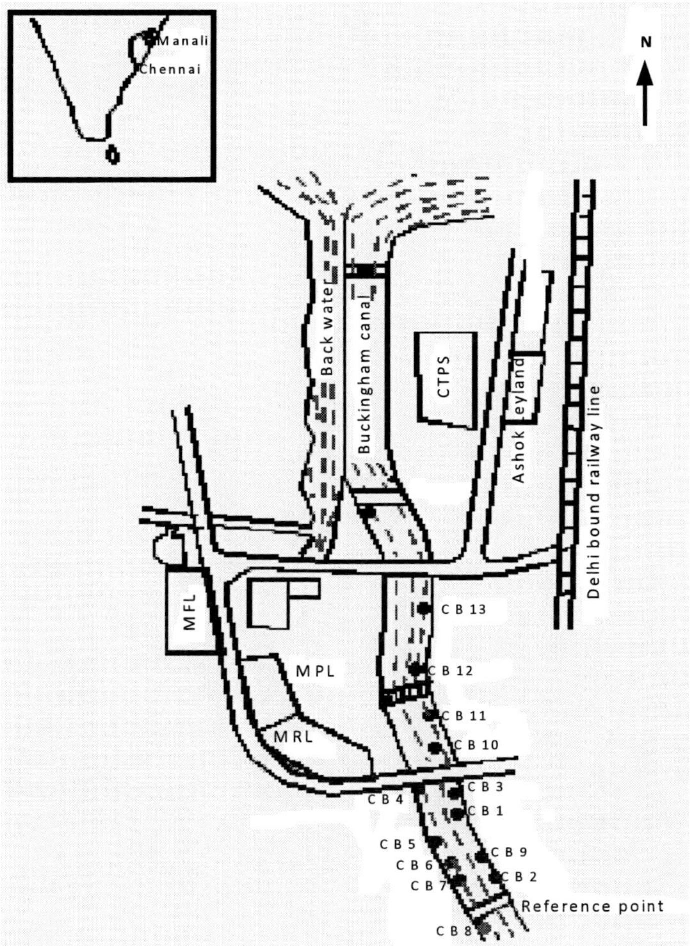

Figure 3.2. Sampling points in the Buckingham canal. All samples were collected from the middle of the canal, the sampling process strictly confirmed with the Standard methods (please see footnote of page 9).

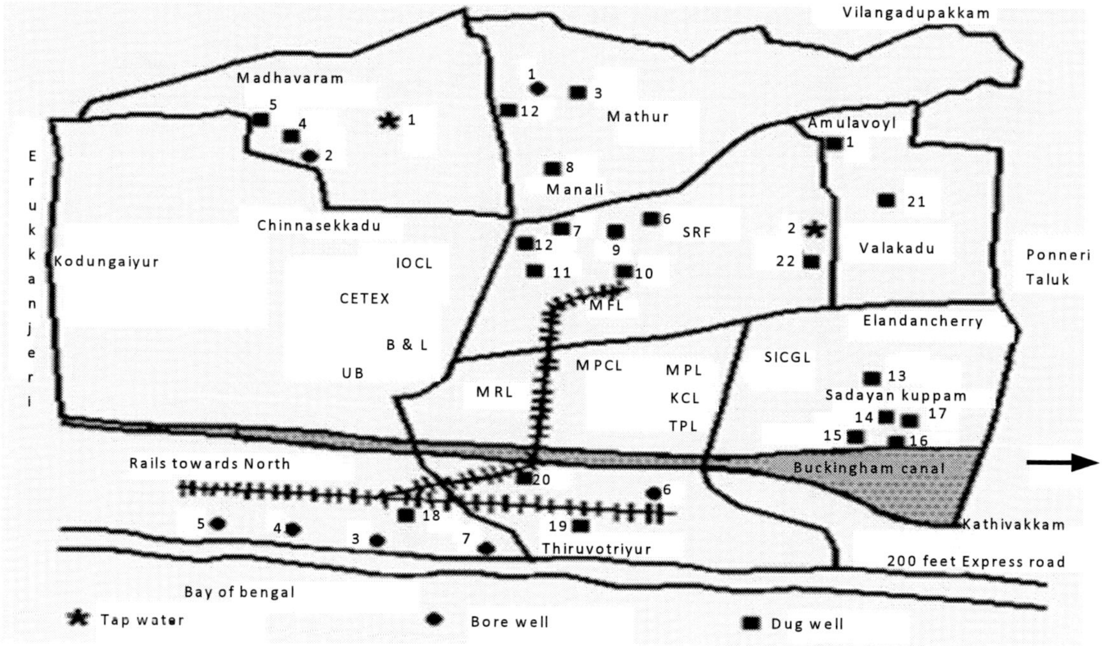

Figure 3.3. Location o ground water monitoring stations.

Plates 3.1 and 3.2. Solid wastes dumped close to a pond (above), and right next to a water body (below).

Plates 3.3. Solid waste dumped all round houses: persistent source of groundwater contamination through leaching.

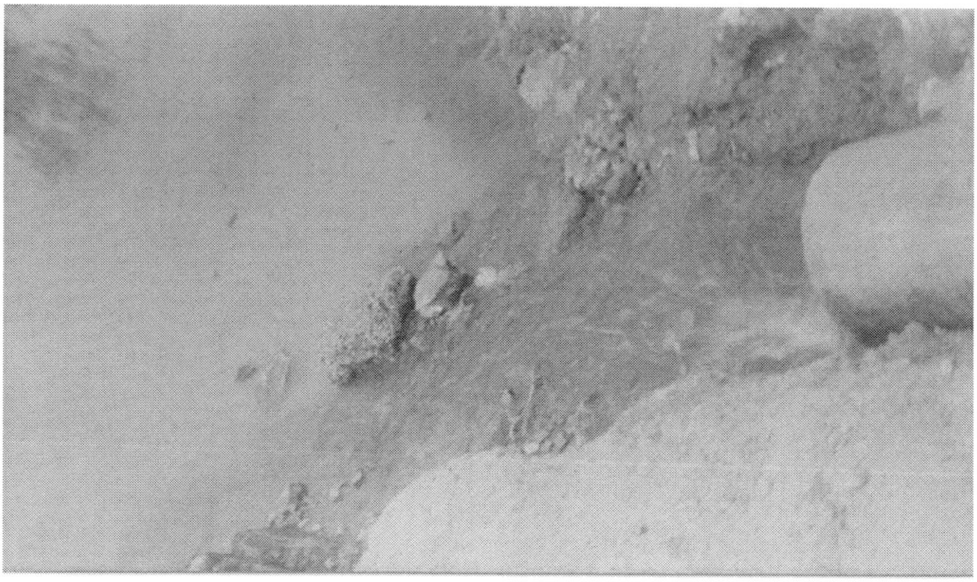

Plate 3.4. Wastewater being released in one of the ponds.

Chapter 4

THE STATUS OF WATER QUALITY

4.1. GENERAL

ABSTRACT

The detailed results of the water quality survey, indicating the levels of various variables at given location and given season, are presented in Appendices A-C. The gist of the findings are summarized in this chapter. The gist of the recheck surveys (2001-2002), which corroborate the earlier studies, is presented at the end of this chapter.

4.2. MAJOR FINDINGS: SURFACE WATER

The results have been discussed in the context of the following water quality standards.

1) Domestic water (For use as raw water for public water supply and bathing ghats) specification: IS 2296 (1974)
2) Drinking water: IS 10500 (1991)
3) Freshwater for fish culture: IS 13891 (1994)

The focus of the study was to ascertain:

1) The present status of the quality of water resources in the context of the typical features of the study area.
2) The extent of resource degradation vis-à-vis different types of water use.

Extensive studies were conducted during 1996-98 after earlier reconnaissance. In order to check the position half a decade later several resurveys were conducted during 2001-2002. The findings indicate that by and large the situation *vis a vis* the water quality or the sources of contamination hasn't improved; it rather has deteriorated. Hence the findings presented in this report are an accurate reflection of the very poor status of water quality in the study area. The findings of the cross-check are presented in Section 4.4.

It may be emphasised that the compliance with respect to standard for fish culture (IS 13891) has been assessed together with the other two standards for the following reasons : a) some of the ponds are used for fishing and b) the fish culture standard is more liberal than the other two standards and non-compliance with this standard is an indicator of the extent of poorness of the given water, and c) compliance with IS 13891 would indicate that the water body is capable of supporting fisheries, in turn reflecting utility value of the water body.

The gist of the findings are presented in this chapter. The raw data is collected in Appendices I-III.

4.2.1. Pre-Monsoon

Most of the samples either had offensive colour, or odour, or both (Table 4.1). Three in every four samples had offensive colour and 41% had odours (Table 4.2). The levels of electrical conductivity, phosphorous and total dissolved solids were above the tolerance limit for domestic water and drinking water in most of the samples (Table 4.3). In more than 75% of the samples the free ammonia levels exceed the tolerance limit for fresh water fish culture. In all samples atleast one parameter is outside the acceptable limits; in several samples more than one parameter is failing to comply (Appendices I-III).

4.2.2. Monsoon

During monsoon, 90% of the samples had offensive colour and 36% had unpleasant odours (Tables 4.4 and 4.5). The levels of electrical conductivity and phosphorus were above the tolerance limit for domestic water in more than 80% of the samples. In more than 60% of the samples the levels of alkalinity, ammonia and nitrate exceed the tolerance level for fresh water fish culture. Total iron exceeded the tolerance limit for drinking water in more than 80% of the samples. The results are summarized in Table 4.6.

4.2.3. Post-Monsoon

During this season, most of the samples (94%) had offensive colour and 47% had odours (Tables 4.7 and 4.8). The electrical conductivity exceeded the tolerance limit for domestic water in all the samples and 69% of the samples exceeded the tolerance limit for the water used for fish culture. The levels of total dissolved solids and phosphorus, too, were above the tolerance limit for domestic water in more than 85% of the samples. In a similarly large fraction (80%) of the samples, the levels of total dissolved solids, total iron and chloride were above the tolerance limit for drinking water. Here again, none of the samples complied with the requirement of water quality for any use (Table 4.9).

4.2.4. Heavy Metals, Arsenic and Boron

The results are summarized in Tables 4.10 and 4.11. The fractions of samples containing higher-than-permissible metal levels vis-a-vis drinking and fish culture were computed. It was seen that in case of surface water, 96% samples had mercury above tolerable level and 40% had excessive cadmium. In nearly all the samples one or more metals were present above permissible limits.

4.3. Major Findings: Ground Water

4.3.1. Pre-Monsoon

Of the ground water samples, 45% were coloured and 13% were odorous (Tables 4.12 and 4.13). In all samples atleast one, but often more than one, water quality parameter was above tolerance limits for drinking water, domestic usage and fresh water for fish culture, thereby making all samples below acceptable quality (Table 4.14). Phosphate was the main pollutant, found to cross the tolerance level for domestic water in 93.5% of the samples.

4.3.2. Monsoon

It was observed that 42% of the samples had colour and 6% had offensive odours during this season (Tables 4.15 and 4.16). More than 50% of the samples had total dissolved solids, total iron and chloride above levels tolerable for drinking water. The levels of electrical conductivity, total dissolved solids and phosphorus exceeded the tolerance level of domestic water in more than 70% cases. All the samples exceeded the tolerance level for ammonia. The results are summarised in Table 4.17. All-in-all none of the samples indicated suitability of water for any use: drinking or domestic.

4.3.3. Post-Monsoon

Among these samples 45% were coloured, and 16% had odours (Tables 4.18 and 4.19). The levels of electrical conductivity, phosphorus and total dissolved solids were above the tolerance limit for domestic water in most of the samples. Total dissolved solids, nitrate and hardness exceeded levels acceptable for drinking water in 77%, 77% and 65% of the samples. In most of the samples the levels of electrical conductivity, alkalinity and ammonia exceeded the tolerance limits (Table 4.20). Overall 100% samples were found to be unfit for use as drinking or domestic water.

4.3.4. Heavy Metals, Arsenic and Boron

Each and every sample contains one or more metals above permissible limits (Tables 4.21 and 4.22).

Therefore in terms of heavy metal and metalloid contamination, too, nearly all the samples are unfit for human use.

4.4. How Relevant the Findings Are as on Date

Resurveys to check whether the situation vis a vis effluent inflows into the water-bodies, location of industrial waste-dumps that contribute pollutants through leachates into ponds and groundwater, and use of the water-bodies by cattle and humans, etc, has significantly improved during the next 5 years, were conducted during 2001 and 2002 (upto January). Nine of the representative sampling stations, covering surface as well as ground water bodies, were also studied. The findings are illustrated in Tables 4.23 and 4.24, plates 3.1-3.4 (chapter 3), and 4.1-4.2. They reveal that as was the situation during 1995-98, in the years 2001-2002 as well, 100% of the water samples of the corresponding stations fail to comply with the water quality standards for drinking water : IS 10500 (1991) as well as with the standard 'for use as raw water and for bathing ghats': IS 2296. In other words the water quality situation is essentially the same as it was during the late 1990s. The solid waste dumps continue to exist near roads, houses, ponds, and grazing sites (Plates 3.1 - 3.4, Chapter 3); the use of ponds by humans and cattle continues (2.1 and 2.2); and waste inflows/spills continue to contaminate water bodies (Plates 4.1-4.2) as it used to be vide our earlier reports.

Table 4.1. Occurrence of colour and odour in the surface water samples during pre-monsoon

Source and Sample number	Colour	Odour
Ponds		
P1	Pale green	Nil
P2	Dark yellow	Nil
P3	Pale yellow	Nil
P4	Grey	Fishy smell
P5	Pale yellow	Nil
P6	Pale yellow	Garlic Smell
P7	Pale green	Garlic smell
P8	Pale green	Nil
P9	Dark green	Nil
P11	Pale yellow	Nil
P12	Nil	Nil
P13	Pale yellow	Nil
Canal		
CA3	Dark grey	Garlic smell
CA4	Pale yellow	Pungent smell
CB1	Pale grey	Oily smell
CB4	Nil	Organic smell
CB6	off-white	Pungent smell
CB7	off-white	Oily smell

Table 4.2. Percentage of surface water samples of unacceptable quality in terms of physical parameters – pre-monsoon

S. No.	Parameter	Percentage of samples unfit as per the limit set for	
		Drinking water	Fish culture
1	Colour*	77%	---
2	Odour[@]	41%	---
3	Turbidity	95%	90%
4	Temp	0%	0%
5	Overall	100%	100%

[*] Samples with colour other than of normal water.
[@] Samples with some or other unpleasant odour.

Table 4.3. Percentage of surface water samples of unacceptable quality in terms of physico-chemical and chemical parameters – pre-monsoon

S No.	Parameter	Percentage of samples unfit as per the limits set for		
		Drinking water %	Domestic use %	Fish culture %
1	pH	23.8	9.5	23.8
2	EC	*	90.5	57.1
3	TDS	81.0	81.0	*
4	Alkalinity	42.9	*	57.1
5	Turbidity	95.2	*	90.5
6	Total hardness	61.9	*	*
7	Total iron	81.0	*	38.1
8	Ammonia	*	*	76.2
9	Nitrate	14.3	*	47.6
10	Chloride	57.1	57.1	*
11	Sulphate	23.8	9.5	*
12	Phosphate	*	95.2	*
13	Copper	30.4	*	30.4
14	Zinc	8.7	*	21.7
15	Mercury	95.7	*	95.7
16	Boron	13.0	*	*
17	Lead	26.1	*	26.1
18	Arsenic	30.4	*	30.4
19	Manganese	47.8	*	*
20	Cadmium	39.1	*	39.1
21	Overall water quality	100.0	100.0	100.0

*Standards not set.

Table 4.4. Percentage of surface water samples of unacceptable quality in terms of physical parameters – monsoon

Sl. No	Parameter	Percent samples unfit as per the limits set for	
		Drinking water	Fish culture
1	Colour *	90%	---
2	Odour @	36%	---
3	Turbidity	100%	86%
4	Temperature	0%	0%
5	Overall	100%	100%

* Samples with colour other than of normal water
@ Samples with some or other unpleasant odour.

Table 4.5. Occurrence of colour and odour in the surface water samples during monsoon

Source and Sample number.	Colour	Odour
Ponds P1	Greyish green	Odourless
P2	Pale yellow	Odourless
P3	Pale grey	Odourless
P4	Pale yellow	Odourless
P5	Greenish	Odourless
P6	Greenish	Odourless
P7	Pale green	Odourless
P8	Pale green	Odourless
P9	Dark green	Odourless
P10	Pale yellow	Odourless
P11	Pale green	Fishy smell
P12	Pale yellow	Muddy smell
P13	Pale yellow	Odourless
P14	Pale green	Odourless
Canal CAM 1	Pale green	Odourless
CAM2	Colourless	Rotten smell
CAM3	Colourless	Oily smell
CAM4	Pale yellow	Odourless
CB1	Black	Sewage smell
CB4	Black	Oily smell
CB6	Black	Sewage smell
CB7	Pale Yellow	Sewage smell

Table 4.6. Percentage of surface water samples of unacceptable quality in terms of physico-chemical and chemical parameters – monsoon

SNo.	Parameter	Percentage of samples unfit as per the limits set for		
		Drinking water %	Domestic use %	Fish culture %
1	PH	40.9	9.0	40.9
2	EC	*	81.8	40.9
3	TDS	59.1	59.1	*
4	Alkalinity	36.4	*	68.2
5	Turbidity	100.0	*	86.4
6	Total hardness	31.8	*	*
7	Total iron	81.8	*	40.9
8	Ammonia	*	*	68.2
9	Nitrate	0.0	*	63.6
10	Chloride	50.0	22.7	*
11	Sulphate	9.1	0.0	*
12	Phosphate	*	90.9	*
13	Overall water quality	100.0	100.0	100.0

Table 4.7. Percentage of surface water samples of unacceptable quality in terms of physical parameters – post-monsoon

Sl. No	Parameter	Percent samples unfit as per the limits set for	
		Drinking water	Fish culture
1	Colour *	94%	---
2	Odour @	47%	---
3	Turbidity	100%	97%
4	Temperature	0%	0%
5	Overall	100%	100%

* Samples with colour other than of normal water

@ Samples with some or other unpleasant odour.

Table 4.8. Occurrence of colour and odour in the surface water samples during post-monsoon

Source and Sample number.	Colour	Odour
Ponds P1	Pale yellow	Odourless
P2	Pale green	Oily smell
P3	Brick red	Fishy smell
P4	Pale yellow	Fishy smell
P5	Grey	Odourless
P6	Pale yellow	Odourless
P7	Pale yellow	Odourless
P8	Pale yellow	Odourless
P9	Pale green	Odourless
P10	Yellowish green	Odourless
P11	Pale green	Fishy smell

Table 4.8. (Continued)

Source and Sample number.	Colour	Odour
P12	Pale yellow	Odourless
P13	Pale yellow	Odourless
P14	Colourless	Odourless
Canal CAM1	Pale yellow	Odourless
CAM2	Pale yellow	Odourless
CAM3	Black	Oily smell
CAM4	Grey	Odourless
CB1	Pale yellow	Oily smell
CB2	Pale grey	Rotten smell
CB3	Dark grey	Rotten smell
CB4	Dark grey	Rotten smell
CB5	Pale grey	Rotten smell
CB6	Dark grey	Rotten smell
CB7	Pale grey	Unpleasant smell
CB8	Pale grey	Odourless
CB9	Pale yellow	Odourless
CB10	Pale yellow	Odourless
CB11	Dark black	Rotten smell
CB12	Dark black	Unpleasant smell
CB13	Pale yellow	Unpleasant smell
CB14	Colourless	Odourless

Table 4.9. Percentage of surface water samples of unacceptable quality in terms of physico-chemical and chemical parameters – post-monsoon

S. No.	Parameter	Percentage of samples unfit as per the limits set for		
		Drinking water %	Domestic use %	Fish culture %
1	PH	3.1	0.0	3.1
2	EC	*	100.0	68.8
3	TDS	87.5	87.5	*
4	Alkalinity	56.3	*	65.6
5	Turbidity	100.0	*	96.9
6	Total hardness	65.6	*	*
7	Total iron	96.9	*	40.6
8	Ammonia	*	*	56.3
9	Nitrate	53.1	*	71.9
10	Chloride	81.3	31.3	*
11	Sulphate	25.0	0.0	*
12	Phosphate	*	87.5	*
13	Overall water quality	100.0	100.0	100.0

Table 4.10. Trace metals* in surface (pond) water samples

SNo.	Cu	Ni	Zn	Hg	B	Co	Pb	As	Mn	Mo	Cd	Be
P2U	0.033	BDL	BDL	0.226	0.367	0.039	BDL	BDL	0.444	BDL	0.009	BDL
P2F	0.031	BDL	BDL	0.170	BDL	0.010	BDL	BDL	0.303	BDL	0.001	BDL
P3U	0.002	0.019	BDL	0.339	1.014	0.010	BDL	BDL	BDL	0.082	0.015	0.004
P3F	BDL	0.009	BDL	0.226	0.663	BDL	BDL	BDL	BDL	0.011	0.009	BDL
P4U	0.025	BDL	BDL	0.226	0.962	0.039	BDL	BDL	BDL	0.100	0.007	BDL
P4F	0.018	BDL	BDL	0.170	0.747	0.039	BDL	BDL	BDL	0.015	0.005	BDL
5U	0.021	BDL	BDL	0.226	BDL	0.010	BDL	BDL	0.677	BDL	0.008	BDL
P5F	BDL	BDL	BDL	0.170	BDL	BDL	BDL	BDL	BDL	BDL	BDL	BDL
P6U	0.023	BDL	BDL	0.339	1.411	0.059	BDL	BDL	2.486	0.272	0.009	BDL
P6F	0.021	BDL	BDL	0.113	BDL	0.029	BDL	BDL	BDL	BDL	0.007	BDL
P8U	0.045	0.028	BDL	0.283	BDL	0.039	BDL	BDL	BDL	BDL	0.021	BDL
P8F	BDL	BDL	BDL	0.113	BDL	0.010	BDL	BDL	BDL	BDL	0.004	BDL
P9U	0.037	BDL	BDL	0.170	0.169	0.049	BDL	BDL	BDL	BDL	0.004	BDL
P9F	BDL	BDL	BDL	BDL	BDL	0.029	BDL	BDL	BDL	BDL	BDL	BDL
P14U	0.033	0.028	BDL	0.396	BDL	0.029	BDL	BDL	BDL	BDL	BDL	BDL
P14F	BDL	BDL	BDL	0.056	BDL	0.010	BDL	BDL	BDL	BDL	BDL	BDL

U-Unfiltered

F-Filtered

* All the values are in mg/L.

Table 4.11. Trace metals* in surface (Buckingham canal) water samples

SNo.	Cu	Ni	Zn	Hg	B	Co	Pb	As	Mn	Mo	Cd	Be
BC1	0.675	0.182	47.51	0.896	1.313	0.177	Nil	0.125	3.670	0.381	0.037	0.054
BC2	0.341	0.228	43.11	0.443	0.222	0.305	0.198	0.125	5.772	0.652	0.034	0.076
BC3	0.165	0.191	BDL	0.010	BDL	0.305	0.159	0.125	7.431	0.519	0.032	0.084
BC4	1.311	0.410	0.253	2.313	BDL	0.483	0.357	0.345	12.81	1.323	0.066	0.113
BC5	3.176	0.255	1.157	0.783	BDL	0.384	0.396	0.125	10.49	0.974	0.059	0.091
BC6	1.422	0.282	2.192	0.103	BDL	0.414	0.555	0.690	7.753	1.260	0.050	0.123
BC7	1.573	0.556	1.357	0.503	BDL	0.799	0.753	0.690	17.58	2.063	0.099	0.179

* All the values are in mg/L.

Table 4.12. Percentage of ground water samples of unacceptable quality in terms of physical parameters – pre-monsoon

Sl. No	Parameter	Percent samples unfit as per the limits set for	
		Drinking water	Fish culture
1	Colour *	45%	---
2	Odour [@]	13%	---
3	Turbidity	35%	29%
4	Temperature	---	0%
5	Overall	100%	100%

*Samples with colour other than of normal water.

[@] Samples with some or other unpleasant odour.

Table 4.13. Occurrence of colour and odour in the ground water samples during pre-monsoon

Source and Sample number	Colour	Odour
Dug well DW1	Colourless	Odourless
Dw2	Colourless	Odourless
Dw3	Pale green	Odourless
Dw4	Pale yellow	Odourless
Dw5	Colourless	Odourless
Dw6	Colourless	Odourless
Dw7	Colourless	Odourless
Dw8	Colourless	Odourless
Dw9	Colourless	Odourless
Dw10	Colourless	Odourless
Dw11	Pale yellow	Odourless
Dw12	Dark yellow	Rust smell
Dw13	Rusty red	Odourless
Dw14	Pale green	Odourless
Dw15	Pale yellow	Odourless
Dw16	Dark green	Odourless
Dw17	Dark green	Odourless
Dw18	Pale yellow	Odourless
Dw19	Colourless	Odourless
Dw20	Colourless	Odourless
Dw21	Pale yellow	Odourless
Dw22	Colourless	Odourless
Bore well Bw1	Colourless	Odourless
Bw2	Colourless	Odourless
Bw3	Dark yellow	Pungent smell
Bw4	Grey	Oily smell
Bw5	Colourless	Odourless
Bw6	Colourless	Odourless
Bw7	Colourless	Odourless
Tap water Tw1	Colourless	Odourless
Tw2	Dark yellow	Fishy smell

Table 4.14. Percentage of ground water samples of unacceptable quality in terms of physico-chemical and chemical parameters – per-monsoon

SNo.	Parameter	Percentage of samples unfit as per the limits set for		
		Drinking water %	Domestic use %	Fish culture %
1	pH	12.9	3.2	12.9
2	EC	*	93.5	54.8
3	TDS	80.6	80.6	*
4	Alkalinity	48.4	*	58.1

SNo.	Parameter	Percentage of samples unfit as per the limits set for		
		Drinking water %	Domestic use %	Fish culture %
5	Turbidity	38.7	*	32.3
6	Total hardness	61.3	*	*
7	Total iron	35.5	*	22.6
8	Ammonia	*	*	51.6
9	Nitrate	0.0	*	64.5
10	Chloride	58.1	22.6	*
11	Sulphate	19.4	6.5	*
12	Phosphate	*	93.5	*
13	Copper	4.2	*	4.2
14	Zinc	0.0	*	0.0
15	Mercury	100.0	*	100.0
16	Boron	12.5	*	*
17	Lead	0.0	*	0.0
18	Arsenic	0.0	*	0.0
19	Manganese	25.0	*	*
20	Cadmium	25.0	*	25.0
21	Overall water quality	100.0	100.0	100.0

Table 4.15. Percentage of ground water samples of unacceptable quality in terms of physical parameters – monsoon

Sl. No	Parameter	Percent samples unfit as per the limits set for	
		Drinking water	Fish culture
1	Colour *	42%	---
2	Odour @	6.4%	---
3	Turbidity	64.5%	51.6%
4	Temperature	---	0%
5	Overall	100%	100%

* Samples with colour other than of normal water

@ Samples with some or other unpleasant odour.

Table 4.16. Occurrence of colour and odour in the ground water samples during monsoon

Source and Sample number	Colour	Odour
Dug well		
DW1	Yellowish	Odourless
DW2	Colourless	Odourless
DW3	Colourless	Odourless
DW4	Colourless	Odourless
DW5	Colourless	Odourless
DW6	Pale yellow	Odourless
DW7	Colourless	Odourless
DW8	Dark yellow	Odourless
DW9	Colourless	Odourless
DW10	Colourless	Ammonia smell

Table 4.16. (Continued)

Source and Sample number	Colour	Odour
DW!!	Greenish	Odourless
DW12	Pale yellow	Odourless
DW13	Colourless	Odourless
DW14	Colourless	Odourless
DW15	Colourless	Odourless
DW16	Pale green	Odourless
DW17	Pale yellow	Odourless
DW18	Pale reddish	Odourless
DW19	Colourless	Odourless
DW20	Pale yellow	Odourless
DW21	Pale yellow	Fishy smell
DW22	Colourless	Odourless
Bore well BW1	Colourless	Odourless
BW2	Colourless	Odourless
BW3	Reddish	Odourless
BW4	Pale yellow	Odourless
BW5	Colourless	Odourless
BW6	Colourless	Odourless
BW7	Pale yellow	Odourless
Tap water TW1	Colourless	4 Odourless
TW2	Colourless	Odourless

Table 4.17. Percentage of ground water samples of unacceptable quality in terms of physico-chemical and chemical parameters – monsoon

Sno.	Parameter	Percentage of samples unfit as per the limits set for		
		Drinking water %	Domestic use %	Fish culture %
1	PH	3.2	0.0	0.0
2	EC	*	87.1	48.4
3	TDS	74.2	74.2	*
4	Alkalinity	45.2	*	54.8
5	Turbidity	64.5	*	51.6
6	Total hardness	38.7	*	*
7	Total iron	64.5	*	25.8
8	Ammonia	*	*	48.4
9	Nitrate	35.5	*	100.0
10	Chloride	54.8	16.1	*
11	Sulphate	29.0	3.2	*
12	Phosphate	*	95.2	*
13	Overall water quality	100.0	100.0	100.0

Table 4.18. Percentage of ground water samples of unacceptable quality in terms of physical parameters – post-monsoon

Sl. No	Parameter	Percent samples unfit as per the limits set for	
		Drinking water	Fish culture
1	Colour *	45%	---
2	Odour @	16%	---
3	Turbidity	51.6%	48.3%
4	Temperature	---	0%
5	Overall	100%	100%

* Samples with colour other than of normal water

@ Samples with some or other unpleasant odour.

Table 4.19. Occurrence of colour and odour in the ground water samples during post-monsoon

Source and Sample number	Colour	Odour
Dug well DW1	Colourless	Odourless
DW2	Colourless	Odourless
DW3	Pale green	Odourless
DW4	Pale yellow	Odourless
DW5	Colourless	Odourless
DW6	Colourless	Odourless
DW7	Colourless	Odourless
DW8	Colourless	Odourless
DW9	Colourless	Odourless
DW10	Colourless	Odourless
DW11	Colourless	Odourless
DW12	Pale yellow	Odourless
DW13	Pale yellow	Odourless
DW14	Pale green	Odourless
DW15	Pale green	Odourless
DW16	Grey	Odourless
DW17	Grey	Rotten smell
DW18	Pale grey	Oily smell
DW19	Colourless	Odourless
DW20	Brick red	Fishy smell
DW21	Pale yellow	Odourless
DW22	Colourless	Odopurless
Bore well BW1	Colourless	Odourless
BW2	Colourless	Odourless
BW3	Black	Oil smell
BW4	Pale yellow	Oily smell
BW5	Colourless	Odourless
BW6	Colourless	Odourless
BW7	Colourless	Odourless
Tap water TW1	Colourless	Odourless
TW2	Pale yellow	Odourless

Table 4.20. Percentage of ground water samples of unacceptable quality in terms of physico-chemical and chemical parameters – post-monsoon

Sl. No.	Parameter	Percentage of samples unfit as per the limits set for		
		Drinking water %	Domestic use %	Fish culture %
1	PH	0.0	0.0	0.0
2	EC	*	100.0	58.1
3	TDS	77.4	77.4	*
4	Alkalinity	41.9	*	64.5
5	Turbidity	51.6	*	48.4
6	Total hardness	64.5	*	*
7	Total iron	51.6	*	22.6
8	Ammonia	*	*	3.2
9	Nitrate	77.4	*	90.3
10	Chloride	54.8	25.8	*
11	Sulphate	48.4	3.2	*
12	Phosphate	*	67.7	*
13	Overall water quality	100.0	100.0	90.0

Table 4.21. Trace metals* in bore well water samples

SNo.	Cu	Ni	Zn	Hg	B	Co	Pb	As	Mn	Mo	Cd	Be
BW3U	0.060	0.028	BDL	0.113	0.335	0.010	BDL	BDL	0.106	BDL	0.007	BDL
BW3F	BDL	BDL	BDL	0.056	0.210	0.010	BDL	BDL	0.081	BDL	0.002	BDL
BW6U	0.023	BDL	BDL	0.339	0.715	0.039	BDL	BDL	0.028	0.056	0.010	BDL
BW6F	BDL	BDL	BDL	0.283	0.549	0.029	BDL	BDL	BDL	0.032	0.001	BDL
BW7U	0.033	BDL	BDL	0.339	3.347	0.088	BDL	BDL	BDL	1.506	0.013	BDL
BW7F	0.002	BDL	BDL	0.170	BDL	0.039	BDL	BDL	BDL	1.393	0.010	BDL

U – Unfiltered; F – Filtered.

* All the values are in mg/L.

Table 4.22. Trace metals* in dug well water samples

SNo.	Cu	Ni	Zn	Hg	B	Co	Pb	As	Mn	Mo	Cd	Be
DW3U	0.037	0.009	BDL	0.056	BDL	0.020	BDL	BDL	10.21	BDL	0.013	BDL
DW3F	BDL	BDL	BDL	0.056	BDL	0.020	BDL	BDL	10.03	BDL	0.003	BDL
DW6U	0.019	BDL	BDL	0.679	BDL	0.079	BDL	BDL	BDL	BDL	0.007	BDL
DW6F	BDL	BDL	BDL	0.226	BDL	0.010	BDL	BDL	BDL	BDL	0.002	BDL
DW10U	0.025	0.019	BDL	0.283	BDL	0.020	BDL	BDL	0.512	BDL	0.009	BDL
DW10F	0.012	BDL	BDL	0.226	BDL	0.020	BDL	BDL	0.489	BDL	0.001	BDL
DW12U	0.002	BDL	BDL	0.170	0.016	0.049	0.119	BDL	0.258	BDL	0.005	BDL
DW12F	BDL	BDL	BDL	0.113	BDL	BDL	BDL	BDL	BDL	BDL	0.005	BDL
DW15U	0.029	BDL	BDL	0.170	BDL	0.039	BDL	BDL	BDL	BDL	0.010	BDL
DW15F	0.027	BDL	BDL	0.113	BDL	0.010	BDL	BDL	BDL	BDL	0.004	BDL
DW16U	0.033	BDL	BDL	0.283	2.165	0.049	BDL	BDL	BDL	0.976	0.005	BDL
DW16F	0.029	BDL	BDL	0.226	BDL	0.020	BDL	BDL	BDL	0.922	BDL	BDL
DW17U	0.021	BDL	BDL	0.226	0.295	0.079	BDL	BDL	BDL	0.244	0.008	BDL
DW17F	BDL	BDL	BDL	0.056	BDL	0.049	BDL	BDL	BDL	0.233	0.005	BDL
DW18U	0.041	BDL	BDL	0.113	0.416	0.148	BDL	BDL	BDL	BDL	0.007	BDL
DW18F	BDL	BDL	BDL	0.056	0.113	0.039	BDL	BDL	BDL	BDL	0.002	BDL
DW22U	0.018	BDL	BDL	0.170	1.249	0.010	BDL	BDL	BDL	0.054	0.006	BDL
DW22F	BDL	BDL	BDL	0.113	0.59	BDL	BDL	BDL	BDL	0.048	0.005	BDL

U-Unfiltered

F-Filtered

* All the values are in mg/L.

Table 4.23. Water quality at Manali: September 2001

Parameters	Station: Thiruvottryur	Station: Vaikkadu	Station: Amulavoil	Station: Sadayan Kuppam	Station: Amulavoil	Station: Amulavoil	Station: Vaikkadu & Elandencheri	Station: Sathankadu	Station: Sadayankuppam
	Bore well BW1	Tap Water TP1	Dug well DW1	Dug well DW2	Pond water PW1	Canal water CW1	Canal water CW2	Canal water (Surface 1 water)	Canal water (Surface water 2)
pH	7.4	8.2	8.3	7.7	5.4	6.3	5.6	6.9	5.5
EC μ mhos	1040	1220	1230	3470	3450	5300	2480	23400	14250
Acidity mg/l	9.6	14.8	4	16	17.2	39.6	66	20	51.6
Alkalinity mg/l	38.16	239.51	294.78	21	28.76	35.53	34.2	88.17	33.99
Total Hardness mg/l	370	62	340	1740	920	1480	430	5700	5200
Ca Hardness mg/l	190	34	180	480	570	482	192	3500	2280
Chloride mg/l	134.14	159.09	171.57	784.70	1575.39	2963.62	717.5	13800	10606.6
Nitrite mg/l	0.2800	0.178	2.406	1.1075	2.558	40.099	71.29	0.764	0.917
Total Phosphate mg/l	0.002	0.032	0.033	0.136	0.169	1.925	1.142	0.534	0.1
Sulfate mg/l	250.9	34.23	98.042	1902.22	47.589	763.6	196.29	1588.5	1909.0
Nitrate mg/l	0.252	0.28	0.52	0.304	0.52	1.276	1.248	6.12	0.484

Table 4.24. Compliance of the recent water samples with water quality standards

Source	*Percent samples complying with IS 10500*	*Percent samples complying with IS 2296*
Bore well	0	0
Dug well	0	0
Pond water	0	0
Canal water	0	0

Plates 4.1. Pond water tainted yellow by the contaminants coming from nearby units.

Plate 4.2. A typical effluent outfall.

Chapter 5

MODELLING OF BUCKINGHAM CANAL WATER QUALITY

5.1. INTRODUCTION

As mentioned in Chapter 2, Buckingham Canal, when commissioned in 1897, used to be a 419 Km long channel running close to the Coromandal coast from Peddagangam town to Marakkanam town. The canal was extensively used for navigation uptill the advant of rail and road traffic in the 20th century. This gradually diminished the canal's importance. Due to the effects of cyclonic storms, disrepair, and disuse, the canal is now silted up for most of its length. One of the existing functional stretches of the canal passes through Manali Industrial Complex (MIC). In this span, the canal receives inflows from point as well as non-point sources.

This chapter describes the modelling of this stretch of Buckingham canal. For this purpose the computer-automated tool QUAL 2E-UNCAS was used. This tool has been developed for the Environmental Protection.

Agency (EPA) of The United States of America and belongs to a series of such tools beginning with QUAL I developed by F.D. Masch and Associates and the Texas Water Development Board (1970, 1971). QUAL 2E-UNCAS can handle upto 15 water quality constituents including DO, BOD, temperature, algae as chlorophyll a, various forms of N, P, coliforms, upto three conservative and an arbitery non-conservative, constituents. It can be used to model dendritic streams that are well mixed. As the Buckingham canal is shallow (maximum water level ~ 2.0 m during rains) and turbulent, the assumption of well-mixed stream is justified in this case. QUAL 2E-UNCAS further assumes that the major transport mechanisms, advection and dispersion, are significant only along the main direction of flow (longitudinal axis of the stream or canal). It allows for multiple waste discharges, withdrawals, tributary flows, and incremental inflow and outflow.

Hydraulically QUAL 2E-UNCAS can be operated in steady state and dynamic modes. In the former it can be used to study the impact of waste loads on stream water quality and to identify the contributions to it of non-point source waste loads. In the dynamic mode the user can study the effects of diurnal variations in meteorological data on water quality (primarily dissolved oxygen and temperature) and diurnal dissolved oxygen variations due to algal growth and respiration.

The UNCAS facility allows the modeler to perform uncertainty analysis on the steady state water quality simulations. Three uncertainty options are available: sensitivity analysis, first order error analysis, and Monte Carlo simulation.

5.2. Hydrological, Including Water Quality Data

The requisite data on canal dimensions, flow, and water quality was generated by us, using standard procedures (Linsely 1975, APHA 1995, Abbasi 1998). Similar set of data earlier generated by an independent agency[1] was used to validate the model after we had calibrated it with our data.

5.3. Model Validation

Typical results validating the model are presented in Figures 5.1 – 5.5. Other validation curves are given at the end of this chapter (Figures 5.8 – 5.21). It may be seen that the observed values of temperature, BOD, nitrate nitrogen, nitrite nitrogen, organic nitrogen, total nitrogen, and TDS match well with the predicted values all the way.

In case of chloride, TDS (total dissolved solids), sulphate, and ammonical nitrogen, the observed values are 50% higher in summer than the predicted value *after about 4.6 km from the reference point* (Figures 5.8 – 5.21). This is due to the tidal influence from the sea.

The model was validated for the canal water quality for four seasons, post-monsoon (December-February), summer (March-May), pre-monsoon (June-August), and monsoon (September-November).

During summer the match between the observed and the predicted values is closest for temperature, BOD, nitrate nitrogen, nitrite nitrogen, organic nitrogen, and total nitrogen. For TDS, chloride, sulphate and ammonical nitrogen, there is a mismatch after about 4.6 km from the reference point, as mentioned above. We attribute this to tidal influence. Such an influence is noticed only during summer because the flow in the canal is at its lowest during summer allowing the tidal influence to penetrate deeper upstream than in other seasons. In some years in peak summer the flow even ceases leaving small puddles of water across the canal length.

During pre-monsoon, monsoon and post monsoon seasons, too, the trend is similar. The predicted values match with the observed ones all the way upto the point about 4.6 km downstream waste discharge point. Beyond this range the water quality is influenced by tidal inflows.

After the model was validated, it was used to simulate different situations of variations in flow, incoming water quality, and waste inputs. The impact of these variations on the assimilative capacity of the canal was then forecast. The salient findings are summarized below.

5.4. IMPORTANT FINDINGS

1) The model was successful in predicting the water quality of Buckingham canal at 90-95% confidence levels across the 7.80 km stretch studied by us, throughout the year. The only exception was the mismatch between the observed and the predicted values of some variables in the third reach, about 4.6 km from the reference point (headwater) - this was due to the impact of tidal inflow which was particularly severe during summer when the flow in the canal was at its leanest.

2) The two variables of greatest concern in managing the water quality are total dissolved solids (TDS) and BOD; these are significantly higher than permissible for irrigation and non-consumptive domestic use throughout the year (except monsoon).

3) Simulations were carried out to forecast the impact of changes in the levels of one or more water quality variables, and of flow augmentation, on the canal water quality during summer, monsoon, and post-monsoon (Figures 5.6 and 5.7).

The aim of the simulations was to identify strategies for improving the water quality of the canal to the extent that it confirms to standards prescribed by the regulatory agencies for use in irrigation and non-consumptive domestic use. The salient findings are:

a) several variables, notably TDS, suspended solids (SS), chloride (Cl), and temperature were not influenced by changes in other variables;
b) increase in temperature caused reduction in BOD, evidently by hastening microbial activity which in turn hastened BOD decay, but increase in headwater DO had no significant impact on BOD.
c) stepwise reduction in the inflows from two point sources (industrial effluents) down to zero, had only marginal impact on the canal water quality.
d) flow augmentation reduced pollutant levels down to permissible limits but the freshwater required to effect such flow augmentation is just not available in that region.

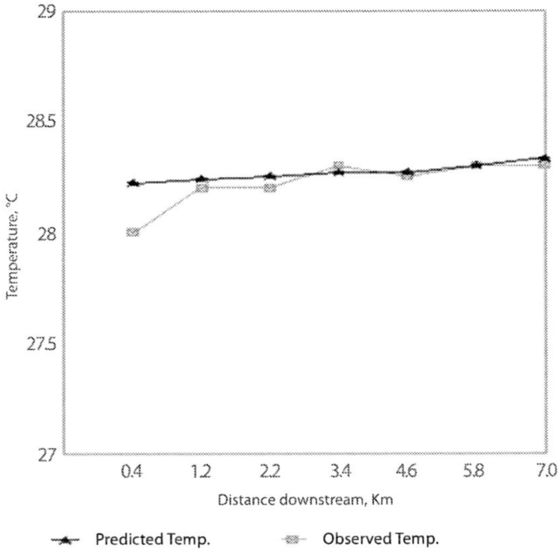

Figure 5.1. Validation curve for temperature during summer.

4) Overall the modelling and simulation indicate that in order to manage the canal water quality the following steps are imperative.

a) sources of BOD and TDS upstream must be identified and steps taken to reduce or eliminate their contributions;

b) The BOD and TDS levels of the effluents from two industries which join the canal in the study area should be reduced drastically by implementing appropriate treatment at the industries before disposal.

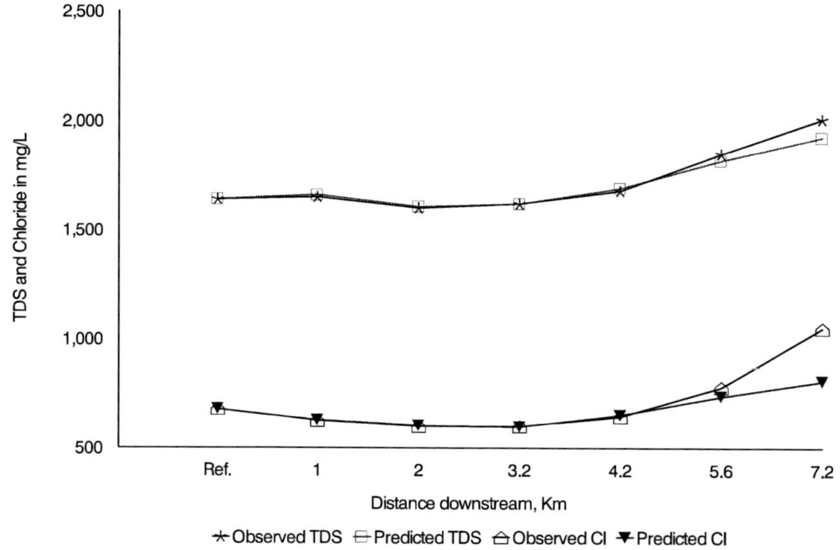

Figure 5.2. Validation curve for TDS and chloride during premonsoon.

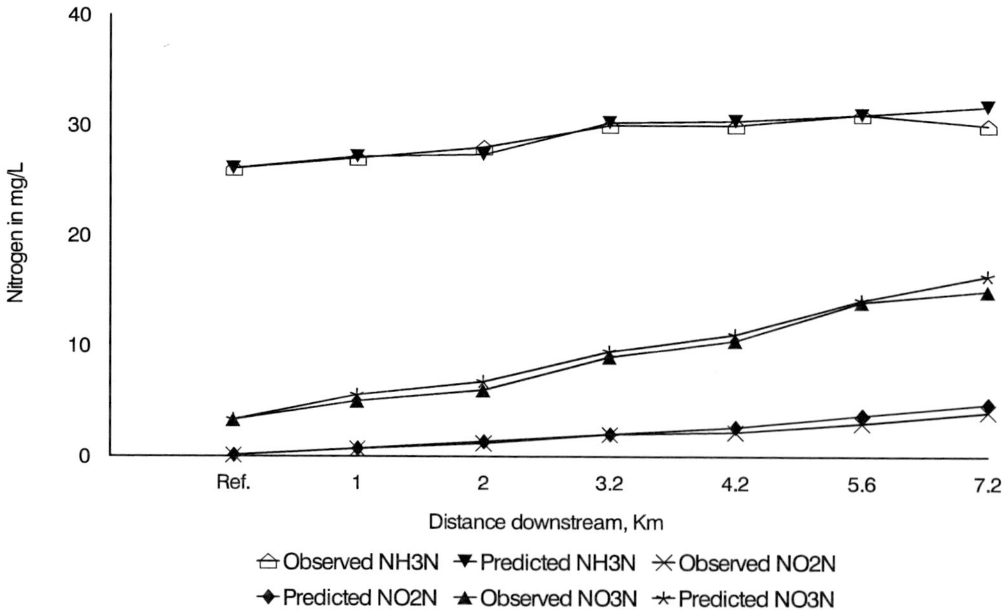

Figure 5.3. Validation curve for Nitrogen (NH_3, NO_2, NO_3) during premonsoon.

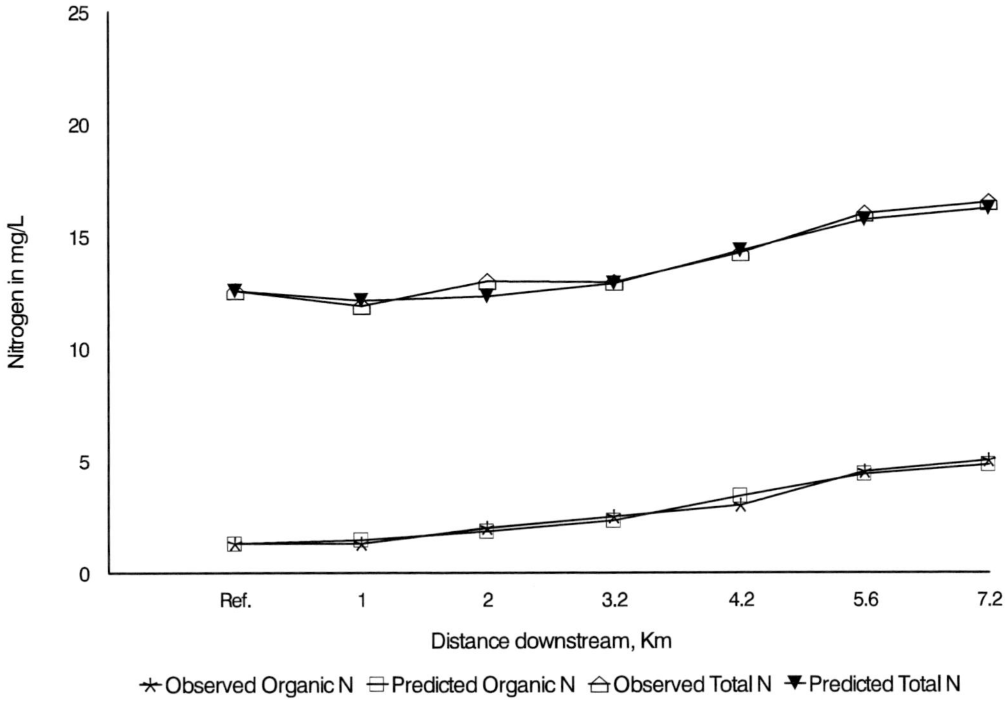

Figure 5.4. Validation curve for organic and total nitrogen during monsoon.

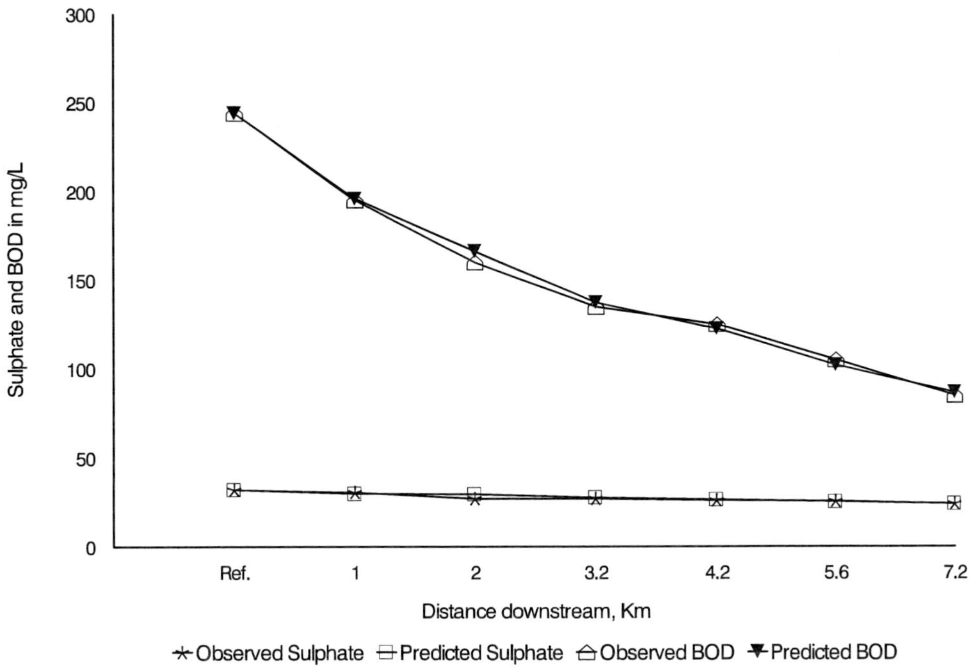

Figure 5.5. Validation curve for sulphate and BOD during postmonsoon.

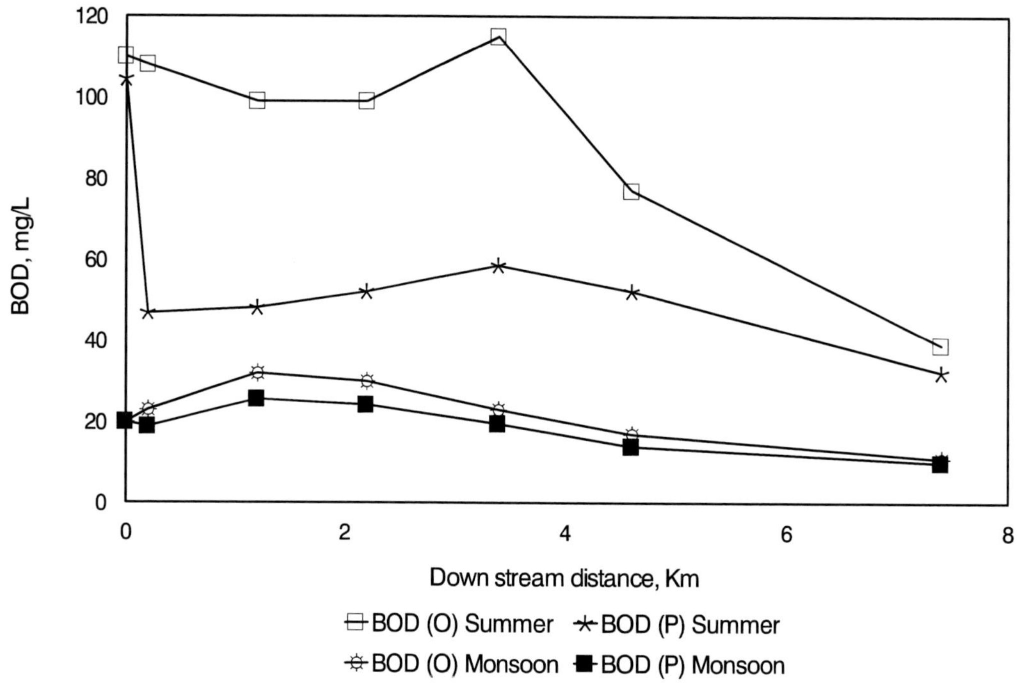

Figure 5.6. impact of addition of fresh water stream ($3.5m^3/s$ and $2.5m^3/s$) and increase in head water DO (4 mg/L) on BOD of the canal during summer and monsoon seasons.

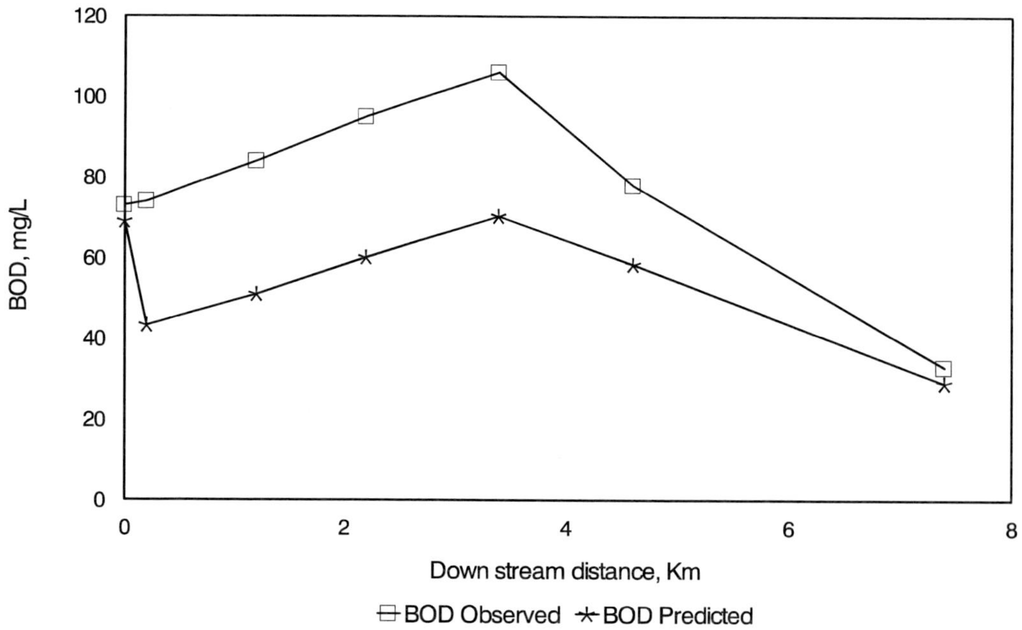

Figure 5.7. Impact of addition of fresh water stream ($3.0m^3/s$) and increase in had water DO (4 mg/L) on BOD of the canal during post monsoon season.

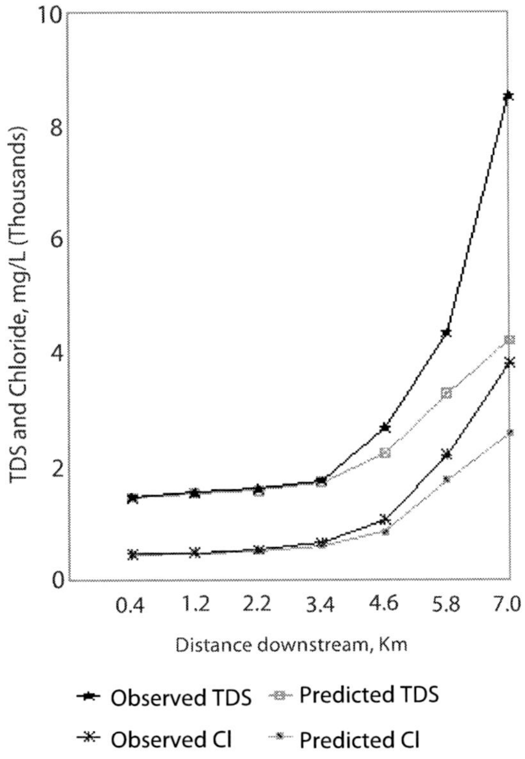

Figure 5.8. Validation curve for TDS and chloride during summer.

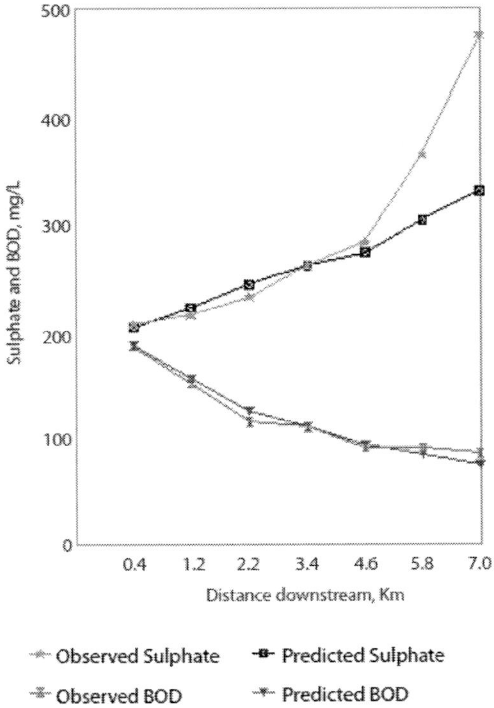

Figure 5.9. Validation curve for Sulphate and BOD during summer.

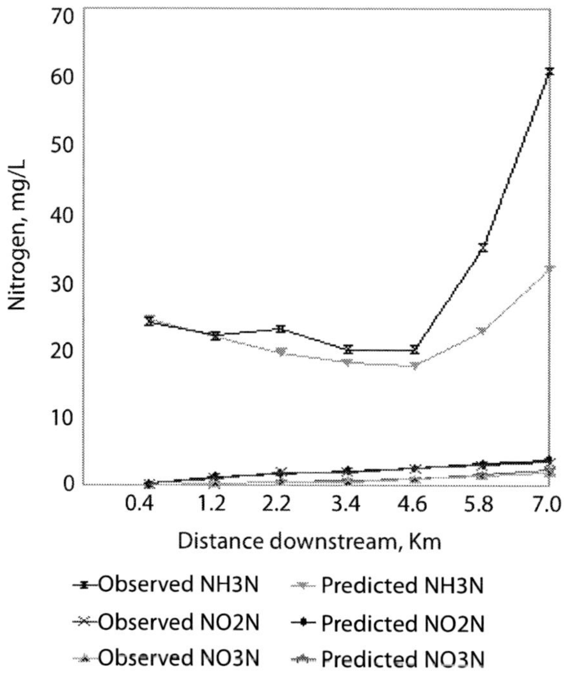

Figure 5.10. Validation curve for Nitrogen (NH_3, NO_2, NO_3) during summer.

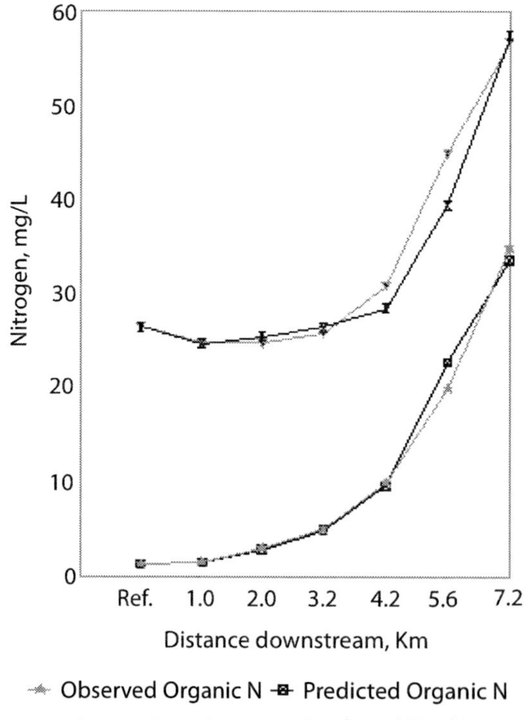

Figure 5.11. Validation curve for Organic and Total Nitrogen during summer.

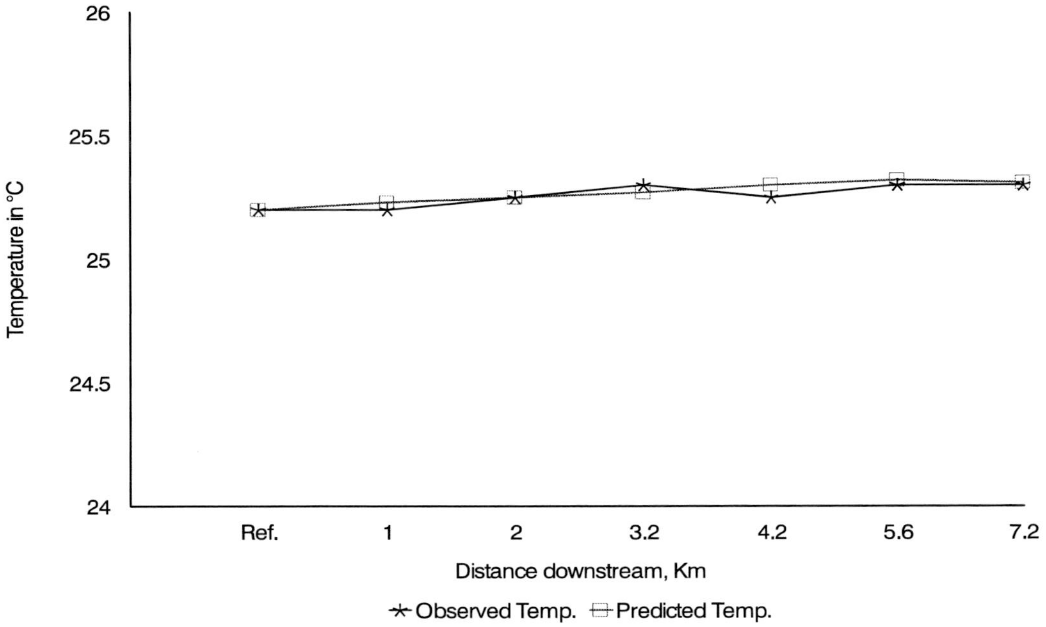

Figure 5.12. Validation curve for temperature during monsoon.

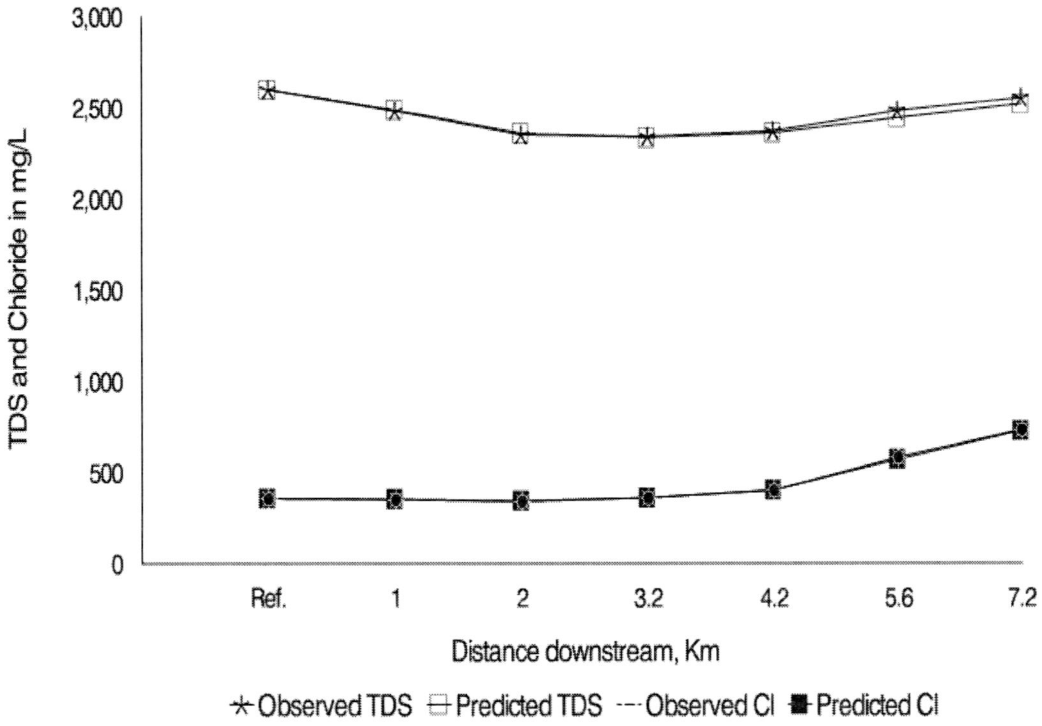

Figure 5.13. Validation curve for TDS and chloride during monsoon (observed curve for chloride totally overlaps the predicted curve).

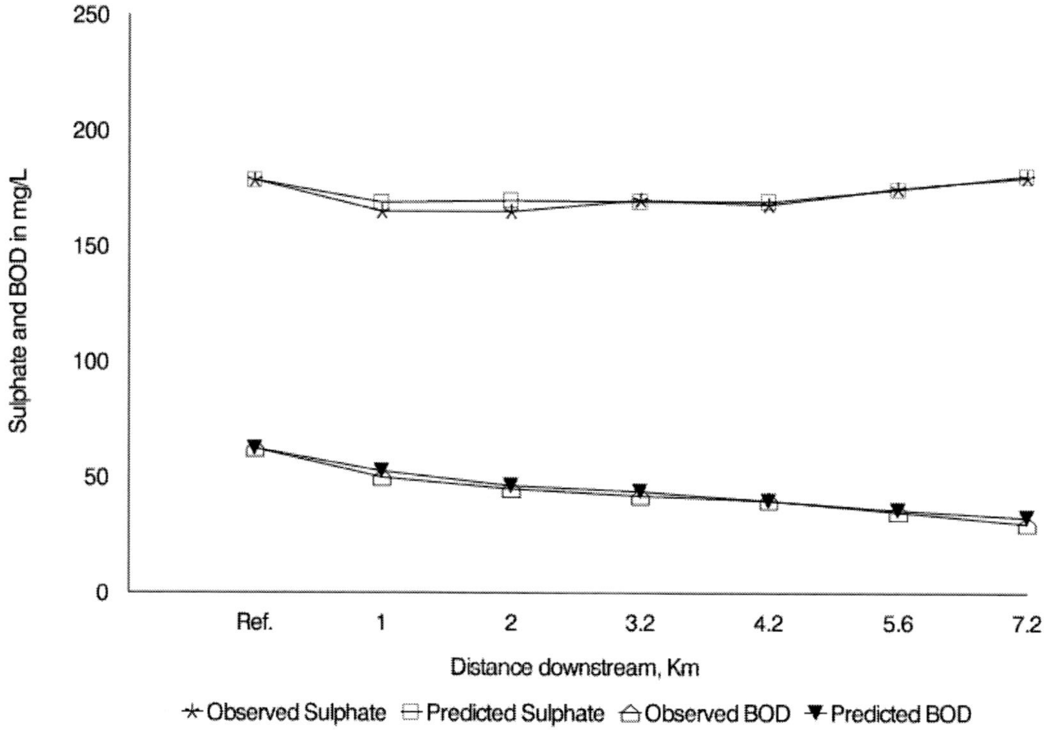

Figure 5.14. Validation curve for sulphate and BOD during monsoon.

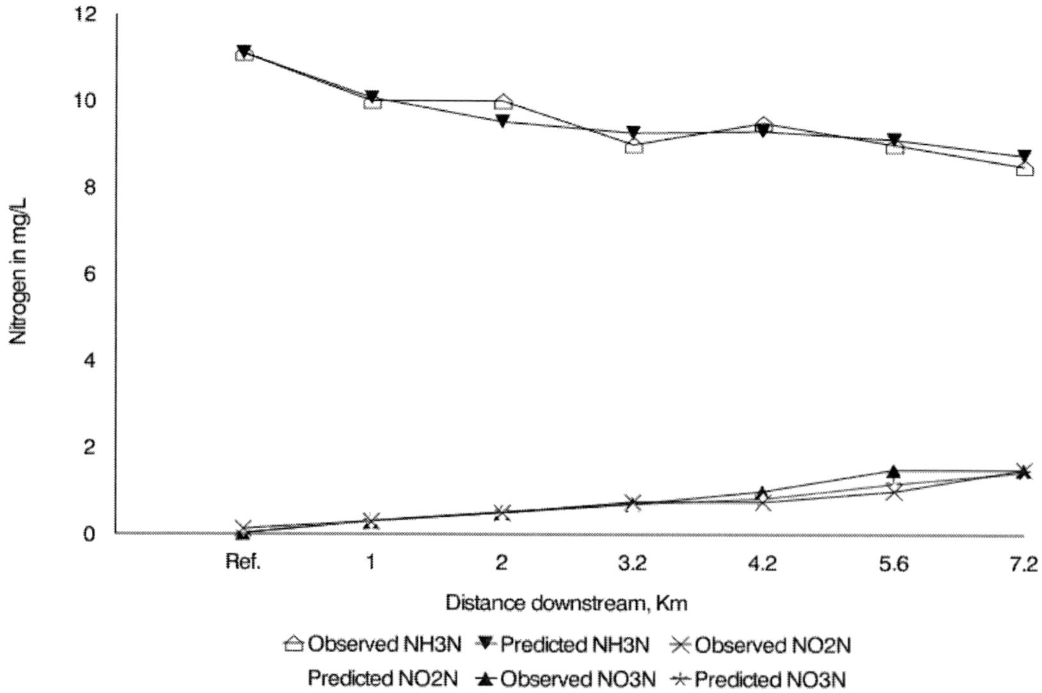

Figure 5.15. Validaiton curve for nitrogen (NH_3, NO_2, NO_3) during monsoon.

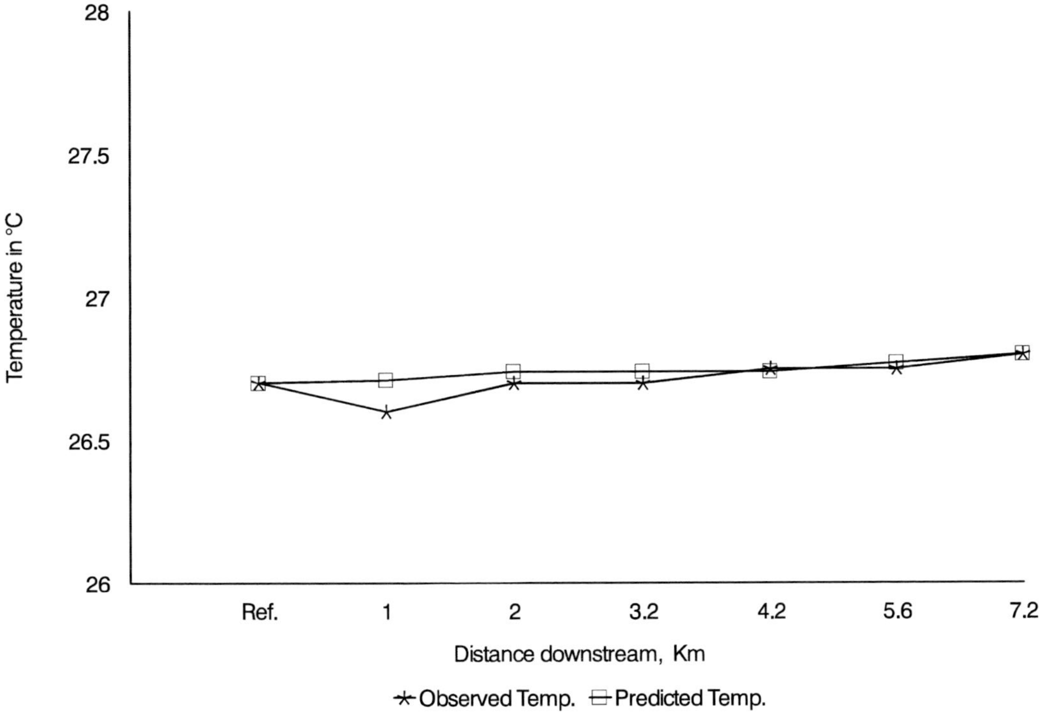

Figure 5.16. Validation curve for temperature during premonsoon.

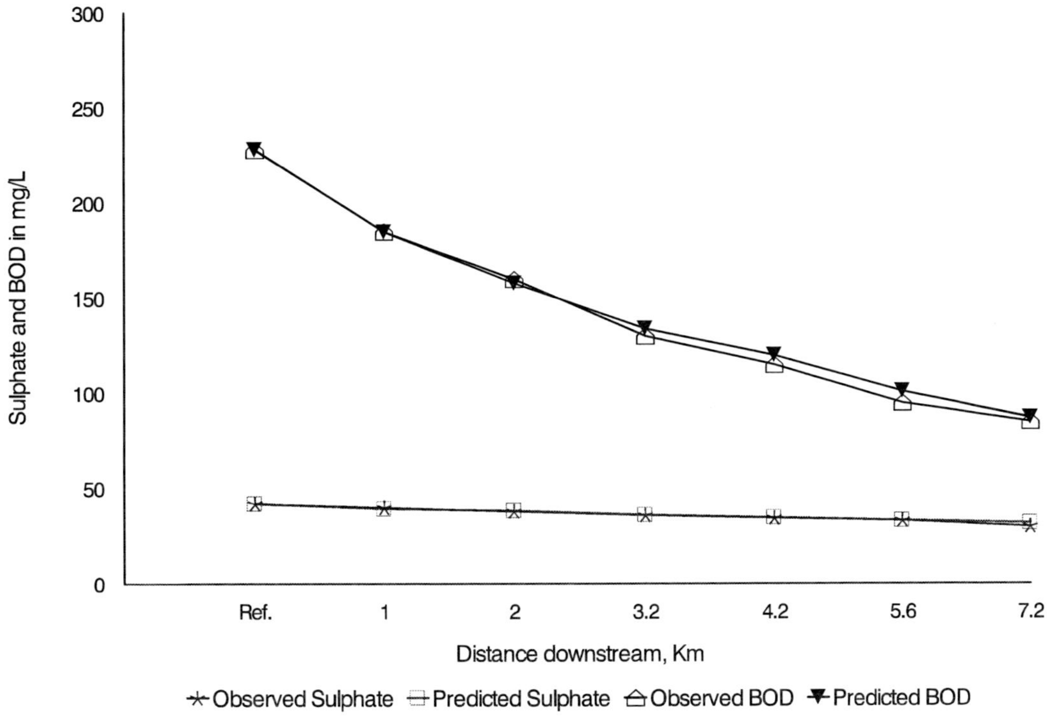

Figure 5.17. Validation curve for sulphate and BOD during premonsoon.

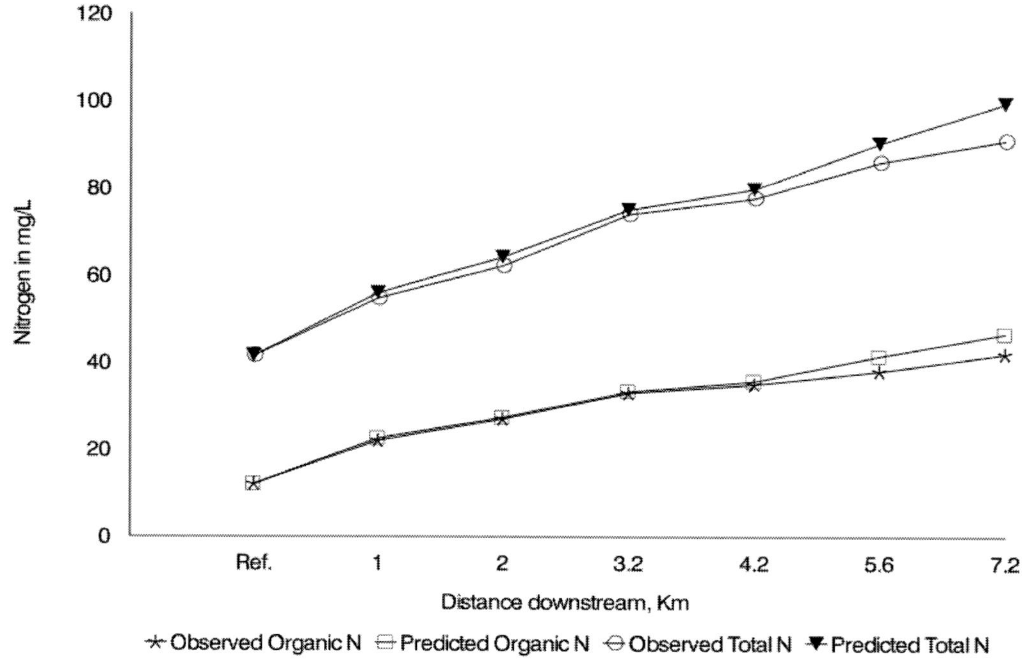

Figure 5.18. Validation curve for organic and total nitrogen during premonsoon.

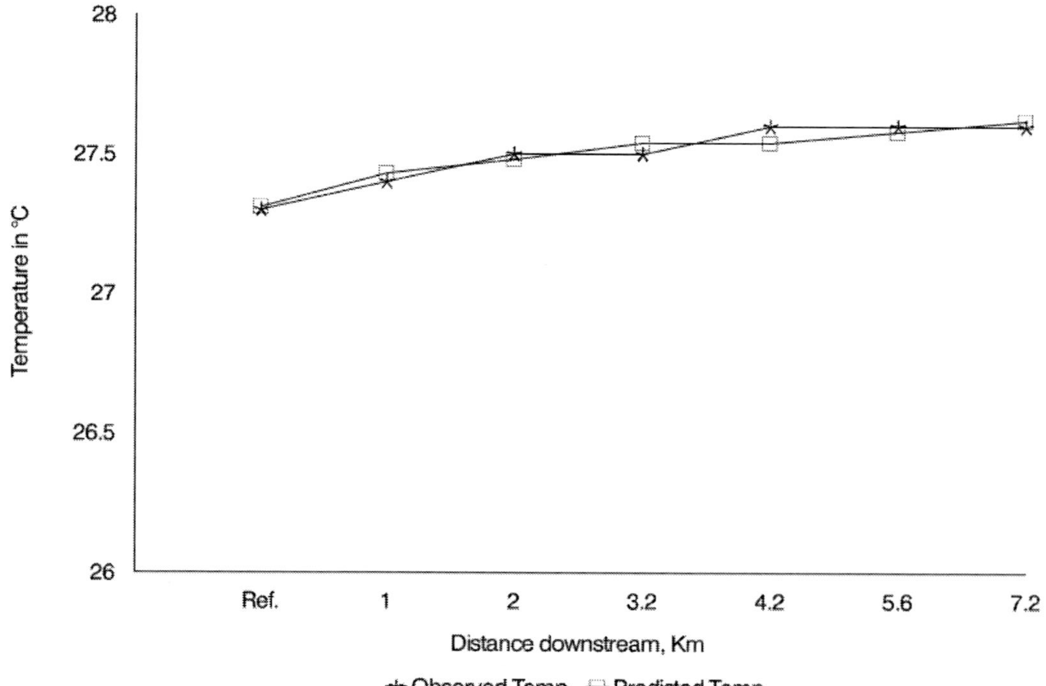

Figure 5.19. Validation curve for temperature during postmonsoon.

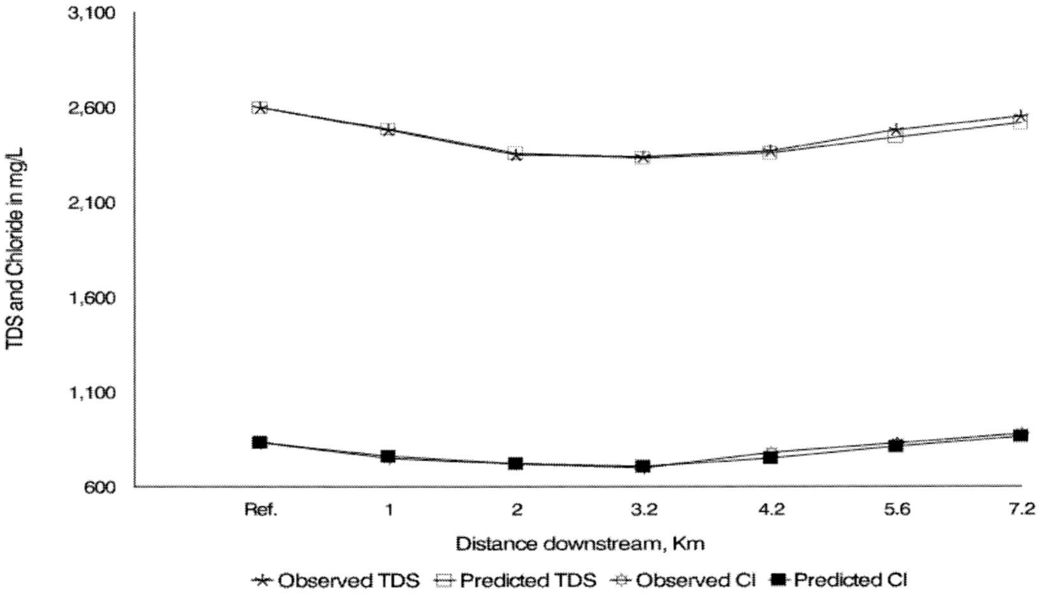

Figure 5.20. Validation curve for TDS and chloride during postmonsoon (the observed and the predicted curves for chloride almost totally overlap).

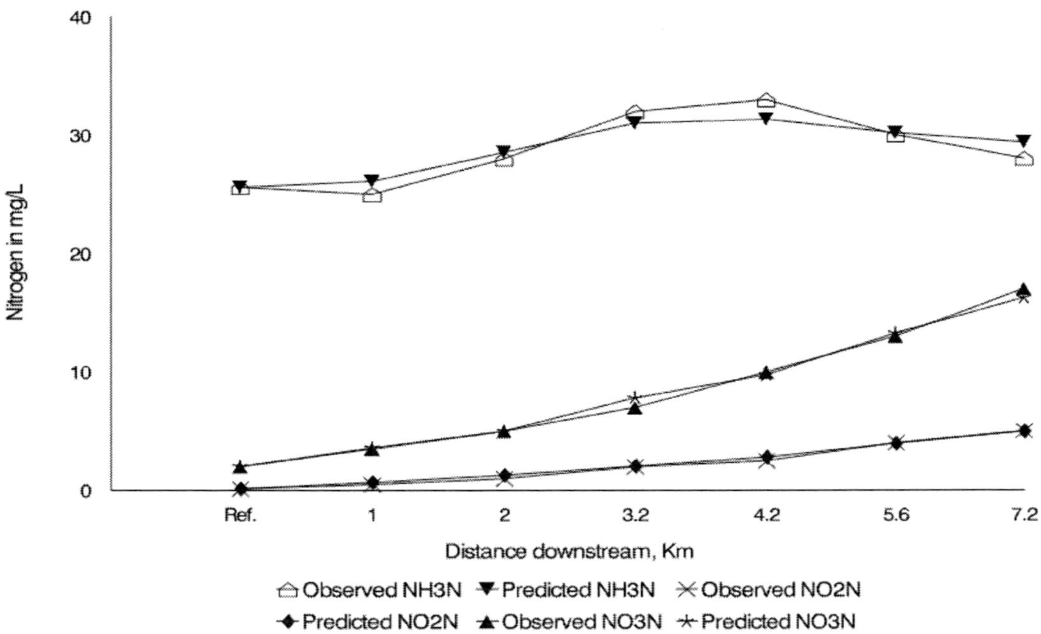

Figure 5.21. Validation curve for nitrogen (NH_3, NO_2, NO_3) during postmonsoon.

Chapter 6

The Airshed at Manali Industrial Complex

6.1. Introduction

Abstract

The study area comprising of Manali Industrial Complex (MIC) and its surrounding is situated very close to the sea on the East Coast of Peninsular India, 20 Km north of downtown chennai (Figure 6.1).

As such the meteorology of the area is subject to coastal effects characterized by rapid changes in wind directions, and high relative humidity. MIC has over ninety stacks (Tables 6.1– 6.5). Some 63% of the stacks have a height of 30-50 m, whereas 16% are 50 m or taller.

6.2. Terrain

The terrain at MIC and surrounding areas is relatively flat at the mean sea level of 3.15 m. The population of the area encompassing 10 Km radius from the reference point (CPCL) is about 40 lakhs. It includes several populous villages. Part of the land is under paddy, maize, and ragi. The terrain is interspersed with ponds, neighbourhoods, canal, and tracts of barren land. A visitor to the area is likely to notice murky plumes of emissions coming out of the stacks (plates 6.1 and 6.2) and may even pass through a plume which might have come close to the ground. The mean roughness factor is 0.3 m.

The region is semi-arid, with average annual precipitation of the order of 100 – 125 cm. There is sparse rainfall during the south-east monsoon (July-September) and more intense during north-east monsoon (October-November). The ambient temperatures are characteristic of humid tropics with very low diurnal variation; peak day temperatures generally hover between 32-37°C, rarely falling below 30°C except in December and January.

6.3. Wind Roses

The wind roses, drawn on the basis of meteorological data of nine year span (1990-98), are shown in Figures 6.1 – 6.4.

Table 6.1. Air emission stacks at *MRL * (now CPCL)*

SNo	Location	Stack	
		Height [m]	Outlet dia [m]
1.	1F1W Atmospheric heater	38.75	1.94
2.	1F1 E Atmospheric heater	38.75	1.94
3.	1F2/1F3 Vacuum heater [common]	55.58	1.20
4.	3F1 Unifier heater	15.8	0.635
5.	3F3/3F4/3F5 Platformer heater	15.2	0.952
6.	4F1 K.HDS heater	37.8	1.172
7.	5F1 HSD.HDS heater	23.01	0.686
8.	6F1 Thermal cracker heater	36.6	1.45
9.	6F2 Visbreaker heater	36.6	1.52
10.	8F1 Raffinate heater	22.17	0.504
11.	8F2 Extract heater	41.46	1.5
12.	10F1 Lube hydrofinisher heater	42.70	0.760
13.	11F1 Hydrogen heater	45.1	1.746
14.	13F1 Vaccum distillation HDS heater	23.01	0.74
15.	14F1 Foots oil heater	32.38	1.146
16.	15F1 Atmospheric heater	63.41	2.6
17.	15F2 Vaccum heater	36.35	1.75
18.	16F1 FCC feed heater	58.27	1.78
19.	Boiler I	30.48	2.438
20.	Boiler II	30.48	2.438
21.	Boiler III	30.48	2.438
22.	Boiler IV	30.00	1.80
23.	15F1B atmospheric heater	60.00	1.80
24.	15F2B atmospheric heater	49.00	1.43
25.	73F101 NMP heater	53.38	1.45
26.	73F102 Raffinate heater	48.85	1.48
27.	9F201 MEK dewaxing Unit	37.93	1.08
28.	9F301 Slack wax heater	36.40	0.68
29.	71F101 PDA heater	45.80	1.165
30.	CPP [common stack]	100.00	1.73

* The data in this and in Tables 2 – 5 is based on information provided to us. Some additions or subtractions are likely on the basis of current changes in the industries' production schedules and expansion / diversification plans.

Table 6.2. Air emission stacks at MFL, IAL and Madras Petrochem Ltd

MFL

SNo	Location	Stack	
		Height[m]	Outlet dia [m]
1.	Utility Boilers I&II	70	4.7
2.	Process condensote boiler in NH3 plant	50	1.83
3.	Primary Reformers	46	1.20
4.	Prill Tower in Urea plant	55	1.82
5.	NPK plants	42	2.5

IAL

SNo	Location	Stack	
		Height[m]	Outlet dia [m]
1.	Boiler	30	0.8
2.	Hot oil system	30	0.4
3.	Incinerator	60	0.4

Madras Petro-Chem Ltd

SNo	Location	Stack	
		Height[m]	Outlet dia [m]
1.	Boiler house	20	1.350
2.	Rotary Kiln	15.5	0.805
3.	ASD plant	45	0.600
4.	Hot air furnace	--	0.200
5.	Oleum plant	45	0.500

Table 6.3. Air emission stacks at TPL, UB Petro Products Ltd and SRF Ltd

TPL

SNo	Location	Stack	
		Height[m]	Outlet dia [m]
1.	Boiler chimney	30	1.0
2.	Hydro treater	30	0.8
3.	Fired heater pacol	30	1.0
4.	Fired heater pacol	56	3.25
5.	Common flare A	70	0.5
6.	Common flare B	70	0.2
7.	Captive power plant [2 Nos]	31.5	1.2
8.	Captive power plant	32.5	1.4

UB Petro Products Ltd

SNo	Location	Stack	
		Height[m]	Outlet dia [m]
1.	Boiler	30	0.85
2.	Flare	30	0.75
3.	Incinerator	30	---
4.	DG Set	12.5	---

SRF Ltd

SNo	Location	Stack	
		Height[m]	Outlet dia [m]
1.	Boiler [2 Nos.]	24	0.900
2.	Boiler [2 Nos.]	8	0.150
3.	Boiler [1 No.]	28	0.500

Table 6.4. Air emission stacks at Balmer Lawrie, Kothari Sugars, and IOCL

Balmer Lawrie

SNo	Location	Stack	
		Height[m]	Outlet dia [m]
1.	Steam boiler	>12	0.254
2.	Scrubber	13	0.153
3.	Thermic Fluid Heater	>12	0.254
4.	D.G. Set	9	0.203

KOTHARI SUGARS & CHEMICALS

SNo	Location	Stack	
		Height[m]	Outlet dia [m]
1.	Boiler	35	0.700
2.	Oil heater	35	0.250
3.	Flare	30	0.350

IOCL

SNo	*Location*	*Stack Detail*	
		Height[m]	*Outlet dia [m]*
1.	Steam boiler [600kg/hr] and liquid heater	25	0.875
2.	Steam boiler 8T/hr	25	0.875
3.	Steam Boiler 32T/hr [2Nos]	25	1.450
4.	Liquid Dow heater KTI	35.9	0.68
5.	Liquid Dow heater Thermax	30.5	0.75
6.	Diesel Generator sets 1100 KVA 3 sets	16.9	1.27
7.	Diesel Generator sets1525 KVA 3 sets	24.0	1.45

Table 6.5. Air emission stacks at Manali Petrochemicals Ltd. and SPIC

Manali Petrochemical Ltd.

SNo	*Location*	Stack	
		Height[m]	*Outlet dia [m]*
1.	Boiler & Incinerator	30	1.6
2.	1100 KVA DG Set	7.5	0.2
3.	1450 KVA DG Set	15	0.3
4.	550 KVA DG Set	7.5	0.15
5.	Flare	--	---

SPIC

SNo	Location	Stack	
		Height[m]	Outlet dia [m]
1.	Fusion plant [concentrate or stack]	20	0.8
2.	Fusion plant [Salt heater]	25	0.38
3.	Ammonium chloride Plant	23	0.38
4.	Ammonium chloride plant [washing tower exit]	23	0.2
5.	Hcl plant [unit 1]	18	0.15
6.	Hcl plant [unit 2]	18	0.15
7.	Hcl plant	18	0.15
8.	Steam boiler	14	0.35
9.	D.G. Set	8	0.15
10. WAD plant	WAD plant	16	0.30
11. Hydrogen Unit	Hydrogen Unit	12	0.2

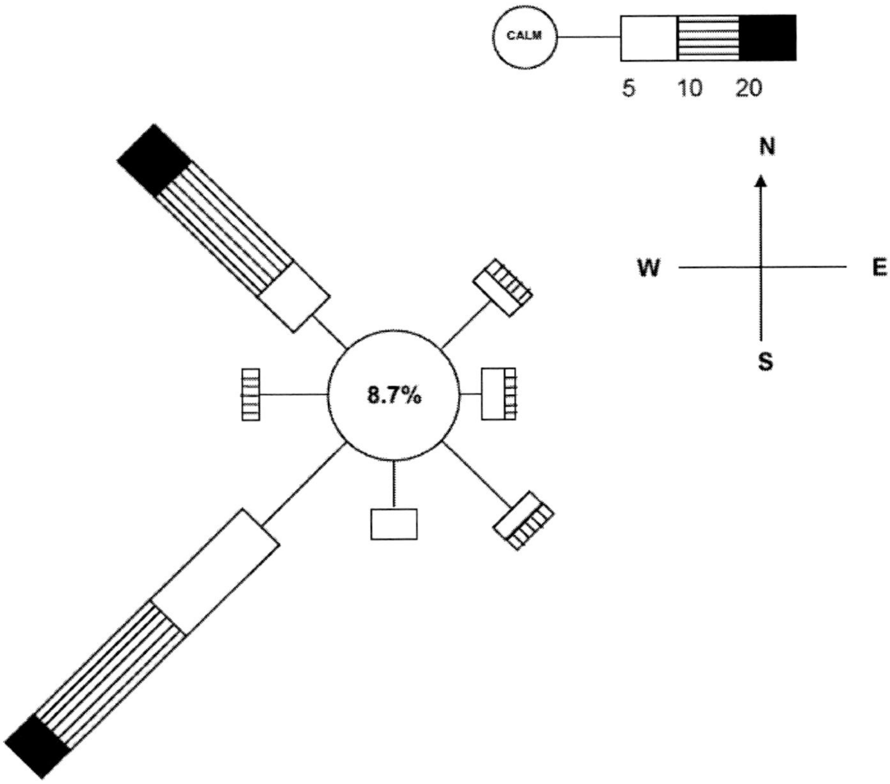

Figure 6.1. Wind roses during winter: 1990 – 1998.

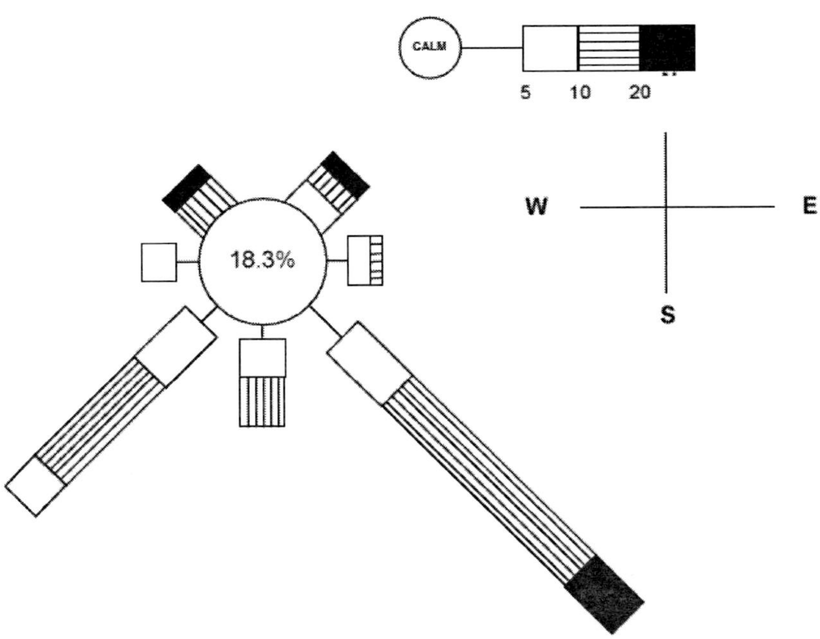

Figure 6.2. Wind roses during summer: 1990 – 1998.

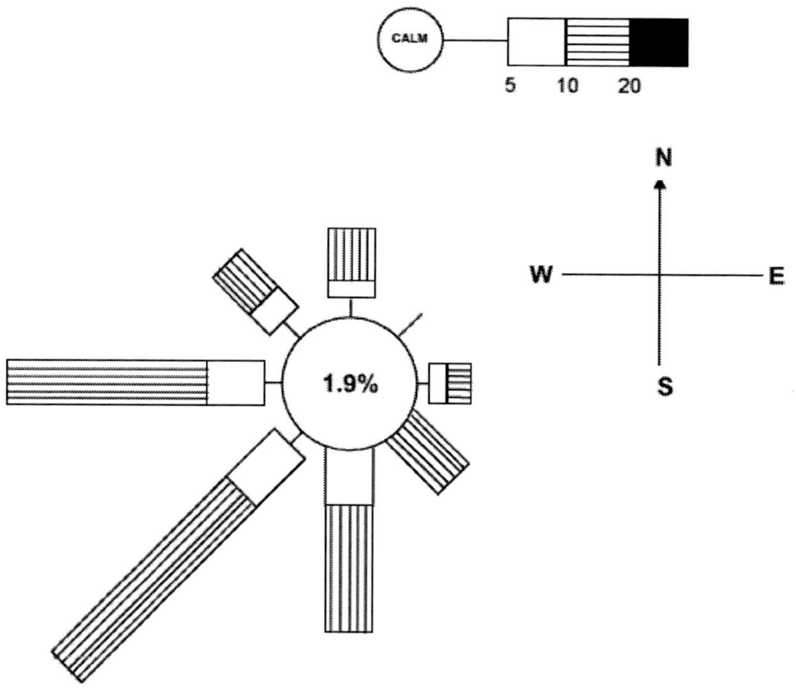

Figure 6.3. Wind roses during pre-monsoons:1990 – 1998.

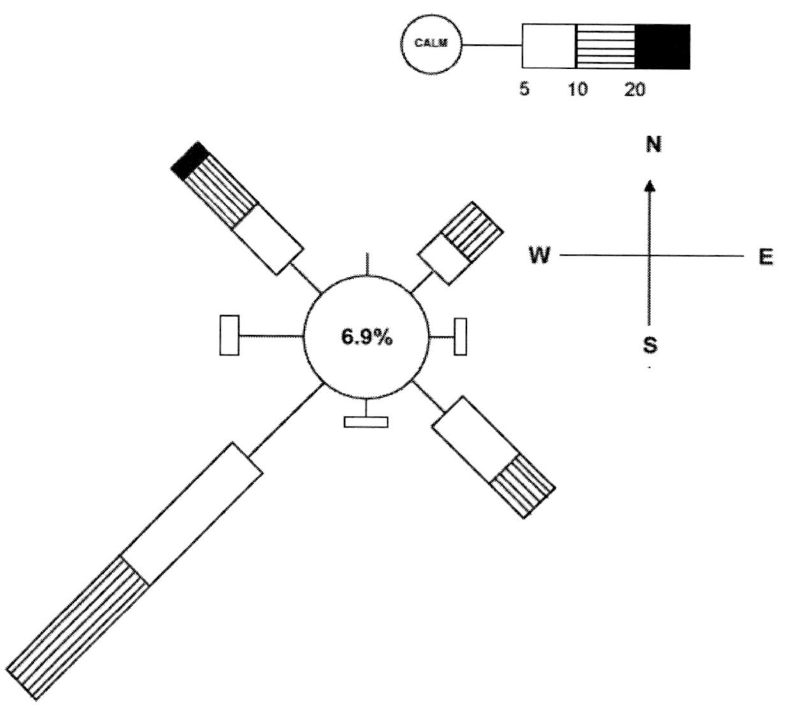

Figure 6.4. Wind roses during post-monsoons: 1990 – 1998.

Plates 6.1 and 6.2. At times dense-gas plumes fall on the ground, posing serious health hazard.

Chapter 7

SETTLING UP OF AIR QUALITY SAMPLING STATIONS

7.1. SAMPLING STATIONS

A network of eleven sampling stations, each equipped with high-volume samplers, and concurrently operated, in and around Manali Industrial complex (MIC) was set-up to monitor air quality. Based on wind roses (Chapter 6) the stations were positioned to representatively monitor ambient air quality covering industrial, residential, and 'sensitive' locations at various predominant wind directions across the year. Sampling was conducted during days as well as nights.

The sampling covered four seasons: summer (March-May), pre-monsoon (June-August), monsoon (September-November) and post-monsoon or 'winter' (December-February) for two successive years. A brief description of the sampling stations is given in Table 7.1. The locations are depicted in Figure 7.1.

A noteworthy feature of our air quality study is that we have monitored chlorine and ammonia besides SPM, SO_X and NO_X. Others have not monitored the first two parameters before, for any significant width or depth of sampling.

7.2. SAMPLING AND ANALYSIS

High volume samplers manufactured by Vayubodhan, New Delhi, were used in accordance with the procedures stipulated by American Public Health Association*, and Bureau of Indian Standards. Each contiguous high volume air sampling spanned 8 hours, and three such samples were collected in succession on every sampling day, in shifts spanning 06-14 hrs, 14-22 hrs, and 22-06 hrs. Overall, some 13900 air samples were collected, each covering 5 parameters.

The mode of analysis was also in strict confirmity with the revelant BIS stipulations as endorsed by Central Pollution Control Board (CPCB).

Very rigorous quality control was exercised to achieve authenticity in the sampling as well as analysis. To further check upon the quality of our monitoring, we had test-run a

---

* Methods of Air Sampling and Analysis, American Public Health Association, Washington; 981 pages.

sampling station by positioning it close to the National Ambiant Air Quality Monitoring (NAAQM) station set by the Central Pollution Control Board (CPCB). Typical data generated by us, in that reconnaissance, together with the NAAQM data for the corresponding periods, is presented in Table 7.2. The agreement indicates that the experimental work of this team compared favourably with the CPCB findings. A note on the findings of other agencies which had done ambient air quality surveys for preparing feasibility reports for some individual industries, is presented below.

7.3. EXISTING DATA-BASE: AN OVERVIEW

In the course of examining the 'dry data' – i.e. data generated in the past by other agencies – we came across results of AAQ (ambient air quality) surveys done by a few consultancy firms in the course of preparing feasibility reports for individual industries. The 'dry data' was carefully assessed. Whereas we do not intend to question the general utility or authenticity of the data, and believe that their findings are quite useful in their context, we nevertheless feel that *vis a vis guaging the impact of MIC industries on ambient air quality*, the said dry data may not serve much purpose. This surmise is based on the following observations: a) quite often sampling stations are located either too close to, or too far from, the emission source; b) often the stations have not been placed in predominant wind directions; c) appropriate standards have not always been used to decide upon the fitness of ambient air. Some illustrative examples are given below.

I. In one of the studies*, out of 14 stations set for monitoring, 3 were placed too close (within 0.2 Km radial distance from the zone of the most dense concentration of stacks) and 10 too far (more than 5 Km) away. As the plumes from the stacks may most often reach the ground within the area between 0.5 to 4 km radius, this positioning would miss the locations where the ambient air quality is really bad. Even then, it is seen that at station A3 10% of samples had SPM concentration more than 245 $\mu g/m^3$, thus exceeding CPCB norms of appropriate air quality (as per which 2% of the samples should not have higher-than-permissible level of any pollutant). Likewise, in another study**, four stations were placed for AAQ monitoring, out of which one was too close and one too far away.

II. Quite often the samplers have not been placed on the basis of wind directions. For example in the study mentioned above@ during September-October the predominant wind direction is SW but only two stations (A1 and A7) out of seven locations monitored during that season are in the down-wind direction. The remaining five stations are either upwind or in the crosswind direction as shown in Figure 7.2.

Similarly in the other study**, only one station is in down-wind direction out of 4 stations monitored during all the four seasons. During pre-monsoon and monsoon seasons the

* EIA study for TPL during 1993

** EIA study for MFL during 1991-92

predominant wind direction is SW but except the location MA5 all other samplers were in either upwind or crosswind directions as shown in the Figure 7.3.

iii) To assess the pollution level most of the agencies have used only the permissible limit for industrial zone and have calculated the percentage of samples polluted solely on that basis. But for AAQ stations located in residential or 'sensitive' locations, appropriate standards set for those types of locations ought to have been used. In a typical study[@] most of the samples exceeded the limit for SPM set for residential areas at locations A2, A3, A4, A5 and A7. The percentage of samples found pullulated were 70% in A2, 16% in A3, 50% in A4, 50% in A5 and 70% in A7 (Table 3-4A; page 3.22) of the said report. In another study[@] it was reported that all the samples were within CPCB limits which is not really so. For example, in the following places during summer the percentage of samples exceeded the prescribed limits for residential areas: Manali village (>90%), Sathangadu (>90%), Redhills (>90%), Kadapakkam (>90%) Thiruvottiyur (>75%), Ennore school (>75%), Sadayankuppam (>50%), Vadaperumbakkam (>50%) and Jyothinagar (>20%).

Therefore, we believe, an extensive ambient air quality survey, as done and now being reported by us, was necessary to accurately guage the impact of the Manali Industrial Complex on the airshed of the study area.

Table 7.1. A brief description of the sampling sites*

Sample No.	Sampling site	Description
MS1	Chinnasekkadu	On top of the house, backside of IOCL
MS2	Periyasekkadu	On top of the house
MS3	Manali fire station	On top of the office, 0.3 km from MFL
MS4	Manali school	On top of the school.
MS5	TNHB (Periyamathur)	On top of the office, 0.4 km from NAPCO.
MS6	Chinnamathur	On top of the house.
MS7	Ammulavoil	On top of the house.
MS8	Madhavaram	Open area on the ground, 3.5 km from IOCL.
MS9	Vaikkadu	On top of the house.
MS10	Sadayankuppam	On top of the house.
MS11	Thiruvottriyur	On top of the house, 1.5 km from CPCL.

* The samplers were placed scrupulously according to the norms, adequately above the ground level, free from lateral obstructions, and local disturbances.

[@] EIA study for TPL during 1993 ; ** EIA study for MRL during 1988-89

Table 7.2. Test runs to compare the ambient air quality data generated by CPET with the NAAQM data

Date	SO_X, μg/m³		NO_X, μg/m³		SPM, μg/m³	
	CPET	NAAQM	CPET	NAAQM	CPET	NAAQM
25/01/96	6.6	8.2	9.7	9.2	102.2	87.6
13/02/96	10.5	11.3	14.5	13.3	126.1	102.6
22/02/96	8.3	7.3	12.7	16.2	106.1	130.0
28/03/96	12.3	4.6	19.4	17.2	210.6	199.3
04/04/96	7.9	5.0	3.6	4.5	61.9	72.6
18/04/96	6.0	5.0	5.0	5.5	133.1	129.6
30/05/96	6.7	9.0	10.6	8.3	225.0	190.0
20/06/96	143.3	126.0	24.3	26.5	142.4	100.0
30/07/96	5.6	3.7	11.9	7.8	217.3	183.6
22/08/96	25.3	26.5	10.6	6.0	188.3	67.6
29/08/96	51.3	55.5	23.3	22.1	114.3	128.0
05/09/96	10.5	10.1	6.1	7.2	109.1	103.3
19/09/96	8.9	9.2	8.3	9.9	106.3	93.3
03/10/96	31.2	34.5	9.1	6.5	120.3	97.0
10/10/96	19.8	13.9	6.6	9.7	78.7	76.3
29/10/96	35.6	34.6	15.2	13.9	177.3	144.3
07/11/96	20.1	17.4	4.4	3.2	175.1	158.5
26/11/96	16.1	15.5	8.7	2.0	25.6	77.5
17/12/96	48.5	50.5	4.5	5.2	83.5	68.0

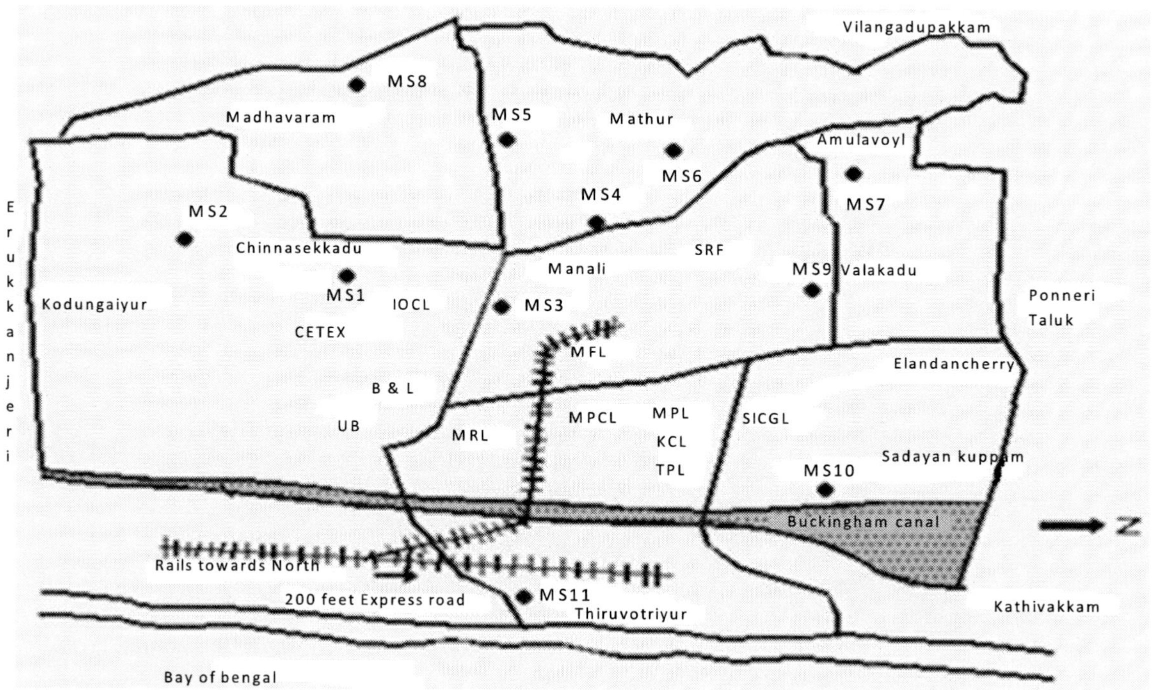

Figure 7.1. Location of air quality monitoring stations [•].

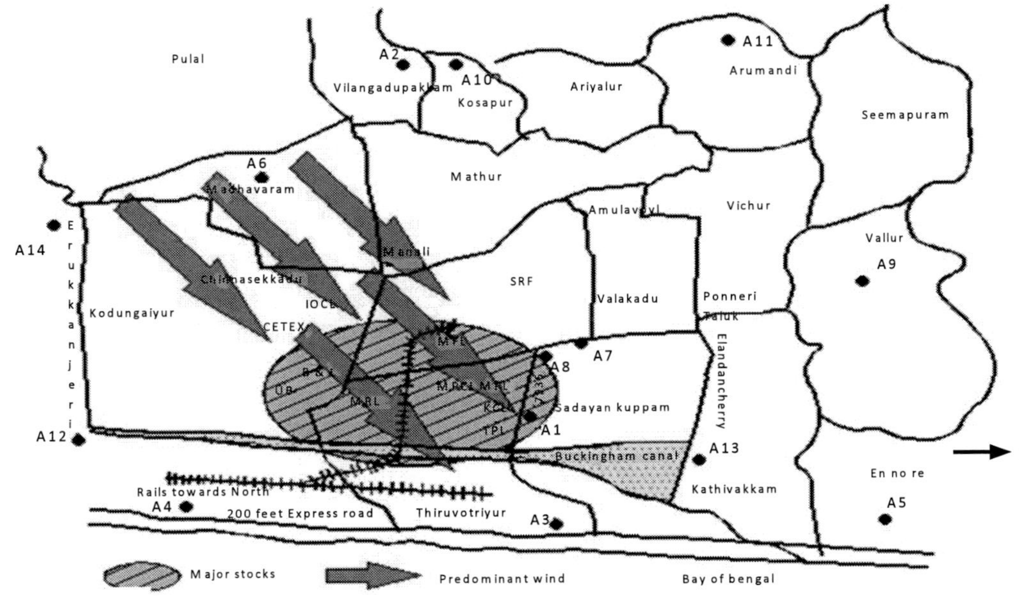

Figure 7.2. Illustrative example of Improper positioning of sampling locations with respect to predominant wind directions during premonsoon and monsoon: example I.

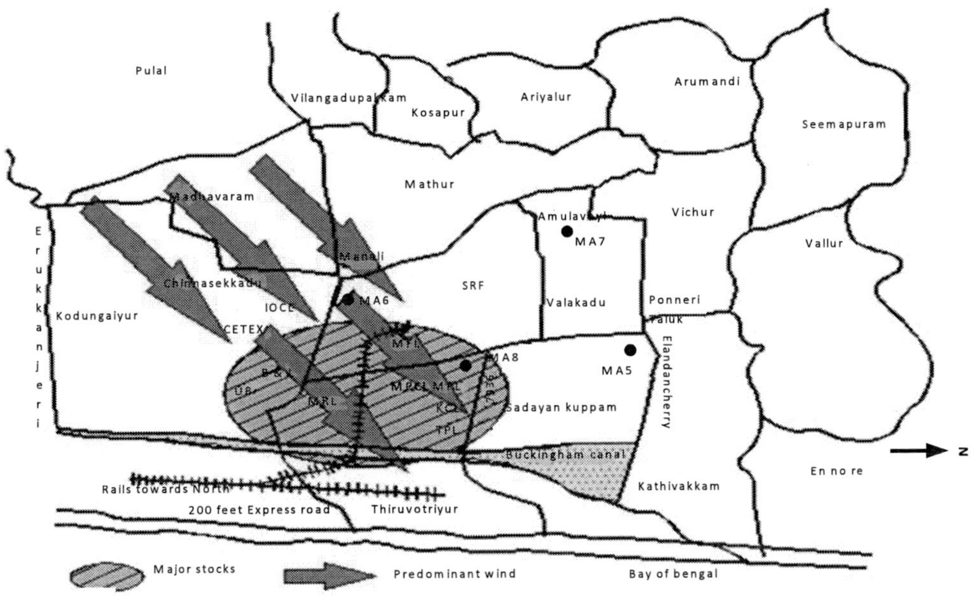

Figure 7.3. Illustrative example of Improper positioning of sampling locations with respect to predominant wind directions during premonsoon and monsoon: example II.

Chapter 8

AMBIENT AIR QUALITY

8.1. GENERAL

The detailed results of the ambient air quality monitoring are presented in Appendix-IV. The gist of the findings is discussed in this chapter.

Appendix-IV details the findings of the ambiant air quality in terms of concentrations of five different parameters in 3 different samples over each 24–hour span, taken during days as well as nights, at eleven different locations and in four different seasons. To resolve this enormous mass of information into an easy-to-comprehend and easy-to-interpret form, all the data has been processed to see the compliance of different individual parameters and different total samples with the relevant national ambiant air quality (NAAQ) standards. The standards are summarized in Table 8.1. As per CPCB (Central Pollution Control Board) norms, a location is considered polluted if 24 hourly / 8 hourly air quality values exceed the relevant standard in more than 2% of the samples [vide SO 384 (E), Air (Prevention and central of pollution) Act 1981 dt 11.4.1994 and EPA Notification GSR 176 (E), 2.4.1996].

8.2. STUDIES ON SO_x

8.2.1. So_x: Pre-Monsoon

As may be seen from Table 8.1, the national ambiant air quality standards for SO_x are: 120 $\mu g/m^3$ for industrial locations, 80 $\mu g/m^3$ for residential or rural locations, and 30 $\mu g/m^3$ for sensitive locations. Further, as per the norms of CPCB, these standards should not exceed in the relevant locations for more than 2% of the samples in a year.

But, as is revealed from Table 8.2, in ten of the eleven stations, the air quality is unacceptable vis a vis SO_x, as in more than 2% of the samples the SO_x levels are higher than the standards applicable to industrial location.

However the study area also encompasses residential locations as also ‘sensitive’ locations such as schools and hospitals (primary health centres). When we apply the air quality standards appropriate to such locations, it is revealed that in all the eleven stations the CPCB norms (of not more than 2% samples exceeding the prescribed limits), are surpassed.

The extent of SO_x pollution in terms of compliance with residential or sensitive locations is most marked at the sampling stations situated at Chinnamathur, TNHB, Madhavaram, and Vaikkadu where more than half the samples exceed the prescribed limits. In terms of the standards for sensitive locations, which are the most stringent, the non-compliance of the air samples is even more marked: 88.7% samples exceed this limit at Periyasekkadu, followed by 85.6% at Chinnamathur, and 81.7% at TNHB.

8.2.2. SO_X: Monsoon

During 'monsoon' (September – November), too, all but one sampling stations has recorded unacceptable air quality as per NAAQ standards of SO_x levels and the CBCB norm of number of above–limit samples tolerable at a given location (Table 8.3). The most liberal of the three standards, applicable to industrial locations, are exceeded to the extent of ~ 38% at Manali Fire Station and Amulavoil. At Vaikkadu and TNHB the non–compliance is in 25% and 21.4% of the samples respectively.

The residential / rural and sensitive locations in the study area receive unacceptable SO_x levels much more frequently, as reflected in the high percentage of samples exceeding the limits set for such locations at all but one of the stations.

8.2.3. SO_X: Post–Monsoon (or 'Winter')

The SOx levels are generally lesser during the post–monsoon months (December – February) due, perhaps, to the generally more unstable atmospheric conditions in these months enabling quicker dispersion of pollutants than in the pre–monsoon and monsoon months of June–November when often the sky is overcast with lesser movements of air. Even then, the fraction of samples found polluted (Table 8.4) as per the standards for industrial locations exceed CPCB norms at nine of the stations. In terms of the standards applicable to residential/rural locations and sensitive locations, the non – compliance is more blatant eventhough lesser in magnitude than in the previous two seasons.

8.2.4. SO_X: Summer

The summer (March–May) in the study area is characterized by increasing ambiant temperatures and correspondingly increasing relative humidity (Table 8.5). Further, the effect of southwest monsoon settling in the west of the western ghats off-an-on spills over to the study area which is situated east of the ghats, resulting in cloudy skies. These factors may combine to create a less favourable climate for the dispersion of air pollutants than was possible in the preceeding months, assuming that all other factors – especially the source strength – had remained constant.

During this study, the air quality with reference to SO_x in summer has, expectedly, deteriorated. All the eleven sampling stations reflect air quality which is unacceptable vis a vis SO_x levels as per the CPCB norms. The impact at Chinnasekkadu, Manali School, and Ammulavoil is particularly severe. In three other stations (Periyasekkadu, TNHB, and

Chinnamathur), too, 20% or more samples have SO_x levels higher – than – permissible for industrial locations. In terms of standards for residential / commercial and sensitive locations, the proportion of polluted samples is even higher; the situation at Chinnasekkadu, Periyasekkadu, Manali School, TNHB, Chinnamathur, and Ammulavoil being particularly bad.

8.3. Studies on NO_x

NO_x is not a pollutant of as serious concern in the study area as is SO_x. In none of the samples NOx levels higher-than-permissible for industrial locations has been recorded. During the monsoon and post-monsoon seasons, NO_x is within acceptable levels for residential/commercial locations as well (Tables 8.2-8.5). Only during summer Chinnasekkadu and Manali School, and during post-monsoon Manali School stations receive NO_x levels higher-than permissible for residential / rural locations.

The possible reason of NO_x levels being significantly lower than SO_x levels is that the former is generally contributed by vehicular exhaust and the traffic density in the study area is not excessively high. Hence the prime contributor to the ambient air pollution are industrial emissions.

8.4. Studies on SPM

At all sampling stations, except one - that too in only one of the four seasons studied- SPM levels are higher-than-permissible for residential / commercial or 'sensitive' locations (Tables 8.2 – 8.5). The sole exception is TNHB during summer. This is an exceedingly undesirable situation as several densely populated villages and other residential areas lie within the impact area studied by us. The area also contains sensitive locations such as schools and primary health centres. Indeed some of the highest SPM levels have been recorded at Manali School, during pre-monsoon (Table 8.2), when more than half the samples exceeded the standards for residential / rural locations and 87.7% samples exceed the standards for 'sensitive' location (which, by definition, Manali School is). The school has also recorded unacceptable SPM levels during all other seasons (Tables 8.2 - 8.5).

Among other stations seriously affected by SPM levels higher-than-permissible for residential / commercial locations are Manali Fire Station (especially during the post-monsoon months), Chinnasekkadu and Periyasekkadu (all year round), Ammulavoil (especially during pre-monsoon and monsoon), and Vaikkadu (more than 22% samples polluted all year round). The sensitive locations receive unacceptable SPM levels even more frequently; over half the time in majority of cases.

8.5. Overall Air Quality

The gist of the entire ambient air quality survey is presented in Table 8.6. This table has been culled from the raw data provided in Appendix IV. It may be seen that only at one of the eleven sampling locations – Manali Fire Station – the air pollution was within the acceptable limits set by CPCB during pre-monsoon and monsoon seasons. At all other sampling locations and in all seasons the air was polluted above the said norms.

The air-shed patterns are further illustrated in Figures 8.1 –8.3. In summary:

a) During pre-monsoon high concentrations of air pollutants are observed in the sampling stations situated in the northeast direction. The iso-concentration profiles for SO_x and SPM (Figures 8.1 and 8.2) indicate pollutant levels far above the limits permitted by CPCB (Central Pollution Control Board), and encompass residential areas such as Sadayankuppam, Ammulavoil and Vaikkadu.
b) Incidence of high concentration of NO_x is limited to a relatively smaller area.
c) In the post-monsoon months the concentration contours for SO_x (Figure 8.3) over the residential areas of Manali, Chinnasekkadu,
d) Periyasekkadu, Madhavaram and Selavayal indicated levels exceeding the prescribed limits of CPCB.
e) During the post-monsoon months, the concentrations of NO_x and SO_X are lower, compared to the pre-monsoon, yet these concentrations are above CPCB's prescribed limits.
f) In monsoon the iso-pleths for SO_x indicate that the pollutant levels exceed the CPCB standards. They envelop wider area than in other seasons, including residential areas of Manali, Chinnasekkadu, Sadayankuppam, Ammulavoil and Vaikkadu.
g) During summer, the rate of aerial dilution appears to be maximum and concentrations of the pollutants are lower compared to any other seasons. Even then the concentrations of SO_x are higher than the CPCB's prescribed limit.

8.6. Ammonia and Chlorine

Ambient air quality standards for SPM, NO_X and SO_X are available with CPCB. But for Cl_2 and NH_3 the ambient air quality standards are not available as yet. We tried hard to procure these from CPCB and USEPA (United States Environmental Protection Agency) but were not successful. For this reason we have not included these two variables in our computations of overall air quality discussed in the preceding section. The discussion in this section is indicative of how the air quality picture *may* look once we take these two variables also into account.

For the present, we have derived tentative standards for these chemicals on the basis of the logic that has gone in the setting up of standards for SO_X and NO_X. We found that ambient standard (AS) for SO_X and NO_X are related to their threshold lethal values (TLVs) by the empirical formulae:

AS1 = 0.025 * TLV (industrial area)
AS2 = 0.015 * TLV (residential area)
AS3 = 0.006 * TLV (sensitive area)

The ambient standards derived on the basis of these formulae for ammonia and chlorine are given in Table 8.7. We have termed them "*Tentative CPET (TCPET) standards*", where CPET represents Centre for Pollution Control and Energy Technology. For interpreting the results we have used these TCPET standards.

8.6.1. Ammonia

The gist of the monitoring is as follows:

a) In a large number of situations (Tables 8.8-8.11) more than 50% samples had ammonia levels exceeding the TCPET limits.
b) The most affected areas due to high NH_3 concentration are Vaikkadu, Chinnasekkadu, Ammulavoil and Manali (Figure 8.4).

All-in-all the ambient air concentration of ammonia was significantly higher than the TCPET standards.

8.6.2. Chlorine

The monitoring reveals that:

a) During monsoon the concentration of Cl_2 exceeds upto 3 times the TCPET limits (Figure 8.5).
b) The areas under the impact of severe pollution load due to Cl_2 are same as under NH_3 concentration (Tables 8.8-8.11).

8.7. Overall Seasonal Pattern

The following patterns emerge from the mass of data generated by us (Appendix – IV)

1) In general the concentrations of various pollutants are at their highest during post-monsoon and lowest during summer.
2) The main pollutants at the residential and industrial area are NH_3, Cl_2, SO_X and SPM.
3) During all four seasons (throughout the year) the air of MIC and the surrounding areas (14 km radius) is severely polluted when assessed on the basis of prevailing CPCB norms. It is not uncommon to final dense-gas plumes very close to the ground (Plates 8.1 and 8.2); one can be literally walking through or driving through such very unhealthy plumes.

Table 8.1. National ambiant air quality standards as per Air Act and EPA Notification GSR 176 E of 2 April 1996

Pollutant *	Concentration in ambiant air, µg / m ^{3}		
	Sensitive location	Residential, rural & other	Industrial
SO_x	30	80	120
NO_x	30	80	120
SPm	100	200	500
Ammonia	@	@	@
Chlorine	@	@	@

* 24 hourly / 8 hourly values should be met 98% of the time in a year. However, 2% of the time it may exceed but not on two consecutive days

@ The standards for these two have not been announced as yet.

Table 8.2. Percentage of samples polluted during pre-monsoon; cf Appendix A 4.1 – A 4.12

Sl. No	Location	Samples during	%of samples polluted											
			SO_X			NO_X			SPM			Over-all quality		
			I	R	S	I	R	S	I	R	S	I	R	S
1	Chinnasekkadu	Day	5.7	24.5	56.6	0.0	0.0	34.0	0.0	17.0	41.5	5.7	34.0	66.0
		Night	7.5	30.2	66.0	0.0	0.0	37.7	0.0	22.6	47.2	7.5	41.5	73.6
		Total	6.6	27.4	61.3	0.0	0.0	35.8	0.0	19.8	44.3	6.6	37.7	69.8
2	Periyasekkadu	Day	7.5	22.6	86.8	0.0	0.0	41.5	0.0	18.9	41.5	7.5	28.3	88.7
		Night	7.5	24.5	90.6	0.0	0.0	50.9	0.0	30.2	52.8	7.5	32.1	94.3
		Total	7.5	23.6	88.7	0.0	0.0	46.2	0.0	24.5	47.2	7.5	30.2	91.5
3	Manali fire station	Day	0.0	7.5	66.0	0.0	7.5	71.7	1.9	52.8	84.9	1.9	56.6	84.9
		Night	0.0	22.6	79.2	0.0	1.9	77.4	0.0	58.5	90.6	0.0	62.3	94.3
		Total	0.0	15.1	72.6	0.0	4.7	74.5	0.9	55.7	87.7	0.9	59.4	89.6
4	Manali school	Day	24.5	37.7	71.7	0.0	0.0	28.3	0.0	24.5	50.9	24.5	56.6	83.0
		Night	18.9	28.3	60 4	0.0	0.0	22.6	0.0	18.9	41.5	18.9	39.6	67.9
		Total	21.7	33.0	66.0	0.0	0.0	25.5	0.0	21.7	46.2	21.7	48.1	75.5
5	TNHB	Day	17.3	53.8	76.9	0.0	0.0	34.6	0.0	25.0	69.2	17.3	61.5	82.7
		Night	23.1	69.2	86.5	0.0	0.0	48.1	0.0	34.6	75.0	23.1	73.1	86.5
		Total	20.2	51.2	81.7	0.0	0.0	41.3	0.0	29.8	72.1	20.2	67.3	84.6

Sl. No	Location	Samples during	%of samples polluted											
			SO_X			NO_X			SPM			Over-all quality		
			I	R	S	I	R	S	I	R	S	I	R	S
6	Chinnamathur	Day	21.2	48.1	82.7	0.0	0.0	50.0	0.0	11.5	78.8	21.2	53.8	86.5
		Night	26.9	57.7	88.4	0.0	0.0	73.1	0.0	25.0	82.7	26.9	59.6	92.3
		Total	24.0	52.9	85.6	0.0	0.0	61.5	0.0	18.3	80.8	24.0	56.7	89.4
7	Amulavoil	Day	9.4	41.5	54.7	0.0	0.0	30.1	0.0	39.6	49.1	9.4	47.2	62.3
		Night	15.1	45.3	62.3	0.0	0.0	35.8	0.0	34.0	47.2	15.1	52.8	73.6
		Total	12.3	43.4	58.5	0.0	0.0	33.0	0.0	36.8	48.1	12.3	50.0	67.9
8	Madhavaram	Day	3.8	48.1	57.7	0.0	0.0	40.4	0.0	23.1	48.1	3.8	61.5	73.1
		Night	11.5	53.8	73.1	0.0	0.0	34.6	0.0	28.8	53,8	11.5	69.2	76.9
		Total	7.7	51.0	65.4	0.0	0.0	37.5	0.0	26.0	51.0	7.7	65.4	75.0
9	Vajakkadu	Day	11.3	49.1	73.6	0.0	0.0	32.1	0.0	34.0	56.6	11.3	54.7	79.2
		Night	17.0	54.7	79.2	0.0	0.0	39.6	0.0	35.8	66.0	17.0	60.4	81.1
		Total	14.2	51.9	76.4	0.0	0.0	35.8	0.0	34.9	61.3	14.2	57.5	80.2
10	Sadayankuppam	Day	5.7	34.0	56.6	0.0	0.0	34.0	0.0	22.6	47.2	5.7	37.7	60.4
		Night	11.3	39.6	66.0	0.0	0.0	35.8	0.0	18.9	52.8	11.3	43.4	67.9
		Total	8.5	36.8	61.3	0.0	0.0	34.9	0.0	20.8	50.0	8.5	45.6	64.2
11	Thiruvolriyur	Day	18.9	43.4	60.4	0.0	0.0	37.7	0.0	28.3	56.6	18.9	52.8	66.0
		Night	13.2	37.7	56.6	0.0	0.0	30.2	0.0	24.5	49.1	13.2	49.1	60.4
		Total	16.0	45.6	58.5	0.0	0.0	34.0	0.0	26.4	52.8	16.0	50.9	63.2

Table 8.3. Percentage of samples polluted during monsoon; cf Appendix A 4.13 – A 4.24

Sl. No	Location	Samples during	%of samples polluted											
			SO_X			NOX			SPM			Overall quality		
			I	R	S	I	R	S	I	R	S	I	R	S
1	Chinnasekkadu	Day	5.8	38.5	84.5	0.0	0.0	38.4	0.0	26.9	73.1	5.8	48.1	90.3
		Night	11.8	47.1	94.1	0.0	0.0	43.1	0.0	35.3	84.3	11.8	54.9	94.1
		Total	8.7	42.7	89.3	0.0	0.0	40.7	0.0	31.1	78.6	8.7	51.4	92.2
2	Periyasekkadu	Day	9.6	42.3	76.9	0.0	0.0	34.6	0.0	23.1	61.5	9.6	53.8	84.6
		Night	15.7	47.1	84.3	0.0	0.0	39.2	0.0	29.4	68.6	15.7	58.8	94.1
		Total	12.6	44.7	80.6	0.0	0.0	36.9	0.0	26.2	65.0	12.6	56.3	89.3
3	Manali fire station	Day	0.0	5.8	36.5	0.0	0.0	17.6	0.0	30.1	57.7	0.0	34.6	76.9
		Night	0.0	17.6	58.8	0.0	0.0	19.6	0.0	49.0	62.7	0.0	58.8	90.2
		Total	0.0	11.7	47.6	0.0	0.0	18.4	0.0	39.8	60.2	0.0	46.6	83.5

Table 8.3. (Continued)

Sl. No	Location	Samples during	%of samples polluted											
			SO_X			NOX			SPM			Overall quality		
			I	R	S	I	R	S	I	R	S	I	R	S
4	Manali school	Day	34.6	38.5	100	0.0	0.0	11.5	0.0	23.1	67.3	34.6	50.0	100
		Night	43.1	58.8	100	0.0	0.0	19.6	0.0	19.6	78.4	58.8	60.8	100
		Total	38.8	48.5	100	0.0	0.0	15.5	0.0	21.4	72.8	46.6	55.3	100
5	TNHB	Day	19.2	69.2	90.4	0.0	0.0	59.6	0.0	34.6	73.1	19.2	73.1	94.2
		Night	23.5	78.4	96.1	0.0	0.0	68.6	0.0	41.2	88.2	23.5	84.3	98.0
		Total	21.4	73.8	93.2	0.0	0.0	64.1	0.0	37.9	80.6	21.4	78.6	96.1
6	Chinnamathur	Day	11.5	19.2	28.8	0.0	0.0	7.7	0.0	21.1	25.0	11.5	28.8	34.6
		Night	15.7	25.5	35.2	0.0	0.0	15.7	0.0	19.6	29.4	15.7	35.3	39.2
		Total	13.5	22.3	32.0	0.0	0.0	11.7	0.0	20.4	27.2	13.5	32.0	36.9
7	Amulavoil	Day	30.8	67.3	84.6	0.0	0.0	15.3	0.0	23.1	86.5	30.8	71.1	92.3
		Night	46.2	92.3	100	0.0	0.0	23.1	0.0	44.2	100	46.2	96.2	100
		Total	38.5	79.8	92.3	0.0	0.0	19.4	0.0	33.7	93.3	38.5	83.7	96.2
8	Madhavaram	Day	3.8	38.5	69.2	0.0	0.0	53.8	0.0	30.8	73.1	3.8	48.1	76.9
		Night	9.8	45.1	78.4	0.0	0.0	58.8	0.0	33.3	82.4	9.8	54.9	88.2
		Total	6.8	41.7	73.8	0.0	0.0	56.3	0.0	32.0	77.7	6.8	51.5	82.5
9	Vajakkadu	Day	23.1	53.8	96.2	0.0	0.0	11.5	0.0	23.1	57.7	23.1	59.6	96.2
		Night	26.9	57.7	100	0.0	0.0	23.1	0.0	34.6	67.3	26.9	71.2	100
		Total	25.0	55.8	98.1	0.0	0.0	17.3	0.0	28.8	62.5	25.0	65.4	98.1
10	Sadayankuppam	Day	9.6	48.1	76.9	0.0	0.0	19.2	0.0	28.8	73.1	9.6	53.8	84.6
		Night	13.5	55.7	86.5	0.0	0.0	15.4	0.0	30.8	76.9	13.5	61.,5	88.5
		Total	11.5	51.9	81.7	0.0	0.0	17.3	0.0	29.8	75.0	11.5	57.6	86.5
11	Thiruvolriyur	Day	7.7	28.8	42.3	0.0	0.0	34.6	0.0	23.1	36.5	7.7	34.6	50.0
		Night	3.8	23.1	38.5	0.0	0.0	28.8	0.0	19.2	38.5	3.8	30.8	42.3
		Total	5.8	26.0	40.4	0.0	0.0	31.7	0.0	37.5	37.5	5.8	32.7	46.2

Table 8.4. Percentage of samples polluted during post-monsoon; cf Appendix A 4.25 – A 4.36

Sl. No	Location	Samples during	%of samples polluted											
			SOX			NOX			SPM			Over-all quality		
			I	R	S	I	R	S	I	R	S	I	R	S
1	Chinnasekkadu	Day	19.2	48.1	57.7	0.0	0.0	38.5	0.0	25.0	42.3	19.2	53.8	61.5
		Night	11.5	57.7	67.3	0.0	0.0	19.2	0.0	28.8	48.1	11.5	67.3	73.1
		Total	15.4	52.9	62.5	0.0	0.0	28.8	0.0	26.9	45.4	15.4	60.6	67.3
2	Periyasekkadu	Day	9.6	44.2	53.8	0.0	0.0	28.8	0.0	24.0	73.1	9.6	50.0	76.9
		Night	15.4	50.0	57.7	0.0	0.0	38.5	0.0	28.8	71.2	15.4	73.1	80.8
		Total	12.5	47.1	55.8	0.0	0.0	33.7	0.0	26.9	72.1	12.5	61.5	78.8
3	Manali fire station	Day	9.6	15.4	34.6	0.0	0.0	19.2	0.0	19.2	57.7	9.6	28.8	80.8
		Night	7.7	9.6	38.5	0.0	0.0	9.6	0.0	21.2	61.5	7.7	23.1	69.2
		Total	8.7	12.7	36.5	0.0	0.0	14.4	0.0	20.2	59.6	8.7	26.0	75.0
4	Manali school	Day	3.8	17.3	75.0	0.0	0.0	38.5	0.0	57.7	100	3.8	65.4	100
		Night	19.2	38.5	80.8	0.0	0.0	23.1	0.0	67.3	96.2	19.2	61.5	96.2
		Total	11.5	27.9	77.9	0.0	0.0	30.8	0.0	62.5	98.1	11.5	63.5	98.1
5	TNHB	Day	0.0	9.6	38.5	0.0	0.0	63.5	0.0	26.9	61.5	0.0	30.5	92.3
		Night	11.5	19.2	48.1	0.0	0.0	67.3	0.0	30.8	69.2	11.5	44.2	84.6
		Total	5.8	14.4	43.3	0.0	0.0	65.4	0.0	28.8	65.4	5.8	37.5	88.5
6	Chinnamathur	Day	11.5	38.5	67.3	0.0	0.0	28.8	0.0	28.8	61.5	11.5	48.1	73.1
		Night	15.4	48.1	76.9	0.0	0.0	30.8	0.0	32.7	76.9	15.4	53.8	86.5
		Total	13.5	43.3	72.1	0.0	0.0	29.8	0.0	30.8	69.2	13.5	51.0	79.8
7	Amulavoil	Day	0.0	23.5	54.9	0.0	0.0	19.6	0.0.	15.7	39.2	0.0	25.5	68.6
		Night	0.0	27.5	58.8	0.0	0.0	29.4	0.0	13.7	49.0	0.0	31.4	78.4
		Total	0.0	25.5	56.9	0.0	0.0	24.5	0.0	14.7	44.1	0.0	28.4	73.5
8	Madhavaram	Day	11.5	48.1	57.6	0.0	0.0	48.1	0.0	5.8	67.3	11.5	51.9	67.3
		Night	9.6	53.8	61.5	0.0	0.0	46.2	0.0	3.8	57.7	9.6	53.8	61.5
		Total	10.6	51.0	59.6	0.0	0.0	47.1	0.0	4.8	62.5	10.6	52.9	64.4
9	Vajakkadu	Day	0.0	27.5	39.2	0.0	0.0	45.1	0.0	21.6	39.2	0.0	31.4	54.9
		Night	0.0	31.4	47.1	0.0	0.0	43.1	0.0	23.5	41.2	0.0	39.2	58.8
		Total	0.0	29.4	43.1	0.0	0.0	44.1	0.0	22.5	40.2	0.0	35.3	56.9
10	Sadayankuppa m	Day	5.9	31.4	58.8	0.0	0.0	29.4	0.0	19.6	39.2	5.9	35.3	62.7
		Night	11.8	39.2	64.7	0.0	0.0	33.3	0.0	23.5	49.0	11.8	39.2	68.7
		Total	8.8	35.3	61.8	0.0	0.0	31.4	0.0	21.6	44.1	8.8	37.3	65.7
11	Thiruvolriyur	Day	2.0	2.0	49.0	0.0	0.0	39.2	0.0	15.7	41.2	2.0	25.5	54.9
		Night	3.9	3.9	35.3	0.0	0.0	29.4	0.0	15.7	35.3	3.9	23.5	39.2
		Total	2.9	2.9	42.2	0.0	0.0	34.3	0.0	15.7	38.2	2.9	24.5	47.1

Table 8.5. Percentage of samples polluted during summer; the columns of I, R, and S represent samples polluted as per norms for industrial, residential, and sensitive locations respectively; cf Appendix A 4.37-A 4.48

Sl. No	Location	Samples during	%of samples polluted											
			SO_X			NOX			SPM			Over-all quality		
			I	R	S	I	R	S	I	R	S	I	R	S
1	Chinnasekkadu	Day	28.3	52.8	100	0.0	3.7	52.8	0.0	22.6	66.0	28.3	60.4	100
		Night	34.6	67.3	100	0.0	7.7	57.7	0.0	25.0	71.2	34.6	73.1	100
		Total	31.4	60.0	100	0.0	5.7	55.2	0.0	23.8	68.6	31.4	66.7	100
2	Periyasekkadu	Day	18.9	37.7	94.3	0.0	0.0	49.1	0.0	22.6	90.6	18.9	41.5	96.2
		Night	23.1	48.1	98.1	0.0	1.9	55.8	0.0	30.8	100	23.1	53.8	100
		Total	21.0	42.9	96.2	0.0	1.0	52.4	0.0	26.7	95.2	21.0	47.6	98.1
3	Manali fire station	Day	37.7	66.0	90.6	0.0	11.5	66.0	0.0	37.7	52.8	37.7	79.2	94.3
		Night	34.6	69.2	86.5	0.0	0.0	53.8	0.0	34.6	50.0	34.6	73.1	90.4
		Total	36.2	67.6	88.6	0.0	5.7	60.0	0.0	36.2	51.4	36.2	84.6	92.4
4	Manali school	Day	7.5	28.3	75.5	0.0	0.0	66.0	0.0	37.7	86.8	7.5	41.5	94.3
		Night	3.8	19.2	73.1	0.0	0.0	57.7	0.0	30.8	80.8	3.8	30.8	92.4
		Total	5.7	23.8	74.3	0.0	0.0	61.9	0.0	34.3	83.8	5.7	36.2	93.3
5	TNHB	Day	26.9	57.7	100	0.0	0.0	50.0	0.0	0.0	76.9	26.9	57.7	100
		Night	15.4	50.0	96.2	0.0	1.9	48.1	0.0	0.0	82.7	15.4	50.0	100
		Total	21.2	53.8	98.1	0.0	1.0	49.0	0.0	0.0	79.8	21.2	53.8	100
6	Chinnamathur	Day	23.1	48.1	92.3	0.0	0.0	38.5	0.0	13.5	50.0	23.1	57.7	100
		Night	26.9	51.9	100	0.0	3.8	50.0	0.0	19.2	67.3	26.9	67.3	100
		Total	25.0	50.0	96.2	0.0	1.9	44.2	0.0	16.3	58.7	25.0	62.5	100
7	Amulavoil	Day	24.5	49.1	84.9	0.0	0.0	47.2	0.0	11.3	50.9	24.5	56.6	92.5
		Night	38.5	73.1	100	0.0	0.0	57.7	0.0	23.1	65.4	38.5	73.1	100
		Total	31.4	61.0	92.4	0.0	0.0	52.4	0.0	17.1	58.1	31.4	64.8	96.2
8	Madhavaram	Day	3.8	38.4	53.8	0.0	0.0	48.1	0.0	9.6	48.1	3.8	42.3	57.7
		Night	5.8	34.6	57.7	0.0	0.0	53,8	0.0	15.4	53.8	5.8	46.2	61.5
		Total	4.8	36.5	55.8	0.0	0.0	51.0	0.0	12.5	51.0	4.8	44.2	59.6
9	Vajakkadu	Day	7.5	39.6	60.4	0.0	0.0	34.0	0.0	22.6	52.8	7.5	45.3	64.2
		Night	11.5	44.2	76.9	0.0	0.0	44.2	0.0	28.8	67.3	11.5	53.8	80.8
		Total	9.5	41.9	68.6	0.0	0.0	39.0	0.0	25.7	60.0	9.5	49.5	72.4
10	Sadayankuppam	Day	3.8	18.9	47.2	0.0	0.0	32.1	0.0	7.5	37.7	3.8	22.6	52.8
		Night	5.8	28.8	50.0	0.0	0.0	34.6	0.0	15.4	48.1	5.8	34.6	57.7
		Total	4.8	23.8	48.6	0.0	0.0	33.3	0.0	11.4	42.9	4.8	28.6	55.7

Sl. No	Location	Samples during	%of samples polluted											
			SO_X			NOX			SPM			Over-all quality		
			I	R	S	I	R	S	I	R	S	I	R	S
11	Thiruvolriyur	Day	9.4	37.7	66.0	0.0	0.0	69.8	0.0	5.7	47.2	9.4	39.6	71.7
		Night	15.4	34.6	57.7	0.0	0.0	61.5	0.0	1.9	46.2	15.4	34.6	67.3
		Total	12.4	36.2	61.9	0.0	0.0	65.7	0.0	3.8	46.7	12.4	37.1	69.5

Table 8.6. Air samples (%) found polluted in and around Manali industrial complex, cf Tables 8.2 – 8.6 and Appendix A 4.1 – A 4.48

Stations	**Season**											
	Post-monsoon, %			*Summer, %*			*Pre-monsoon, %*			*Monsoon, %*		
	Day	*Night*	*Total*	*Day*	*Night*	*Total*	*Day*	*Night*	*Total*	*Day*	*Night*	*Total*
Chinnasekkadu	53.8	67.3	60.6	60.4	73.1	66.7	34.0	41.5	37.7	48.1	54.9	51.4
Periyasekkadu	50.0	73.1	61.5	41.5	53.8	47.6	28.3	32.1	30.2	53.8	58.8	56.3
Manali school	100.0	96.2	98.1	94.3	90.4	92.4	83.0	67.9	75.5	100.0	100.0	100.0
Manali fire station	9.6	7.7	8.7	7.5	3.8	5.7	1.9	0.0	0.9	0.0	0.0	0.0
TNHB	30.8	44.2	37.5	57.7	50.0	53.8	61.5	73.1	67.3	73.1	84.3	78.6
Chinnamathur	48.1	53.8	51.0	57.7	67.3	62.5	53.8	59.6	56.7	28.8	35.3	32.0
Ammulavoil	25.5	31.4	28.4	56.6	73.1	64.8	47.2	52.8	50.0	71.1	96.2	83.7
Madhavaram	51.9	53.8	52.9	42.3	46.2	44.2	61.5	69.2	65.4	48.1	54.9	51.5
Vaikkadu	31.4	39.2	35.3	45.3	53.8	49.5	54.7	60.4	57.5	59.6	71.2	65.4
Sadayankuppam	35.3	39.2	37.3	22.6	34.6	28.6	37.7	43.4	45.6	53.8	61.5	57.6
Thiruvottriyur	25.5	23.5	24.5	39.6	34.6	37.1	52.8	49.1	50.9	34.6	30.8	32.7

Table 8.7. Tentative CPET standards for chlorine and ammonia

Area	Cl_2 (ppm)	NH_3 (μg/m^3)
Industrial and mixed use	0.025	375
Residential and rural area	0.015	225
Sensitive area	0.006	90

Table 8.8. Ambient air samples exceeding TCPET limits for chlorine and ammonia: pre-monsoon, cf Appendix A 4.1 – A 4.12

SNo.	Season/ months	Location	Samples during	% of samples exceeding TCPET limits					
				Cl_2			NH_3		
				I	R	S	I	R	S
1	Pre-monsoon June - August	Chinnasekkadu	Day	23	29	41	33	39	46
			Night	29	33	44	41	45	50
			Total	26	31	42	37	42	48
2		**Periyasekkadu**	Day	5	13	17	10	18	21
			Night	9	19	28	14	14	19
			Total	7	16	22.5	12	16	20
3		Manali school	Day	52	64	63	71	75	76
			Night	60	69	69	75	79	80
			Total	56	66.5	66	73	77	78
4		Manali fire station	Day	30	36	40	34	37	40
			Night	24	30	32	42	44	44
			Total	27	33	36	38	40	42
5		TNHB	Day	10	29	36	10	19	27
			Night	6	16	29	14	24	32
			Total	8	22	32.5	12	21.5	29.5
6		Chinnamathur	Day	4	13	39	13	16	34
			Night	6	17	38	19	21	39
			Total	5	15	38.5	16	18.5	36.5
7		Amulavoil	Day	53	55	76	70	73	75
			Night	58	60	69	74	77	80
			Total	55.5	57.5	72.5	72	75	77.5
8		Madhavaram	Day	31	37	38	38	46	47
			Night	27	33	32	34	42	42
			Total	29	35	35	36	44	45
9		Vaikkadu	Day	59	60	63	70	73	75
			Night	63	64	67	74	77	79
			Total	61	62	65	72	75	77
10		Sadayankuppam	Day	60	65	69	73	74	76
			Night	53	60	65	70	70	71
			Total	56.5	62.5	67	71.5	72	73.5

SNo.	Season/ months	Location	Samples during	% of samples exceeding TCPET limits					
				Cl_2			NH_3		
				I	R	S	I	R	S
11		Thiruvottriyur	Day	54	68	67	71	71	75
			Night	59	64	71	77	75	79
			Total	56.5	66	69	74	73	77

I - Industrial zone.

R - Residential zone.

S - Sensitive zone.

Table 8.9. Ambient air samples exceeding TCPET limits for chlorine and ammonia: monsoon, cf Appendix A 4.13 – A 4.24

SNo.	Season/ months	Location	Samples during	% of samples exceeding TCPET limits					
				Cl_2			NH_3		
				I	R	S	I	R	S
1	Post-monsoon December-February	Chinnasekkadu	Day	56	75	97	78	86	95
			Night	53	72	93	74	81	91
			Total	54.5	73.5	95	76	83.5	93
2		**Periyasekkadu**	Day	10	17	34	12	22	51
			Night	6	13	30	16	26	57
			Total	8	15	32	14	24	54
3		Manali school	Day	58	76	94	77	87	96
			Night	54	72	90	73	83	92
			Total	56	74	92	75	85	94
4		Manali fire station	Day	20	32	41	38	48	63
			Night	24	36	45	42	52	67
			Total	22	34	43	40	50	65
5		TNHB	Day	0	10	16	18	30	48
			Night	4	14	20	14	34	52
			Total	2	12	18	16	32	50
6		Chinnamathur	Day	20	40	60	30	46	64
			Night	23	36	64	34	50	60
			Total	21.5	38	62	32	48	62

Table 8.9. (Continued)

SNo.	Season/ months	Location	Samples during	% of samples exceeding TCPET limits					
				Cl_2			NH_3		
				I	R	S	I	R	S
7		Amulavoil	Day	9	22	33	18	30	42
			Night	5	18	29	16	27	38
			Total	7	20	31	17	28.5	40
8		Madhavaram	Day	15	28	40	24	38	56
			Night	18	33	41	20	36	50
			Total	16.5	30.5	39.5	22	37	53
9		Vaikkadu	Day	57	73	89	73	81	94
			Night	60	76	94	75	87	98
			Total	58.5	74.5	91.5	74	84	96
10		Sadayankuppam	Day	52	72	98	75	90	97
			Night	58	76	94	71	87	93
			Total	55	74	96	73	88.5	95
11		Thiruvottriyur	Day	57	77	95	78	88	93
			Night	53	73	91	72	84	90
			Total	55	75	93	75	86	91.5

I - Industrial zone.
R.- Residential zone.
S.- Sensitive zone.

Table 8.10. Ambient air samples exceeding TCPET limits for chlorine and ammonia: post monsoon, cf Appendix A 4.25 – A 4.36

SNo.	Season/ months	Location	Samples during	% of samples exceeding TCPET limits					
				Cl_2			NH_3		
				I	R	S	I	R	S
1	Monsoon September- November	Chinnasekkadu	Day	56	74	89	75	87	95
			Night	59	78	94	79	90	98
			Total	57.5	76	91.5	77	88.5	96.5
2		**Periyasekkadu**	Day	9	18	40	17	35	53
			Night	7	14	35	12	29	49
			Total	8	16	37.5	14.5	32	51

SNo.	Season/ months	Location	Samples during	% of samples exceeding TCPET limits					
				Cl_2			NH_3		
				I	R	S	I	R	S
3		Manali school	Day	58	78	92	78	90	98
			Night	54	74	94	73	87	92
			Total	56	76	93	75.5	88.5	95
4		Manali fire station	Day	20	34	50	35	55	62
			Night	25	39	57	40	58	66
			Total	22.5	35.5	53.5	37.5	56.2	64
5		TNHB	Day	5	10	35	12	28	48
			Night	9	14	39	18	34	52
			Total	7	12	37	15	31	50
6		Chinnamathur	Day	22	40	61	30	48	59
			Night	26	44	64	35	52	61
			Total	24	42	62.5	32.5	50	60
7		Amulavoil	Day	7	19	40	22	39	61
			Night	6	15	36	18	34	62
			Total	6.5	17	38	20	36.5	61.5
8		Madhavaram	Day	16	18	39	27	44	67
			Night	12	22	43	31	49	70
			Total	14	20	41	29	46.5	68.5
9		Vaikkadu	Day	58	74	89	78	87	95
			Night	55	78	93	74	82	99
			Total	56.5	76	91	76	84.5	97
10		Sadayankuppam	Day	53	75	89	74	86	92
			Night	58	79	93	78	89	97
			Total	55.5	77	91	76	87.5	94.5
11		Thiruvottriyur	Day	53	78	89	72	72	98
			Night	56	82	92	75	76	95
			Total	54.5	80	91.5	73	74	96.5

I - Industrial zone.
R - Residential zone.
S - Sensitive zone.

Table 8.11. Ambient air samples exceeding TCPET limits for chlorine and ammonia: summer, cf Appendix A 4.37 – A 4.48

SNo.	Season/ months	Location	Samples during	% of samples exceeding TCPET limits					
				Cl_2			NH_3		
				I	R	S	I	R	S
1	Summer March - May	Chinnasekkadu	Day	9	25	42	14	38	51
			Night	5	29	48	18	34	55
			Total	7	27	45	16	36	53
2		**Periyasekkadu**	Day	14	35	44	25	38	50
			Night	18	31	48	21	34	53
			Total	16	33	46	23	36	51.5
3		Manali school	Day	63	73	85	74	85	99
			Night	67	79	89	78	81	95
			Total	65	76	87	76	83	97
4		Manali fire station	Day	68	74	84	71	81	93
			Night	64	78	88	75	85	97
			Total	66	76	86	73	83	95
5		TNHB	Day	60	79	89	68	83	95
			Night	64	75	85	72	89	99
			Total	62	77	87	70	86	97
6		Chinnamathur	Day	62	75	94	73	83	95
			Night	66	79	90	76	87	99
			Total	64	77	92	74.5	85	97
7		Amulavoil	Day	62	78	91	71	85	91
			Night	66	75	95	75	81	95
			Total	64	76.5	93	73	83	93
8		Madhavaram	Day	20	43	56	30	52	60
			Night	24	40	52	33	56	63
			Total	22	41.5	54	31.5	54	61.5
9		Vaikkadu	Day	62	73	92	78	89	95
			Night	66	76	96	74	85	92
			Total	64	74.5	94	76	87	93.5

SNo.	Season/ months	Location	Samples during	% of samples exceeding TCPET limits					
				Cl_2			NH_3		
				I	R	S	I	R	S
10		Sadayankuppam	Day	61	78	96	74	87	98
			Night	58	74	92	70	83	94
			Total	59.5	76	94	72	85	96
11		Thiruvottriyur	Day	12	21	40	12	30	52
			Night	8	25	45	16	34	56
			Total	10	23	42.5	14	32	54

I - Industrial zone.

R - Residential zone.

S - Sensitive zone.

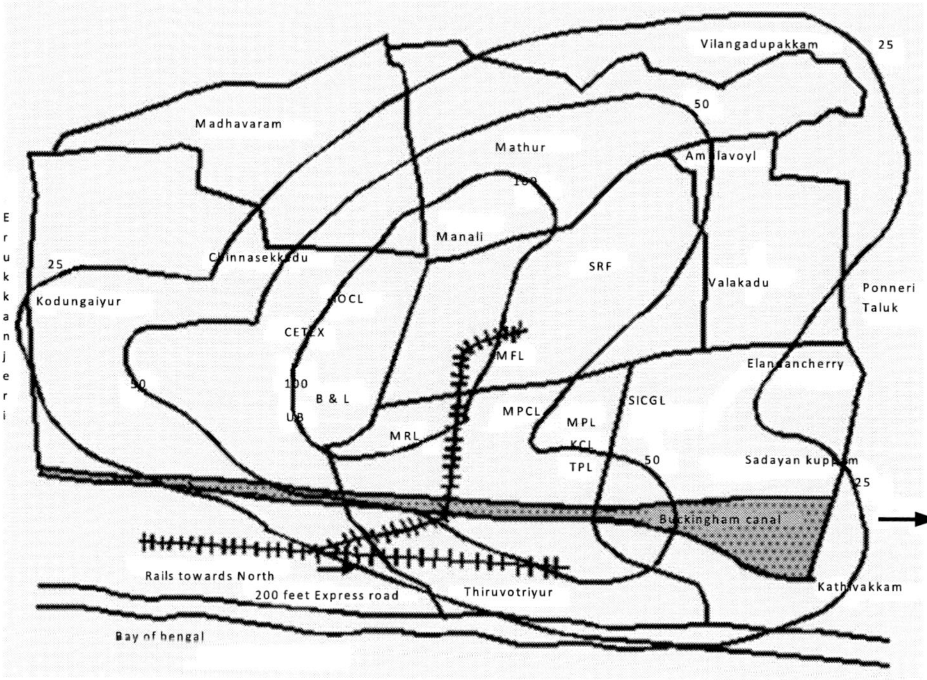

Figure 8.1. Isopleths of So_x ($\mu g/m^3$) during pre-monsoon.

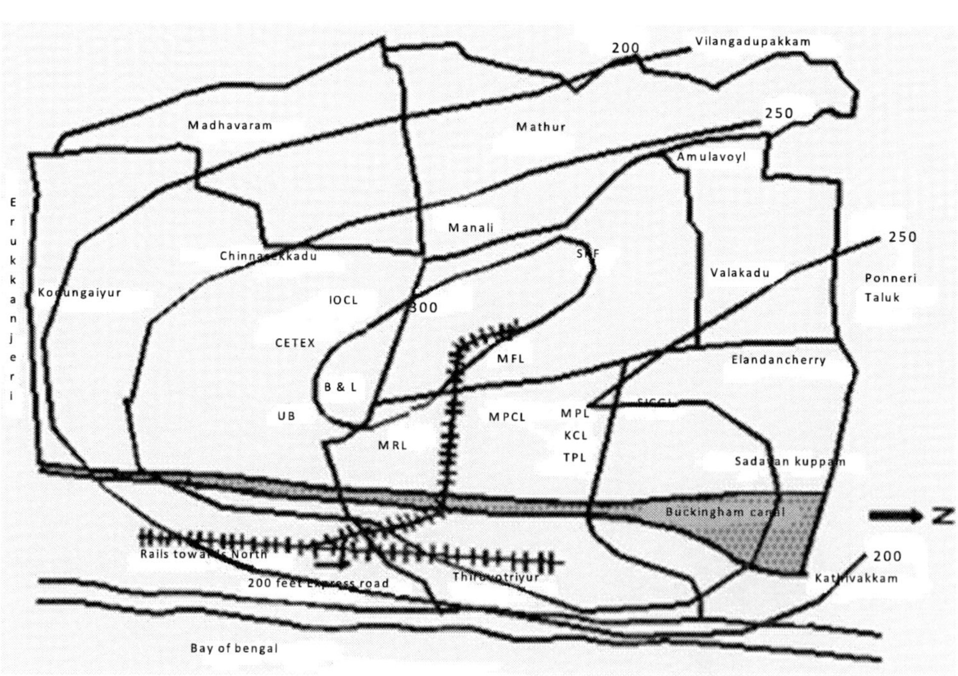

Figure 8.2. Isopleths of SPM ($\mu g/m^3$) during pre-monsoon.

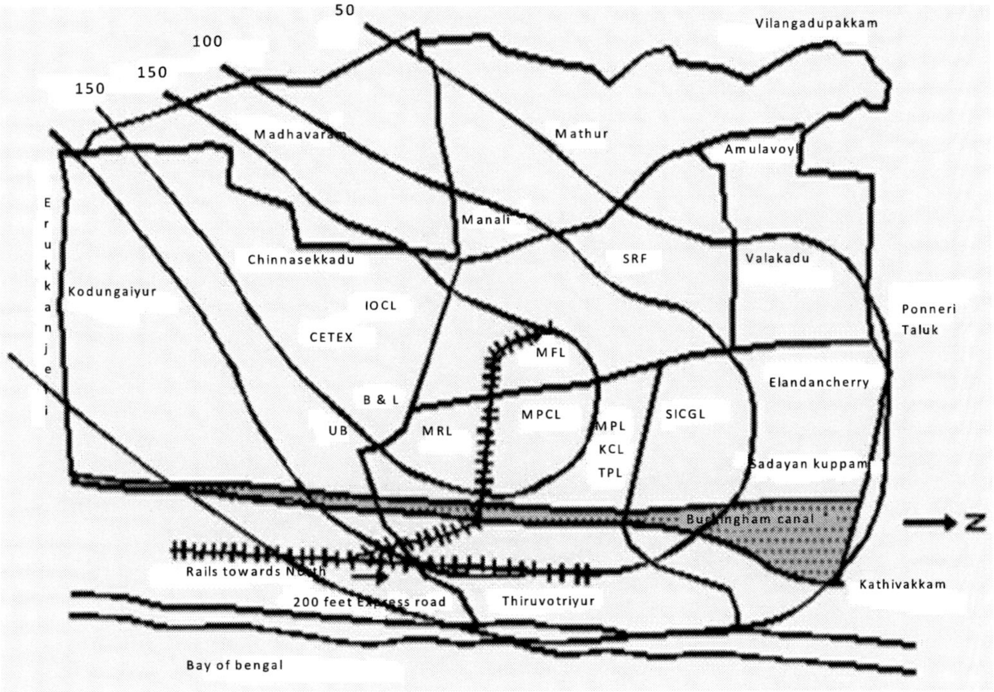

Figure 8.3. Isopleths of SO_x ($\mu g/m^3$) during post-monsoon.

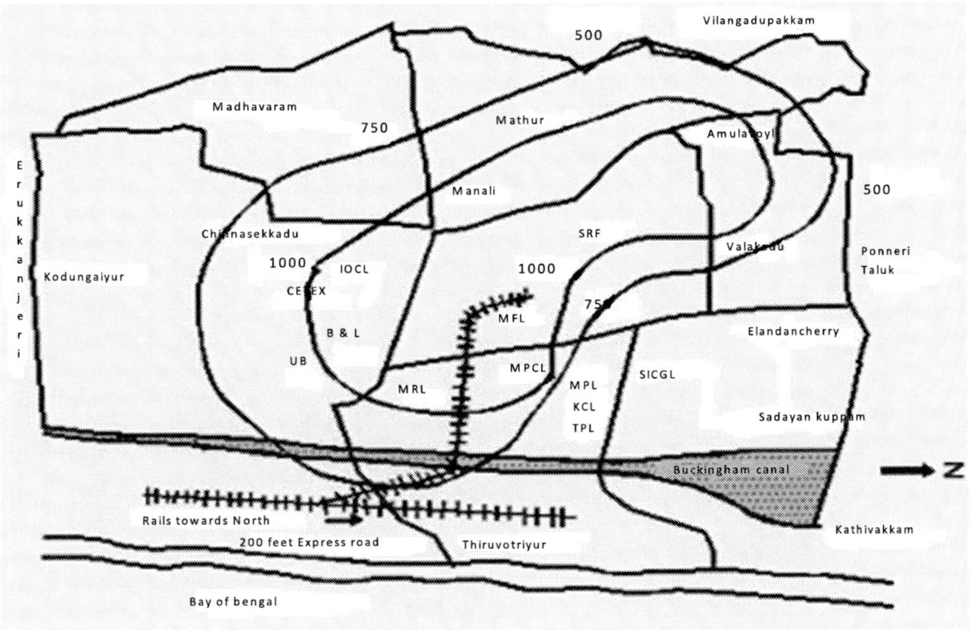

Figure 8.4. Isopleths of NH_3 ($\mu g/m^3$) during monsoon.

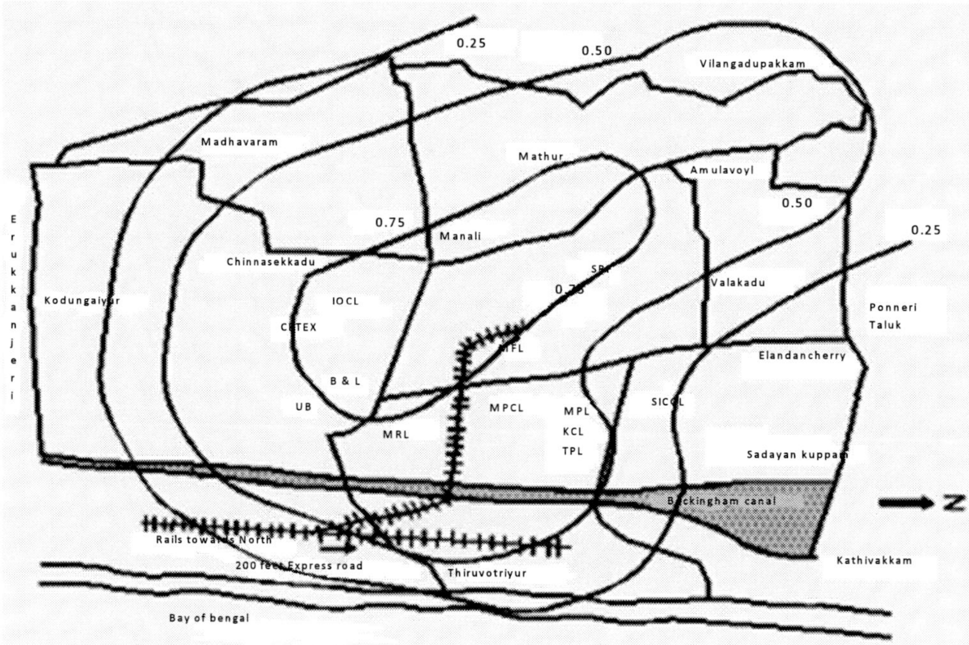

Figure 8.5. Isopleths of Cl_2 (ppm) during monsoon.

(Continued on next page).

Plates 8.1 and 8.2. At times dense-gas plumes fall on the ground, posing serious health hazard.

Chapter 9

AIR QUALITY MODELLING

9.1. INTRODUCTION

Air quality modeling is necessary to identify the relationship between ambient air pollution, the source of the pollution, and the variables such as meteorology and terrain characteristics which influence the pattern of dispersion of the pollutants. Once such relationship is identified and validated, the model becomes usable in forecasting the impacts of decrease or increase in source strength on AAQ as a function of other influencing parameters mentioned above.

When setting out to model the air quality in Manali Industrial Complex (MIC), the following important factors were considered, *inter alia:*

1) The conventional approaches to atmospheric stability classification such as the ones named after Pasquill-Gifford and Turner do not enable precise computations; they are particularly unsuitable under the conditions of coastal meteorology prevailing in the Manali area (Please see the following para). Therefore, we have adopted the most recently modified version of Mohn-Obukhov model for atmospheric stability classification[1, 2].
2) The industrial complex is close to the East coast. As such the classical stability classifications adequate for inland locations cannot be used because the coastal meteorology is different from the inland one, especially in terms of frequent changes in wind directions and speed.
3) Particular care was needed to handle heavy gas dispersion (i.e. dispersion of gases heavier-than-air) as all the five major air pollutants in Manali are 'heavy'. Most conventional models are able to handle only lighter-than-air and as-dense-as-air gases[3].
4) As most of the gases are released continuously for long durations, it is necessary to incorporate this effect in dispersion calculations. However, most of the classical models for dispersion do not respond to this call. We have incorporated this effect by modifying the models. In summary, we have used Mohn – Obukhov stability classification with appropriate modifications to handle coastal effects. In this system the stability is classified using Mohn – Obukhov lengths and Mohn – Obukhov

coefficient. For heavy gas dispersion the latest model based on Plume Path Theory[3,4,5] has been employed after extensive validation and calibration.

The model results have been validated against the field data generated by an exhaustive network of air quality monitoring stations (as detailed in Chapters 2 and 3). Typical results of validation are presented in Figures 9.1-9.4.

9.2. Air Quality Simulations for Impact Forecasting

Using the model validated as above, the air quality of Manali industrial area was simulated as a function of varying source characteristics, atmospheric characteristics and terrain characteristics. The initial setting of various parameters used in the simulations are given in Table 9.1. Present increase/decrease in various parameters needed to bring down the pollutant levels to acceptable limit during various seasons in Manali residential area are given in tables 9.2 – 9.5. Presented below are illustrative examples of simulations carried out by us.

9.2.1. Impact of Changes in Source Characteristics (Flow Rate and/or Source Strength) on Ambient Air Quality in Manali Residential Area

Simulations to assess the impact of changes in source characteristics on ambient air quality reveal the following:

Post-Monsoon

When the flow rate is increased four times, the SO_x, and SPM, and NO_x concentrations increase by 80.5%, 37%, and 25.5% respectively. When the flow rate is reduced by 30%, the concentration of SO_x is reduced by 17.5%, SPM by 9.8% and NO_x by 8%. A 57.25% reduction is needed to bring down the SPM level to acceptable limit (Table 9.2). The simulation results are summarized in Table 9.6.

Changes in the source strength effects ambient SO_x concentration to a much greater degree than it does other pollutants; when the concentration of SO_x in the source is increased by 200%, it leads to an increase of ambient SO_x concentration by 515%. When the source strength is doubled, the SO_x, NO_x and SPM concentrations increase by 214.5%, 13.8% and 18% respectively. To bring down SO_x and SPM to acceptable limits the source strength must be reduced by 20% and 64.25% respectively (Table 9.7).

---

[3] S.A. Abbasi and F.I. Khan, (2000). Computer-aided Environmental Management Discovery Publishing House, New Delhi; viii+ 364 p

[4] F.I. Khan and S.A. Abbasi, (2000). Modelling and simulation of heavy gas dispersion, Journal of Hazardous Materials (U.K), A 80, 15-30.

[5] F.I. Khan and S.A. Abbasi, (1999). HAZDIG : a new software package for assessing the risks of toxic release, Journal of Loss Prevention (U.K.), 12 167-181.

The references 2 and 3 are reproduced in Appendix B. Ref1, a book, is available in the market.

Summer

The impacts in summer differ markedly in degrees compared to the previous season; an increase in flow rate by 200% leads to increase in ambient SO_X by 154.5%, NO_X by 139% and SPM by 17.4%. On the other hand, a 30% reduction of SO_X and NO_X at source reduces their ambient concentrations by 35% and 21.5% respectively (Table 9.7). A minimum of 75.5% reduction in flow rate is needed to bring the ambient pollutant concentrations down to acceptable limits.

The ambient SO_x increases to 278%, when the source concentration of SO_x is increased by 200% (Table 9.7). There is less significant change in the ambient SPM concentration with the increase in source SPM levels: the former increases by only 9.5% when source concentration is increased by 100%. But a similar NO_X at source increases the ambient NO_X concentration by 60.5%; a 30% reduction at source decreases the ambient concentration by 21.5%. During this season, a reduction of atleast 80% at source is needed to bring down the ambient SPM level to the permissible limit.

Pre-Monsoon

When the flow rate is doubled, the ambient concentrations of SO_x, NO_x and SPM increase by 55.5%, 33% and 27% respectively (Table 9.6).

When the source concentration of SO_x, NO_x and SPM are doubled, their ambient concentrations increase by 419%, 20% and 21% respectively (Table 9.7).

Monsoon

An increase in the flow rate by 200%, would lead to 100%, 34%, and 62% increase in the ambient of SO_x, NO_x and SPM levels respectively (Table 9.6).

An increase in source strength by 100% would lead to increase in ambient SO_x by 240%, NO_x by 17% and SPM by 21% (Table 9.7).

The impacts of flow rate and source strength on ambient air quality of the locations where the pollutant levels are the highest, are presented in Tables 9.8 and 9.9. The extent of decrease in flow rate, source strength and both needed to bring down the pollutant concentrations to acceptable levels in these worst case situations' is presented in Table 9.10.

9.2.2. Impact of Changes in Terrain Characteristics on Ambient Air Quality in Manali Residential Area

The terrain characteristic, friction factor, of the area is simulated to find out the impact of their changes on ambient air quality.

Post-Monsoon

As the friction factor increases by 100%, the ambient SO_x, NO_x and SPM concentrations increase by 28.5%, 6% and 30% respectively. When the friction factor is reduced by 30%, it leads to a decrease in ground level concentration of SO_x by 12%, NO_x by 3% and SPM by 12.5%. A 48% reduction in friction factor brings down the SPM concentration to 199.74$\mu g/m^3$.

Summer

When the friction factor is increased by 100%, the ground level concentrations of SO_x increases by 28%, NO_x by 25.5% and SPM by 20%. The reduction in pollutants concentration is found to be 12% in SO_X, 11% in NO_x and 10% in SPM with a reduction of friction factor by 30%.

Pre-Monsoon

There is an increase of 35% in SO_x, 30% in NO_x and 50.5% in SPM with 100% increase in the friction factor. When the friction factor is reduced by 30%, the pollutant concentrations of SO_x, NO_x and SPM reduce by 14%, 13% and 19% respectively.

Monsoon

The most significant increase in the pollutant concentration is seen to occur during monsoon as the increase in SO_x is 37.5%, NO_x is 40% and SPM is 56% for an increase of 100% in friction factor. A 30% reduction in the friction factor causes the ambient concentrations to drop by 15% (SO_x), 16% (NO_x) and 20.5% (SPM), again indicating the sensitivity of pollutant attenuation to the friction factor. Reduction of 40% and 36.5% in friction factor is needed to bring down the SO_x and SPM concentrations to acceptable limits.

9.2.3. Impact of Changes in Atmospheric Conditions on the Ambient Air Quality in Manali Residential Area

The pollutant dispersion under different conditions of wind velocity and atmospheric stability was simulated to forecast the impacts.

Post-Monsoon

There is a reduction of 36% in SO_x, 15% in NO_x and 23% in SPM concentrations with a two-fold increase in the wind velocity. A 77% increase in wind velocity would bring the SO_X concentration down to permissible level and a 90.5% increase would bring down the SPM to permissible level.

During post-monsoon the atmosphere was generally neutral. The simulations reveal that all other factors remaining the same, a shift in atmospheric condition to unstable would cause a decrease in the ambient SO_x by 24%, NO_x by 7% and SPM by 25%. A similar shift to stable conditions would cause increase in SO_x by 65%, NO_x by 12% and SPM by 68.5%.

Summer

When the wind velocity increases by 100% there is a reduction of SO_x by 61%, NO_x by 54% and SPM by 16%. Only a 160% increase in wind velocity would bring down SPM concentration to permissible levels.

Instability conditions had prevailed during summer in both the years during which the air quality was sampled. An increase of SO_x by 31%, NO_x by 28.5% and SPM by 22% would occur in neutral atmospheric condition and an increase of 115%, 102.5% and 62% in these parameters would occur if stable conditions prevail.

Pre-Monsoon

If wind velocity is increased by 6%, there is a decrease of SO_x concentration to acceptable level but as much as 57.5% increase in wind velocity is needed to bring the ambient SPM level down to acceptable limit.

These observations pertain to unstable atmospheric condition. For neutral condition there would be an increase of SO_x by 39%, NO_x by 34% and SPM by 57% and in stable atmospheric condition there would be increase in these concentrations to the extent of 153% for SO_x, 127% for NO_x and 257% for SPM.

Monsoon

In this season an increase in wind velocity by 100% would depress the ambient concentration of SO_x, NO_x and SPM by 11%, 3% and 18% respectively. A 290% increase in wind velocity is needed to bring down the SO_x level to acceptable limit and 175% increase to achieve the same result with SPM.

When atmospheric conditions are highly unstable, there is a reduction of SO_x by 30%, NO_x by 31% and SPM by 39%. But when atmosphere is stable, the ambient concentrations would rise sharply to 89%, 96% and 143.5% for SO_x, NO_X and SPM respectively.

Table 9.1. The initial setting of various parameters used in the simulation studies

SNo.	Parameter	Post-monsoon	Summer	Pre-monsoon	Monsoon
1	Location	Manali	Manali	Manali	Manali
2	Flow rate, m^3/d	102344232	102344232	102344232	102344232
3	Source strength				
	SO_X, mg/m^3	350	320	285	280
	NO_X, mg/m^3	7.0	6.125	9.5	5.0
	SPM, mg/m^3	400	380	250	246.7
4	Friction factor, m	1.5	1.5	1.5	1.5
5	Distance, m	2500	2500	2500	2500
6	Wind velocity, km/h	4.8	6.5	7.4	7.5
7	Stability class	Neutral	Unstable	Unstable	Neutral
8	Ambient temperature, °K	298.3	305.5	303.7	301.5
9	Stack details				
	Height, m	31.68	31.68	31.68	31.68
	Diameter, m	1.094	1.094	1.094	1.094
	Exit velocity, m/s	9.02	9.02	9.02	9.02
	Exit- temperature, °K	498.5	498.5	498.5	498.5
10	Ambient concentration				
	SO_X, µg/m^3	115.65	128.52	84.47	99.39
	NO_X, µg/m^3	38.38	39.80	49.65	34.88
	SPM, µg/m^3	225.56	257.46	265.16	266.73

Table 9.2. Percent change* in forcing functions needed to bring down the pollutant level at Manali to acceptable limit during monsoon

SNo.	Parameter	Percent increase/decrease in case of the pollutant		
		SO_X	NO_X	SPM
1	Flow rate, Q	-29.5	Nil	-48.25
2	Source strength, S	-12.0	Nil	-64.5
3	Both Q&S	-9.0	Nil	-33.5
4	Ambient temperature, Ta	63.0	Nil	57.0
5	Exit temperature, Te	10.0	Nil	24.5
6	Wind velocity, Vw	290.0	Nil	175.0
7	Friction factor, Uf	-40.0	Nil	-36.5

* Negative sign indicates the need of reduction.

Table 9.3. Percent change* in forcing functions needed to bring down the pollutant level at Manali to acceptable limit during summer

SNo.	Parameter	Percent increase/decrease in case of the pollutant		
		SO_X	NO_X	SPM
1	Flow rate, Q	-43.0	Nil	-75.5
2	Source strength, S	-32.5	Nil	-80.0
3	Both Q&S	-21.0	Nil	-52.75
4	Ambient temperature, Ta	280.0	Nil	117.5
5	Exit temperature, Te	24.0	Nil	75.0
6	Wind velocity, Vw	42.5	Nil	160.0
7	Friction factor, Uf	-74.0	Nil	-54.0

* Negative sign indicates the need of reduction.

Table 9.4. Percent change* in forcing functions needed to bring down the pollutant level at Manali to acceptable limit during pre-monsoon

SNo.	Parameter	Percent increase/decrease in case of the pollutant		
		SO_X	NO_X	SPM
1	Flow rate, Q	-8.25	Nil	-55.75
2	Source strength, S	-2.5	Nil	-64.0
3	Both Q&S	-2.0	Nil	-36.5
4	Ambient temperature, Ta	15.5	Nil	62.0
5	Exit temperature, Te	3.0	Nil	18.25
6	Wind velocity, Vw	6.0	Nil	57.5
7	Friction factor, Uf	-12.5	Nil	-38.0

* Negative sign indicates the need of reduction.

Table 9.5. Percent change* in forcing functions needed to bring down the pollutant level at Manali to acceptable limit during post-monsoon

SNo.	Parameter	Percent increase/decrease in case of the pollutant		
		SO_X	NO_X	SPM
1	Flow rate, Q	-49.75	Nil	-57.25
2	Source strength, S	-20.0	Nil	-64.25
3	Both Q&S	-15.5	Nil	-37.25
4	Ambient temperature, Ta	177.5	Nil	91.75
5	Exit temperature, Te	77.75	Nil	305.0
6	Wind velocity, Vw	77.0	Nil	90.5
7	Friction factor, Uf	-64.0	Nil	-48.5

* Negative sign indicates the need of reduction.

Table 9.6. Gist of simulations – I: Impact of flow rate on ambient air quality of Manali; the rows containing enclosed figures indicate the situation where air quality is brought under permissible levels

SNo.	Season	Percent increase / decrease in flow rate, %	Ambient Concentration			
			SO_X μg/m^3	NO_X μg/m^3	SPM μg/m^3	Overall quality polluted (P) or unpolluted (U)
1	Monsoon	200	198.85	46.77	431.43	P
		100	153.94	41.97	361.27	P
		50	128.38	38.87	318.53	P
		25	114.42	37.02	294.10	P
		0	99.39	34.88	266.73	P
		-29.50	79.71	31.77	228.89	P
		-48.25	65.57	29.26	199.92	U
2	Summer	200	327.02	95.14	302.34	P
		100	231.67	68.97	285.78	P
		50	181.41	54.90	274.02	P
		25	155.36	47.50	266.57	P
		0	128.52	39.80	257.46	P
		-43.0	79.69	25.48	234.49	P
		-75.5	38.88	13.04	199.99	U
3	Pre-monsoon	200	170.12	77.95	388.71	P
		100	131.39	66.00	337.53	P
		50	109.38	58.64	305.36	P
		25	97.38	54.41	286.58	P
		0	84.47	49.65	265.16	P
		-8.25	79.96	47.92	257.33	P
		-55.75	50.24	35.52	199.63	U
4	Post-monsoon	200	208.80	48.16	350.98	P
		100	167.89	44.55	312.20	P
		50	143.83	41.99	287.31	P
		25	130.40	40.37	272.57	P
		0	115.65	38.38	255.56	P
		-49.75	79.88	32.25	209.50	P
		-57.25	73.23	30.81	199.95	U

Table 9.7. Gist of simulations – II: Impact of source strength on ambient air quality of Manali; the rows containing enclosed figures indicate the situation where air quality is brought under permissible levels

SNo.	Season	Percent increase / decrease in source strength, %	Ambient concentration			
			SO_X µg/m³	NO_X µg/m³	SPM µg/m³	Overall quality polluted (P) or unpolluted (U)
1	Monsoon	200	692.04	44.83	362.72	P
		100	338.13	40.86	323.82	P
		50	203.42	38.26	298.77	P
		25	147.41	36.7	283.92	P
		0	99.39	34.88	266.73	P
		-12.0	79.30	33.88	257.36	P
		-64.5	15.95	27.54	199.63	U
2	Summer	200	485.96	84.4	296.94	P
		100	297.45	63.95	282.37	P
		50	209.97	52.52	272.03	P
		25	168.38	46.36	265.48	P
		0	128.52	39.80	257.46	P
		-32.5	79.86	30.41	243.33	P
		-80.0	18.31	13.23	199.62	U
3	Pre-monsoon	200	1149.6	66.06	360.15	P
		100	438.63	59.45	321.67	P
		50	221.41	55.17	296.88	P
		25	143.56	52.61	282.18	P
		0	84.47	49.65	265.16	P
		-2.5	79.54	49.32	263.30	P
		-64.5	7.45	38.07	-199.46	U
4	Post-monsoon	200	711.03	46.76	332.09	P
		100	363.74	43.66	301.49	P
		50	226.07	41.47	281.50	P
		25	167.25	40.08	269.53	P
		0	115.65	38.38	255.56	P
		-20.0	79.97	36.68	242.32	P
		-64.25	21.12	30.54	199.98	U

Table 9.8. Gist of simulations – III:
Impact of flow rate on ambient air quality in the worst situations; the rows containing enclosed figures indicate the situation where air quality is brought under permissible levels

S.No	Season	Location	Percent increase/decrease inflow rate, %	Ambient concentration			
				SO_x µg/m^3	NO_x µg/m^3	SPM µg/m^3	Overall quality polluted (P) or unpolluted (U)
1	Monsoon	TNHB	200	202.62	47.71	442.88	P
			100	156.86	42.82	370.86	P
			50	130.81	39.65	326.99	P
			25	116.59	37.77	301.91	P
			0	101.27	35.58	273.82	P
			-31.25	79.94	32.19	232.40	P
			-51.25	64.35	29.37	199.93	U
2	Summer	Chinna mathur	200	347.98	100.75	315.37	P
			100	246.52	73.03	298.80	P
			50	193.04	58.13	287.05	P
			25	165.32	50.30	279.60	P
			0	136.75	42.14	270.48	P
			-47.0	79.72	25.46	244.54	P
			-82.25	31.45	10.69	199.85	U
3	Pre monsoon	Chinna mathur	200	183.44	83.31	430.93	P
			100	141.67	70.53	374.19	P
			50	117.94	62.67	338.53	P
			25	105.00	58.15	317.71	P
			0	91.09	53.06	293.96	P
			-18.5	79.95	48.78	273.75	P
			-67.0	44.94	33.65	199.83	U
4	Post monsoon	Chinna sekkadu	200	226.33	48.93	381.78	P
			100	181.99	45.32	339.59	P
			50	155.91	42.76	312.52	P
			25	141.35	41.13	296.49	P
			0	125.37	39.15	277.98	P
			-56.75	79.88	31.68	218.22	P
			-68.25	67.65	28.93	199.58	U

Table 9.9. Gist of simulations – IV: Impact of source strength on ambient air quality in the worst situations; the rows containing enclosed figures indicate the situation where air quality is brought under permissible levels

S.No	Season	Location	Percent increase/ decrease in source strength, %	Ambient concentration			
				SO_x µg/m^3	NO_x µg/m^3	SPM µg/m^3	Overall quality polluted (P) or unpolluted (U)
1	Monsoon	TNHB	200	705.17	45.73	372.35	P
			100	344.54	41.68	332.42	P
			50	207.27	39.03	306.71	P
			25	150.2	37.44	291.46	P
			0	101.27	35.58	273.82	P
			-12.5	79.99	34.51	263.77	P
			-67.5	13.91	27.53	199.93	U
2	Summer	Chinna mathur	200	517.11	89.37	309.96	P
			100	316.51	67.71	295.39	P
			50	223.42	55.61	285.05	P
			25	179.17	49.09	278.50	P
			0	136.75	42.14	270.48	P
			-36.0	79.67	31.05	254.44	P
			-86.0	12.65	10.97	199.83	U
3	Pre monsoon	Chinna mathur	200	1239.60	70.60	399.27	P
			100	472.96	63.53	356.60	P
			50	238.74	58.96	329.13	P
			25	154.79	56.23	312.82	P
			0	91.09	53.06	293.96	P
			-5.5	79.63	52.28	289.36	P
			-75.0	3.38	37.00	199.75	U
4	Post monsoon	Chinna sekkadu	200	770.75	47.52	361.23	P
			100	394.29	44.43	327.94	P
			50	245.06	42.24	306.20	P
			25	181.29	40.85	293.17	P
			0	125.37	39.15	277.98	P
			-24.0	79.64	37.05	260.38	P
			-75.0	12.67	28.58	199.74	U

Table 9.10. Percent decrease in the flow rate, source strength and both needed to bring down the pollutants level to acceptable limits in worst situations occurring in the four seasons

SNo.	Season	**Parameter**	Location recording maximum air pollution (i.e. worst situation)	Percent decrease in the parameters needed to bring down the pollutants level to acceptable limits, %	
				for SO_X	for SPM
1	Monsoon	Flow rate	TNHB	-31.25	-51.25
		Source strength		-12.5	-67.50
		Both		-9.5	-35.50
2	Summer	Flow rate	Chinnamathur	-47.0	-82.25
		Source strength		-36.0	-86.00
		Both		-23.0	-60.25
3	Pre monsoon	Flow rate	Chinnamathur	-18.5	-67.00
		Source strength		-5.5	-75.00
		Both		-4.25	-46.00
4	Post monsoon	Flow rate	Chinnasekkadu	-56.75	-68.25
		Source strength		-24.0	-75.00
		Both		-18.75	-46.50

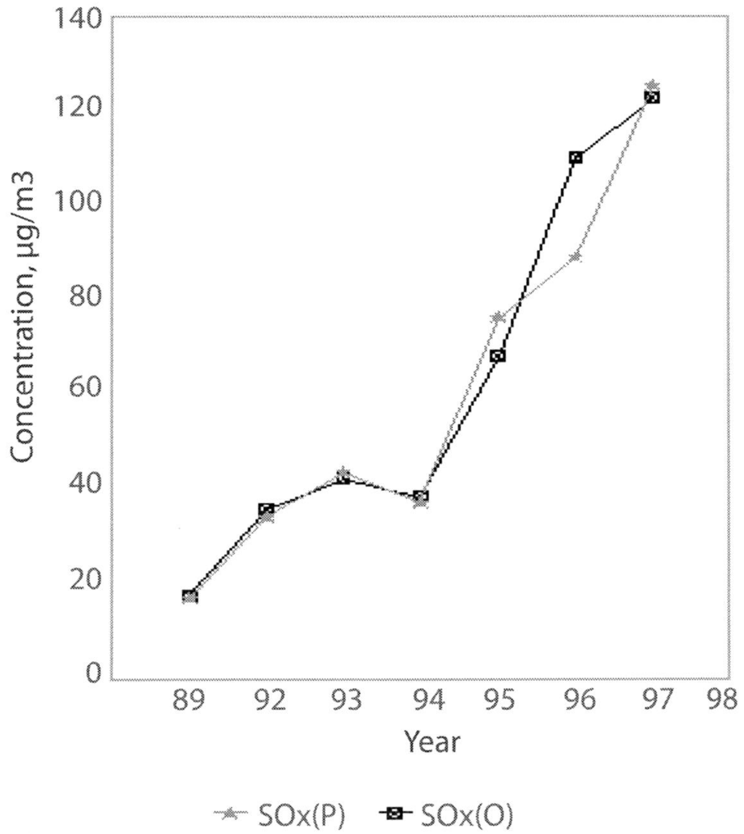

Figure 9.1. Comparison of SO_x (predicted and observed) during post-monsoon.

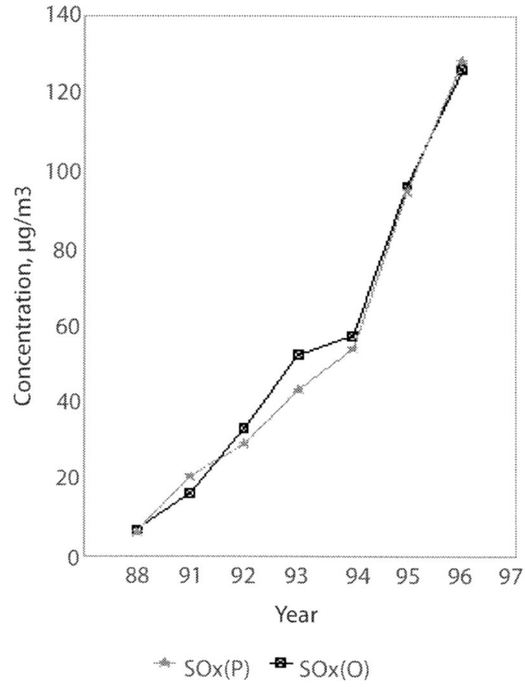

Figure 9.2. Comparison of SO_x (predicted and observed) during summer.

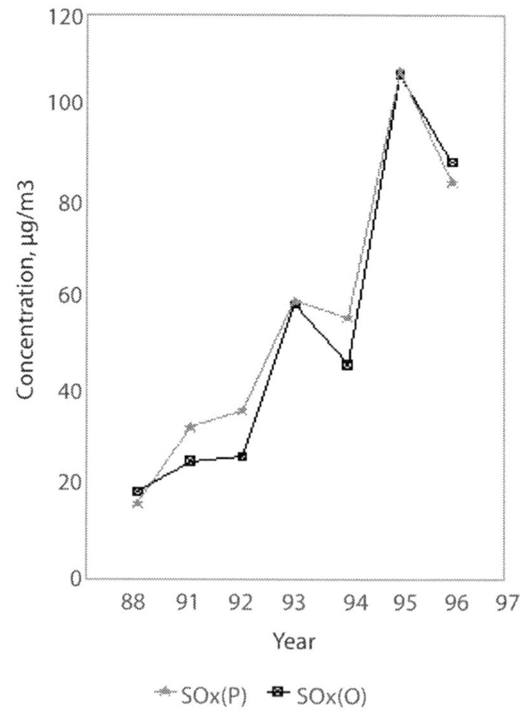

Figure 9.3. Comparison of SO_x (predicted and observed) during pre-monsoon.

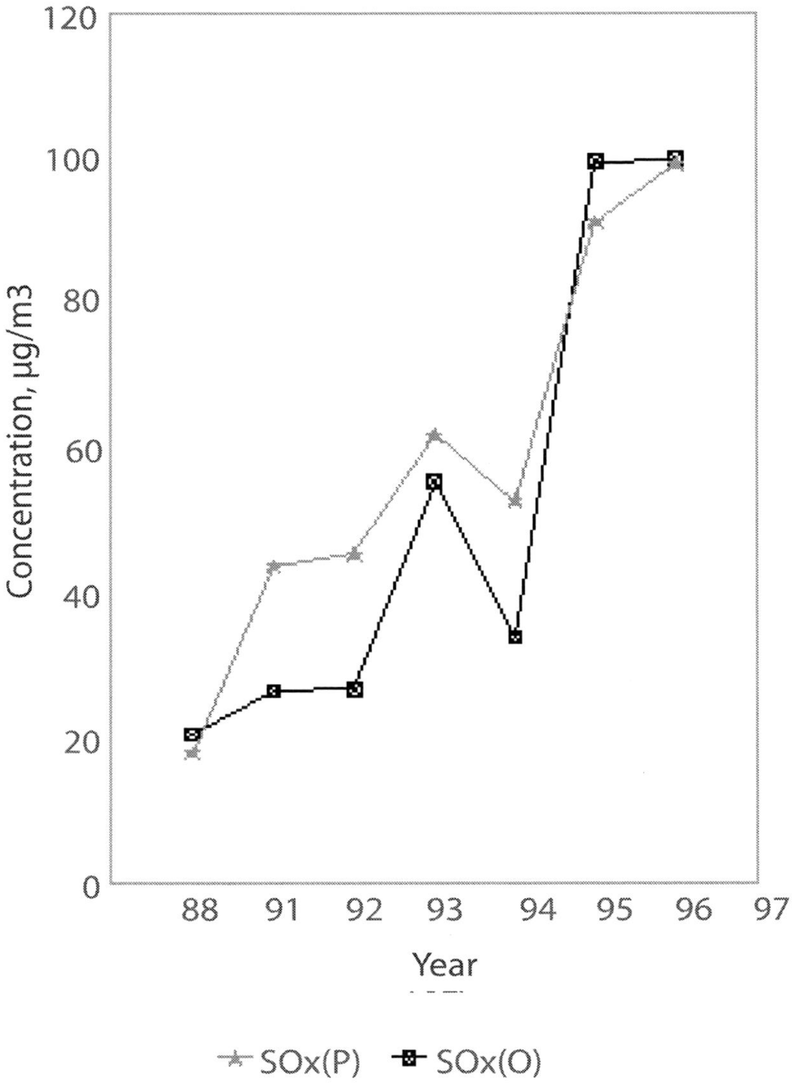

Figure 9.4. Comparison of SO_x (predicted and observed) during monsoon.

Chapter 10

NOISE ENVIRONMENT

10.1. GENERAL

An assessment of ambient noise levels at different times of the day, and at night, was considered important because sensitive locations such as hospitals, besides residential locations, occur in the study area.

10.2. SAMPLING NETWORKS

The ambient noise levels and the frequency of vehicular traffic were monitored during different seasons at 8 different locations (Figure 10.1) covering the entire study zone. The sampling points were selected to represent both industrial and residential areas of the region.

The noise levels were measured as dB(A) using a hand-held noise level meter at each location at five different times of the day (early morning, late morning, mid-day, evening and night). These studies were conducted during near-absence as well as during peak movement of vehicular traffic. A representative set of results is given in Tables 10.1-10.8.

The frequency of vehicular movement was monitored on the highways as well as the interior roads. The vehicular traffic was classified into three groups: (i) heavy vehicles (trucks, lorries and buses), (ii) light vehicles (jeeps and cars) and (iii) two/three wheelers.

10.3. IMPORTANT FINDINGS

10.3.1. Pre-Monsoon

Baring the locations MN2 (Vayakkadu) and MN3 (Aulavoil), the ambient noise levels at other stations are often higher than the limits set for industrial (75 dB)/commercial (65 dB)/residential (55 dB) localities (Table 10.9). But the levels do not exceed the limits by very wide margins (Tables 10.1-10.8). Only in 9 out of 280 samples the noise levels were higher than the USEPA criteria of 'maximum allowable doze' (i,e the doze above which harmful effects are certain to occur on persistent exposure : 90 dB). And in 6 of these 9 samples the noise levels were just a fraction of a dB above 90.

The noise levels at night (after 20 hrs) were within the limit of 75 dB (A) in all but one instance. The zonation of noise levels in illustrated in Figure 10.2.

The traffic density was at its peak during 9-11 and 16-19 hours but the highest average noise levels didn't necessarily occur during these hours.

10.3.2. Monsoon

The noise levels near the residential areas were in the range of 50 - 65 dB(A) with night levels on the lower side. The noise levels near the industrial areas were typically in the range of 70 - 85 dB(A). In this season, too, the majority of samples exceeded the ambient standards, but not by a very wide margin.

10.3.3. Post-Monsoon

The pattern was similar to the one observed in other seasons. The ambient noise levels often crossed the permissible limits but were mostly within the USEPA limit of 90 dB for 'maximum allowable noise levels.' The hours of peak noise level did not always coincide with the hours of peak traffic movement.

All-in-all ambient noise is not a very serious concern at present in the study area eventhough the noise levels are often higher than the statutory limits, because the margins by which the permissible noise levels are crossed is not alarming. But continued industrial expansion, with consequent increase in traffic flow, may change this picture, as noted below.

10.3.4. The Trends

Reconnoitery surveys conducted during 2000-2002 indicate a rising trend in the ambient noise levels. For example, a survey done on 26.1.2002 during 12 hrs -- 16 hrs, reveals that the margin with which the standards are being violated, is going up:

Location	Sampling duration, hrs	Average of seven readings taken at 4 min intervals, dBA
MFL Gate	12-12:30	90.4 ± 4.2
Chakra Foam India Ltd	12:30-13	86.2 ± 1.8
Ponneri Ennur Nedunsalai	13-13:30	95.8 ± 2.2
Rajaji Nagar	13:30-14	96.2 ± 4.4
CBCS Industries	14-14:30	92.0 ± 4.2
Nendunchezhian Salai	14:30-15	92.2 ± 2.8
Manali Highway	--	110.8 ± 11.8

Table 10.1. Noise levels, dBA, at MRL Gate (MN1)

Time,	Noise level at location							Minimum	Maximum	Average
hrs	1	2	3	4	5	6	7			
6 – 7	75.3	80.2	78.2	70.25	66.8	57.9	63.1	57.9	80.2	70.25
9 – 10	77.4	83.1	90.5	76.1	64.6	72.8	68.2	64.6	90.5	76.1
13 – 14	62.7	58.3	67.2	70.75	76.4	78.0	81.9	58.3	81.9	70.75
16 – 17	81.3	84.1	78.9	77.58	72.3	68.9	80.0	68.9	84.1	77.58
20 – 21	62.1	56.3	61.8	60.15	58.9	67.3	54.5	54.5	67.3	60.15

Table 10.2. Noise levels, dBA, at Vayakkadu (MN2)

Time,	Noise level at location							Minimum	Maximum	Average
hrs	1	2	3	4	5	6	7			
6 – 7	56.5	58.65	56.9	63.2	68.6	51.4	55.3	51.4	68.6	58.65
9 – 10	59.3	61.6	54.8	68.6	59.8	61.3	65.8	54.8	68.6	61.6
13 – 14	51.3	53.87	64.1	56.9	50.8	49.9	50.2	50.2	64.1	53.87
16 – 17	62.6	60.73	58.9	64.5	60.8	56.3	61.3	56.3	64.5	60.73
20 – 21	54.2	54.65	60.8	58.3	51.3	53.2	50.1	50.1	60.8	54.65

Table 10.3. Noise levels, dBA, at Amulavoil (MN3)

Time,	Noise level at location							Minimum	Maximum	Average
hrs	1	2	3	4	5	6	7			
6 – 7	53.0	50.1	58.9	62.3	55.82	53.9	56.7	50.1	62.3	55.82
9 – 10	60.3	58.4	63.5	59.0	61.98	66.8	63.9	58.4	66.8	61.98
13 – 14	58.1	52.6	59.2	65.0	60.27	66.4	60.3	52.6	66.4	60.27
16 – 17	54.5	58.9	64.0	59.1	60.45	62.8	63.4	54.5	64.0	60.45
20 – 21	51.3	52.8	60.1	58.5	54.17	50.8	51.5	50.8	60.1	54.17

Table 10.4. Noise levels, dBA, near TPL (MN4)

Time,	Noise level at location							Minimum	Maximum	Average
hrs	1	2	3	4	5	6	7			
6 – 7	90.7	83.2	86.3	68	54.0	79.68	95.9	54.0	95.9	79.68
9 – 10	72.9	90.7	87.9	72.1	80.1	81.90	87.7	72.1	90.7	81.90
13 – 14	74.6	82.6	65.2	79.3	60.7	71.80	68.4	60.7	82.6	71.80
16 – 17	71.4	82.5	72.9	68	90.7	76.15	71.4	68	90.7	76.15
20 – 21	60.8	75.6	68.4	65.2	73.2	68.25	66.3	60.8	75.6	68.25

Table 10.5. Noise levels, dBA, at IOCL Gate (MN5)

Time,	Noise level at location							Minimum	Maximum	Average
Hrs	1	2	3	4	5	6	7			
6 – 7	73.1	62.8	72.9	88.4	75.6	89.2	60.7	60.7	89.2	74.67
9 – 10	67.0	68.9	83.1	78.8	84.3	88.3	89.0	67.0	89.0	79.91
13 – 14	65.2	63.6	67.8	69.2	73.0	80.3	74.9	63.6	80.3	70.57
16 – 17	69.9	81.0	78.3	70.8	74.2	60.3	70.5	60.3	81.0	72.14
20 – 21	56.3	58.9	64.1	68.2	59.1	66.9	63.2	56.3	68.2	62.39

Table 10.6. Noise levels, dBA, at MFL Gate (MN6)

Time,	Noise level at location							Minimum	Maximum	Average
hrs	1	2	3	4	5	6	7			
6 – 7	64.3	69.2	82.3	66.8	71.9	68.8	58.3	58.3	82.3	68.8
9 – 10	67.6	90.4	83.7	72.0	84.6	78.43	72.3	67.6	90.4	78.43
13 – 14	75.1	62.3	72.1	65.9	66.4	66.85	59.3	59.3	75.1	66.85
16 – 17	63.2	87.5	81.9	73.2	61.2	72.13	65.8	61.2	87.5	72.13
20 – 21	52.9	58.3	55.1	63.2	55.7	57.73	61.2	52.9	63.2	57.73

Table 10.7. Noise levels, dBA, at TNEB substation (MN7)

Time,	Noise level at location							Minimum	Maximum	Average
hrs	1	2	3	4	5	6	7			
6 – 7	65.3	77.8	85.4	72.62	60.7	70.2	76.3	60.7	85.4	72.62
9 – 10	77.9	84.3	90.3	80.82	88.8	61.5	82.1	61.5	90.3	80.82
13 – 14	65.3	66.4	80.2	70.4	71.6	70.4	68.5	65.3	80.2	70.4
16 – 17	72.1	74.5	88.1	77.95	89.3	77.6	66.1	66.1	89.3	77.95
20 – 21	62.1	65.3	53.8	61.57	66.9	63.2	58.1	53.8	66.9	61.57

Table 10.8. Noise levels, dBA, at Manali market (MN8)

Time,	Noise level at location							Minimum	Maximum	Average
hrs	1	2	3	4	5	6	7			
6 – 7	85.3	70.3	73.7	94.1	86.4	81.2	76.4	70.3	94.1	81.06
9 – 10	74.4	70.1	68.1	72.1	88.3	79.2	80.3	68.1	88.3	76.07
13 – 14	73.3	80.1	91.2	80.7	67.4	97.4	69.4	67.4	97.4	79.93
16 – 17	70.2	78.1	83.2	81.6	89.2	76.2	72.1	70.2	89.2	78.66
20 – 21	69.0	72.3	60.2	77.5	69.3	74.8	67.9	60.2	77.5	70.14

Table 10.9. Percentage of observations of noise levels when the levels were beyond acceptable limits#: pre-monsoon

Time,	Location	Percentage of samples crossing the limit # for		
hrs		Industrial area	Commercial area	Residential area
6 – 7	Near MRL gate (MN1)	42.86	71.4	100
9 – 10	"	57.14	85.71	100
13 – 14	"	42.86	71.42	100
16 – 17	"	71.42	100	100
20 – 21	"	0	14.29	85.71
6 – 7	Vayakadu (MN2)	0	14.29	85.71
9 – 10	"	0	28.57	85.71
13 – 14	"	0	0	28.57
16 – 17	"	0	0	100
20 – 21	"	0	0	28.57
6 – 7	Amulavoil (MN3)	0	0	57.1
9 – 10	"	0	14.2	100
13 – 14	"	0	14.2	85.7
16 – 17	"	0	0	85.7
20 – 21	"	0	0	28.5
6 – 7	TPL (MN4)	71.4	85.7	85.7
9 – 10	"	71.4	100	100
13 – 14	"	28.5	85.7	100
16 – 17	"	42.8	100	100
20 – 21	"	14.2	85.7	100
6 – 7	IOCL Gate (MN5)	42.8	71.4	100
9 – 10	"	71.4	100	100
13 – 14	"	14.2	85.7	100
16 – 17	"	28.5	85.7	100
20 – 21	"	0	28.5	100

Table 10.9. (Continued)

Time, hrs	Location	Percentage of samples crossing the limit # for		
		Industrial area	Commercial area	Residential area
6 – 7	MFL Gate (MN 6)	14.29	71.42	100
9 – 10	"	57.14	100	100
13 – 14	"	14.29	71.42	100
16 – 17	"	28.57	71.42	100
20 – 21	"	0	0	85.71
6 – 7	TNEB Substation (MN7)	42.8	85.7	100
9 – 10	"	85.7	85.7	100
13 – 14	"	14.2	100	100
16 – 17	"	57.1	100	100
20 – 21	"	0	28.5	85.7
6 – 7	Manali Market (MN8)	71.43	100	100
9 – 10	"	42.86	100	100
13 – 14	"	57.14	100	100
16 – 17	"	71.43	100	100
20 – 21	"	14.29	85.71	100

Ambient air quality standards [Amendment of Environment (protection) act, 1986 (29 of 1986)].

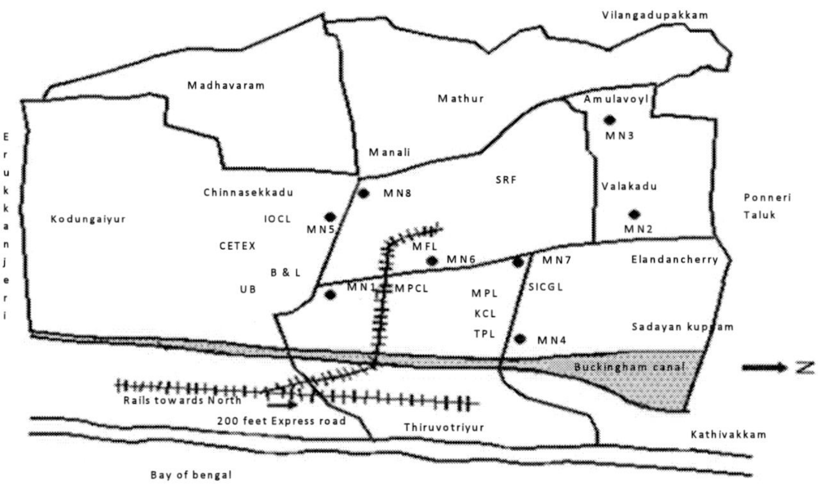

Figure 10.1. Location of noise level monitoring stations (MN 1 – MN 8; cf Tables 10.1–10.8).

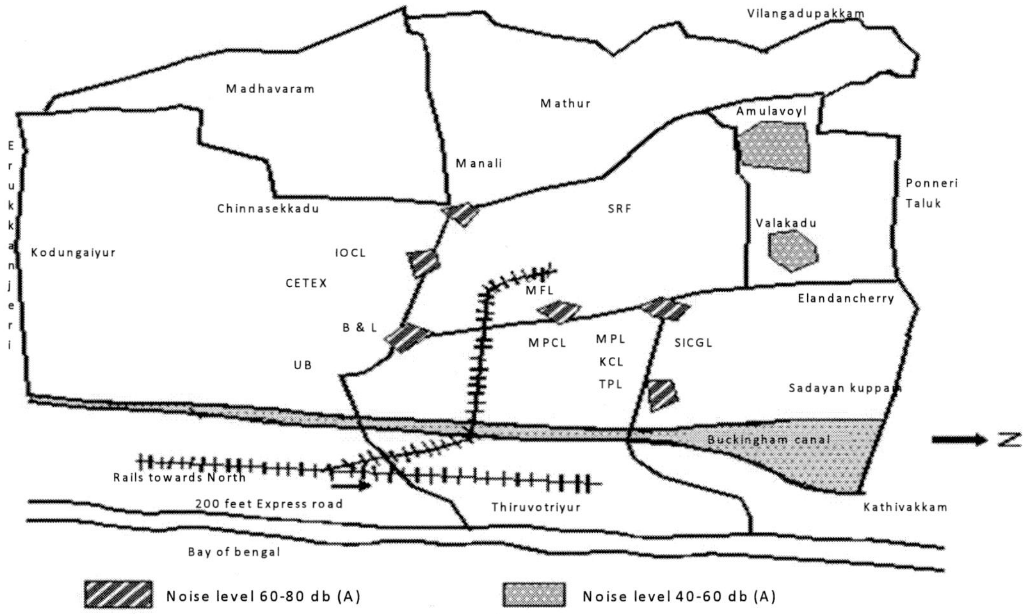

Figure 10.2. Ambient noise level at various locations during pre-monsoon.

Chapter 11

LAND ENVIRONMENT

11.1. LAND-USE PATTERN

The Ponneri and Saidapet taluks, and other parts, of the Chennai District relevant to the study zone, situated within ~10 Km radius from the centre of Manali Industrial Complex (MIC) were studied for land use pattern.

The broad land use pattern (Figure 11.1) in the study area, which has 107 villages, is as follows:

Forest area - Nil
Irrigated agricultural area - 3,772.31 ha
Non-irrigated agricultural area - 2,653.46 ha
Cultivable waste land - 1,747.55 ha
Area not available for cultivation - 2,038.31 ha

The study area comprises of three distinct sections: (i) urban suburban zone (ii) rural zone, and (iii) sea. The urban + suburban zone accounts for 38% of the study area, rural area accounts for 35%; the remaining 27% being the sea. Over half of the available land in the study area is put to non-agricultural use, and only 20% of the land is presently cultivated. The major crop is paddy. In some zones, sugarcane, banana and vegetables are also grown.

11.2. STUDIES ON SOLID WASTES

One of the highly disturbing feature of the study area is solid waste dumps occurring near houses, by the roadside, and near water bodies (Plates 2.1, 2.2 and 11.1-11.4, Figure 11.2). Periodic surveys, including the recent ones indicate that neither the sizes, nor the number of solid waste dumps has gone down across the years. These waste dumps are anonymous; industries just offload their solid waste here and there in a surreptitious manner, with no effort whatsoever to prevent the waste form polluting the environment. The current photographs (Plates 11.1-11.4), as well as the photographs taken during the last few years indicate that dairy animals graze on the vegetation growing near the solid waste dumps and the nearby ponds are used by economically weak section of the people for fishing.

11.2.1. Leachability Studies

The effect of rainfall on the solid wastes and the chemicals that would leach out from the wastes into the soil, underground water, and surface water was simulated. Time- varied simulation for three successive spells of rainfall was done. The studies indicate the very substantial impact the solid wastes are exerting on the soil and water environment of the region. Representative results of the leachability studies are presented in Figures 11.3 and 11.4.

11.2.2. Dry Data on Solid Wastes

In spite of continued probing we have not so far been able to reach to any existing (past) data on the extent, quality, and leachability of solid wastes in Manali Industrial Complex. In all probability no such data exists. The information generated by us, therefore, is highly significant.

11.2.3. Important Findings

The quality of the leachate from the solid waste collected from all the places is found to exceed the limits of water discharged into inland surface waters, discharged on land for irrigation, and water used for domestic purpose. Typical results of the leachability studies are presented in Table 11.1.

Except the leachate from the solid waste collected near Thiruvottriyur railway gate, all other solid waste samples produce alkaline leachate. In general, the pH ranged from 8.12-12.04. The EC of the leachate is as great as 10 times the standard for domestic water. The leachate from the solid waste collected from SW7 had EC in the range of 32700-60500 μhmo/cm^2.

The alkalinity of the leachate from solid wastes collected from locations 2, 4 and 7 (Figure 11.2) was found to be 5 times the drinking water standard, and leachates from location 3 & 5 had alkalinity 10 times higher than drinking water standard. Hardness of the leachate ranged from 1810-20940 mg/L ($CaCO_3$); in case of location 8 (about 500 m away from MFL towards Ennore), the hardness was as high as 10480-20940 mg/L ($CaCO_3$).

The chloride levels in the leachate exceeded the limit for industrial effluent discharged on land for irrigation in all the cases. The highest levels of chloride - 18507-48888 mg/L – occurred in the leachate of the solid waste collected from location 7. Total Nitrogen in the leachate ranged from 5.6-58.8 mg/L. The leachate of solid waste collected from location 7 had total nitrogen in the range of 3542-6199 mg/L.

Total phosphorus levels in the leachate were 10 times greater than the standard for the water for domestic use. The thickest leachate in terms of phosphorous content came from location 7.

In general the leachetes from the solid waste damps were highly polluted and undoubtedly contribute to the deterioration of the already poor ground water quality.

Table 11.1. Leachability studies of solid waste collected at various locations

Location	pH	E C µmohs	Alkal-inity (mg/L as $CaCo_3$)	Hard-ness (mg/L as $CaCo_3$)	Chloride (mg/L)	Total Nitro-gen (mg/L)	Total phosp-horus (mg/L)
Near MRL Bridge	8.56-9.54	4630-57300	40-60	1960-24846	2539-29890	5.6-19.6	1.11-4.17
About 50m away from MRL Bridge	9.77-11.79	39500-4980	82-1396	1310-13020	1310-20800	8.4-44.8	1.26-5.16
On the way to Sadayankuppam	11.43-12.03	10100-30100	1200-2390	1860-14280	737-15723	14- 25.2	1.56-5.72
Near MPL	10.12-11.76	13290-38000	80-1128	5364-17440	6715-20544	11.2-36.4	2.42-5.72
Below Ennore bridge	9.18-12.04	4690-10330	104-2344	1381- 2207	819- 3194	11.2-58.8	3.18-8.59
Near KSCL	8.12-8.5	8680-27400	36-56	2996-10118	4258-14330	8.4- 4.2	1.55-3.28
Near Thiruvottriyur Railway Gate	6.12-6.33	32700-60500	312-1764	4108- 9813	18507-48888	3542-6199.2	13.5-92.8
½km away from MFL towards Ennore	10.89-11.57	23400-37700	602-748	10480-20940	12365-16541	8.4-11.2	1.9- 2.62
½ km before Bhagawandas metals Sadayankuppam	7.71-7.86	16120-18760	20-52	3760- 4240	8762-10400	8.4-14.0	11.16-11.36

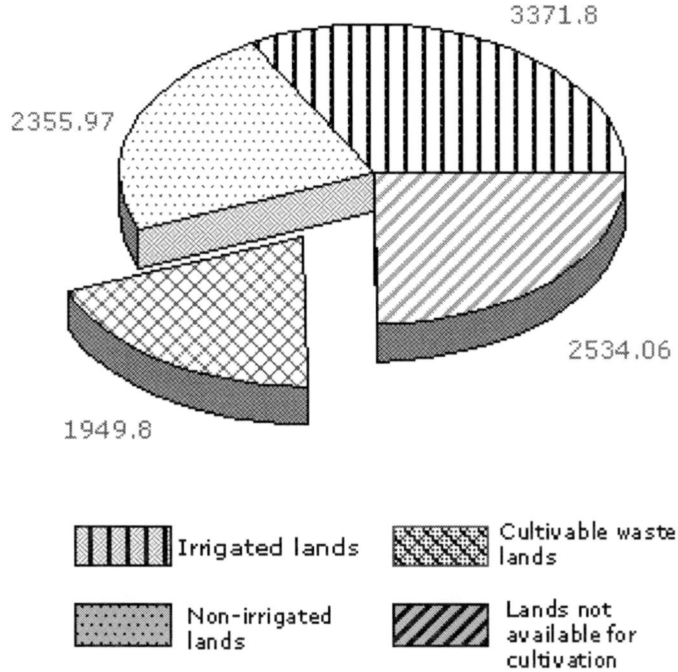

Figure 11.1. The broad land use pattern in the study area.

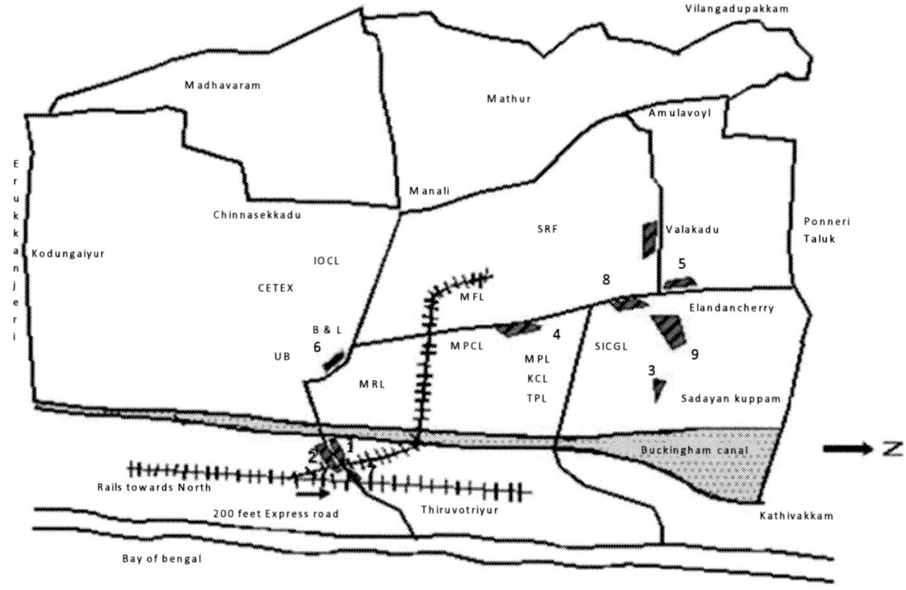

Figure 11.2. Location of Solid waste disposal sites.

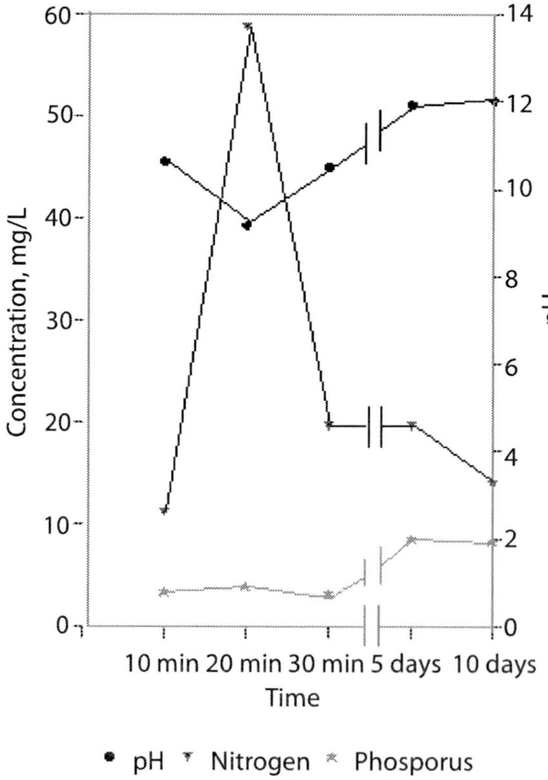

Figure 11.3. The leachate characteristics of the solid waste collected near Ennore bridge (pH, total nitrogen, and total phosphorus).

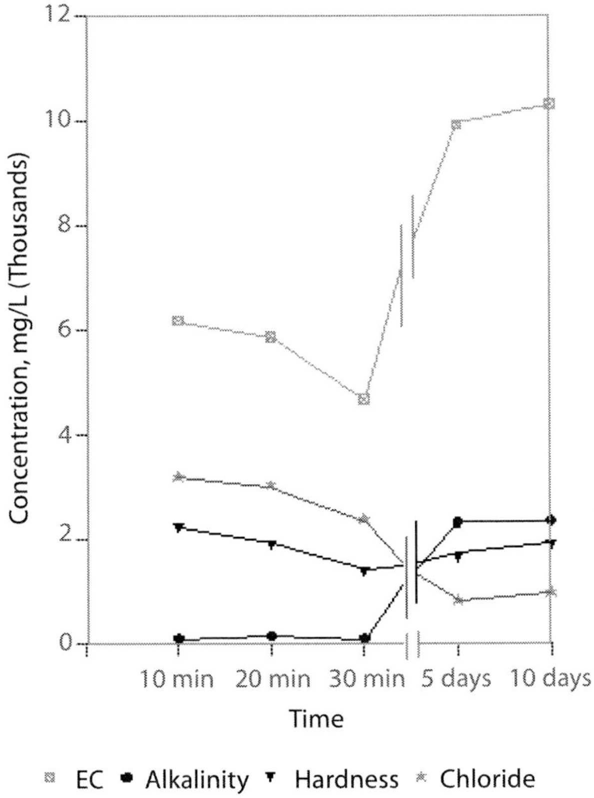

Figure 11.4. The leachate characteristic of the solid waste collected near Ennore bridge (EC, alkalinity, hardness, and chloride).

Plates 11.1 and 11.2. Industrial solid waste dumps in the study area – on land (above) and near human dwellings .

Plates 11.3. Industrial solid waste next to a large pond.

Plate 11.4. Water in the midst of industrial solid waste: major source of ground water and surface water pollution.

Chapter 12

Socio-Economic Environment

12.1. General

Socio-economic environment around any industrial complex is strongly influenced by the beneficial as well as adverse effects arising out of the industrial activity. Industrialization leads to job opportunities, new commerce, migration, creation of infrastructure as well as strain on the existing basic amenities, and impact on the environment quality.

The villages included in the socio-economic survey are represented in Table 12.1 (Figure 12.1). Demographically, parts of the study area fall under zones definable as urban, suburban and rural.

12.2. Population and Personal Details

The urban populations are within the Chennai Metropolitan zones; the suburban populations are from certain parts of Saidapet taluk; and the rural populations are from other parts of Saidapet taluk and Ponneri taluk. Land-area wise, the break-up of urban, suburban and rural zones is 19:33:48 (Figure 12.2). In terms of populations, the break-up is 62:30:8 (Figure 12.3). Thus the urban area is nearly twenty times as densely populated as the rural area. The ratio of male to female population varies from 50:50 to 70:50 in the study area (Figure 12.4).

Hinduism is the predominant religion in the villages surveyed (Figure 12.5). In some villages (V3, V9, V10 and V13) the SC/ST population is more than half of the total (Figure 12.6). In the study area the predominant language spoken is Tamil, more than 75% of the people speak this language (Figure 12.7). More dwellings have joint families than individual families (Figure 12.8).

12.3. Education and Employment

Owing to the predominance of urban / suburban populations this region has a high literacy rate of over 65%, reflected in the high density of primary schools, secondary schools and professional colleges in the area (Figure 12.9). For the same reason, the major means of employment is industrial or office work. The percentage of population engaged in traditional work like farming, fishing etc is relatively low (Figure 12.10). The majority of the population has a declared monthly income of less than Rs 1000 (Figure 12.11).

12.4. Agriculture and Fisheries

12.4.1. Agriculture

It is estimated that only 20% of the land in the study are is cultivated. And this segment of land falls almost predominantly under the rural area. The pattern is somewhat typical of the Chengalpat district.

12.4.2. Fisheries

Outside the industrial sector, fishing is one of the important commercial activities in this region. It is predominantly marine and estuarine, but inland fishing is also done. There are about 40 fishing villages with over 50, 500 fishermen, out of which only about 8500 are known to be active. Both mechanized and non-mechanized (catamarans) fishing craft are used. Over 92% of the total catch is through mechanised craft. The principal varieties caught are sharks, skates, catfish, sardines, perches, red mullets, caranx, seer and prawns.

12.5. Electricity

All the villages are electrified but in the perception of the villagers (Figure 12.12) the actual availability of the supply is not satisfactory.

12.6. Industrial Development

The core of the study zone, Manali - Ennore is a highly developed industrial area. An illustrative list of major industries in the study zone, and an indication of the nature of their activities and their pollution potential are given in Table 12.2. In the perception of over 90% of the villagers, the study area is polluted (Figure 12.13) and they attribute it to the existing industries.

12.7. Transportation

The study area has good transportation facilities (Figure 12.14). The Chennai Airport is within 38 km, the Madras Central Railway Station within 14 km, the Ennore Suburban Railway Station within 7 km, and the Chennai Harbour within 16 km Manali Industrial

Complex (MIC). The access to MIC downtown Chennai city is by road via Chennai-Ennore Express Highway (beach road) followed by Manali Express Highway. Both these highways are in good physical condition and permit smooth vehicular flow.

12.8. Public Health

All diseases, which are common to other parts of the state, are also common to the study zone (Figure 12.15). Filaria is seen mostly in the urban and suburban areas. Malaria in this area has established its endemicity. Leprosy and tuberculosis are also common. Among the other water-borne diseases, acute gastro-enteritis, cholera, diarrhoea, jaundice and typhoid are the most common. Nutritional deficiency diseases like anaemia, bleeding-gums, and eye diseases like maresamus are prevalent among poorer sections of the society. Hereditary diseases like hypertension and diabetes are also seen in these areas.

According to the available statistics the top ten death-causing diseases in the study area are : [i] heart diseases, [ii] tuberculosis, [iii] paralysis, [iv] respiratory diseases other than tuberculosis, [v] diarrhoea, [vi] jaundice, [vii] cancer, [viii] ulcer, [ix] diabetes and [x] dysentery.

The public perception of medical facilities in the study area is summarised in Figure 12.16. Interestingly whereas the majority of people in nine out of thirteen villages believe the medical facilities to be good, the overwhelming majority in the same villages consider environmental pollution status rather bad (Figure 12.13). It follows that given the available medical facilities, improvement in the environmental quality can significantly bring down the incidence of diseases.

12.9. Water Supply

Except in one village (V1) the people are largely dependent on hand pumps and dug wells for water supply (Figure 12.17). The pollution of groundwater is, thus, a very serious concern in the study area.

12.10. Sensitive Areas

There are no areas or aspects within the study zone, which may be considered historically, or culturally sensitive to any exceptional degree. Nevertheless, the pollution of inland surface waters, groundwater, and of the marine environment is an ecologically sensitive issue. In addition, public safety against industrial hazards like release of toxic substances into the atmosphere, and fire and explosion may also be treated as sensitive issues, given the high density of major and hazardous industries in this region.

12.11. Aesthetic Aspects

12.11.1. Land

As the land in Manali Industrial Complex has been developed as an industrial area, it has been levelled. The flat topography is marked by industrial buildings and tall stacks. Waste dumps by the roadside, near residential houses, and along water-bodies seriously mar the presentability of the landscape (Plates 11.1-11.4). There are stretches of barren lands near industries (Plates 12.1 & 12.2) in the study area which are presently infested by weeds. Greenbelts can be developed on these lands to improve the micrometeorology of the area and cushion the effect of air pollution.

12.11.2. Atmosphere

The atmosphere around the complex is often murky. One can see gaseous plumes trying to disperse (Plates 12.3 and 12.4) and, at times, coming close to the ground (Plates 8.1 and 8.2). Unpleasant odours are encountered every now and then. The air quality is described in detail in Chapters 6-9.

12.11.3. Water-Bodies

As detailed in Chapters 2-5, the area is well-endowed in terms of surface water resources but the water bodies are badly affected by pollution and degradation. As the waters of the ponds and the Buckingam canal are used by the cattle, and by the people for laundry, fishing etc (as illustrated in Chapters 2-5), this situation calls for urgent corrective measures.

Table 12.1. The Villages surveyed for studies on socio economic environment

Sl.No	Name of the Village	Distance from MIC (km)	Direction
V1	Manali	1.0	W
V2	Amulavoil	2.5	NW
V3	Vayakkadu	2.0	NW
V4	Chinnasekkadu	1.5	SW
V5	Periyasekkadu	2.0	SW
V6	Sathangadu	1.5	S
V7	Thiruvottriyur	3.5	E
V8	Mathur	2.0	W
V9	Madhavaram	3.5	SW
V10	Sadayankuppam	1.5	NE
V11	Elandanecherry	2.0	NE
V12	Kannadasan Nagar	3.5	S
V13	Kosapur	3.5	NW

Table 12.2. Types of major Industries, products manufactured and processes in brief*

Sl.No	Name of the Industry	Products manufactured in T/Month	Manufacturing process in brief
1.	Madras Fertilizers Limited.	**Products** Urea: 38,500 NPK: 70,00 Intermediate Products Ammonia: 28,875 **By products** CO_2 24,100 Nm^3/hr	Ammonia is produced through Naphtha steam reformation process, Urea through total recycle process, NPK through DORR-Oliver NPK granulation process
2.	Indian Additives Limited.	**Products** Finished lubricating oil additives, 15,000 Intermediate products Succinimides 6,300 Phenates 3,000 Pibsa 4,200	The process units planned for additives manufacture are Succinimide plant Phenate plant Pibsa plant Blending plant
3.	Manali Petrochemicals Limited.	**Products** Propylene oxide: 1,000 Propylene glycol 520 Polyols: 500 By products Di-Chloropropane: 120 Di-propylene glycol: 40	Propylene and chloride are reacted in water to form propylene chlorohydrin. The propylene chlorohydrin is saponified with lime to form propylene oxide. Propylene oxide is reacted with water to form mixture of mono, di and tri-propylene glycols. The mixture is concentrated by triple effect evaporators and vacuum distilled to produce pure pharmaceu-ticals grade propylene glycol and di-propylene glycol
4.	Madras Petro Chemicals Limited	**Products** Transformer oil: 900 Petroleum specialities: 200 Oleum-sulphuric acid: 1200 Petroleum Jellies: 40	Process involves following stages Raw oil storage -Retreat settler - Neutraliser -Oil still - Alcohol recovery -Percolation - Spent brine - Batch still –Drain – desalter – Sulfonate still –Sulfarate – Miley Retort –SO_2 puriification – oleum plant – oleum -Sulphuric acid
5.	Madras refineries Limited.	**Products** LPG Petrol Diesel Kerosene Asphalt Match wax Par. Wax HSD LDO etc. Naphtha Industrial fuel Lube base stocks1	Manufacture process in volves following stages : Crude Tank – Atmospheric Distillation unit – Amine Treating Unit – Merox Treating Unit – Vacuum distillation Unit – Fluid Catalytic Cracking unit – final Products.
6.	Celex Petro Chemicals Limited.	**Products** Methyl Ethyl Ketone 4000T/Annum By Products Secondary Butyl: 49 Heavy Fraction: 66 Sulphuric Acid: 1097	Feed preparation, Secondary Butyl alcohol Synthesis and Secondary Butyl Ether recovery and Methyl Ethyl Ketone Synthesis.

Table 12.2. (Continued)

Sl.No	Name of the Industry	Products manufactured in T/Month	Manufacturing process in brief
7.	Sri Ram Fibers ltd.	**Products** Nylone moulding Powder : 2000 T/A Nylone – 6/Nylone 6, 6 Polymer compounded and natural: 2000 T/A	The manufacturing process consists of mainly melting, prepolymerisation granulation, extractions, drying and weaving
8.	Tamil Nadu Petro products Limited.	**Products** Linear Alkyl benzene : 6667 Heavy normal paraffin : 1667 Heavy Alkylate : 600 Intermediate products Normal paraffin : 5417	The manufacturing unit consists of prefractionation, Hydrotreator, molex, Pacol and Detergent Alkylation.
9.	Indian Organic Chemicals Ltd.	**Products** Polyester staple Fibre: 2500 T/M. Ployester filament yarn: 1250 T/M. **By products** Methanol : 80 T/M.	The manufacturing process involves melting, reaction, fractionation, polycondensation, extrusion, & spinning.
10.	Spic (heavy Chemicals Division) Limited.	**Products** Caustic Soda - 5500 T/m Liquid Ammonia -3333 T/M Hydrochloric acid – 412 T/M Ammonium Chloride – 1788 T/M **By Products** Hydrogen – 137.5 T/M	Brine preparation, cell operation, chlorine preparation, caustic Flakes preparation, Hydrochloric Acid Preparation, Ammonium Chloride preparation.
11.	Kothari Sugar and Chemicals Ltd.	**Products** Poly Butene & Light : 425 T/M Polmers : 120 T/M	Feed preparation, polymerisation of isobutylene, fractionation and storage.
12.	U.B. Petro products Ltd.	**Products** Polyols : 1000 T/M Propylene Glycol : 659 T/M Tripropylene glycol : 10 T/M Propylene oxide : 1000 T/M **By Products** Propylene Dichloride : 175 T/M Dipropylene glycol : 80 T/M	Hypochlorination, Neutralisation, dehy- drochlorination, purification, reaction Concentration and Distillation, reaction Neutralisation and Filteration.
13.	South India Carbonic Gas Industries ltd.	**Products** Carbon-di-oxide liquid : 600 T/M Solid : 300 T/M	The manufacturing process consists of absorption, compression, cooling, separation drying, liquefaction and storing.
14.	Balmer Lawrie & Co.Ltd.	Drum/Barrel – 700 T/M Lubricating grease Specialities – 417 T/M Bituminous compounds – 83 T/m Di Tertiary Butyl para cresol – 12.5 T/M Para tertiary Butyl phenol – 25 T/M Leather Auxidemies 180 T/M HCL – 120 T/M	Deceiting, shearing, levelling. cutting Grinding, Welding, Heating, agitation saponification cooling direction Heating distillation, filtration, washing, drying & packing. Primary reaction, secondary reaction Tertiary reaction, & blending operation.

* This is only an indicative table; the product range and capacities are also illustrative.

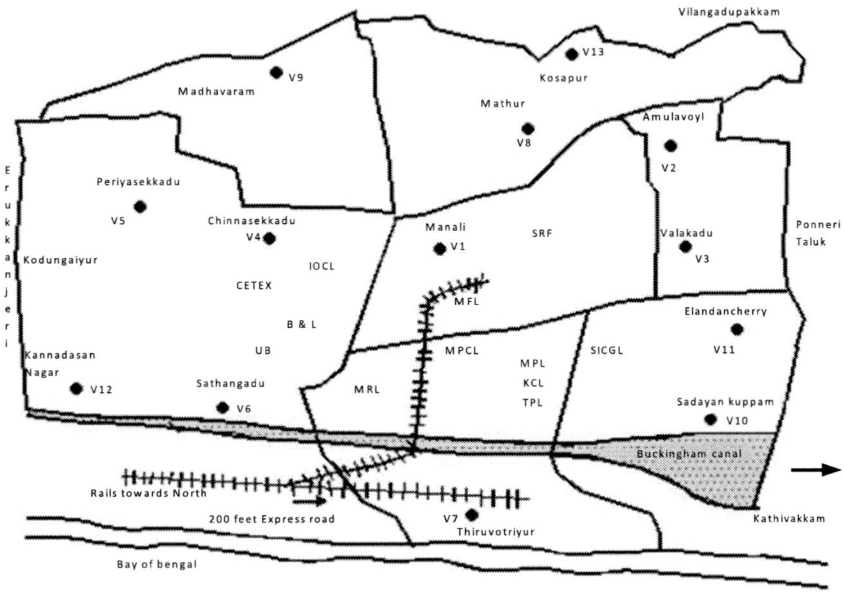

Figure 12.1. Villages included for survey of socio-economic status [•].

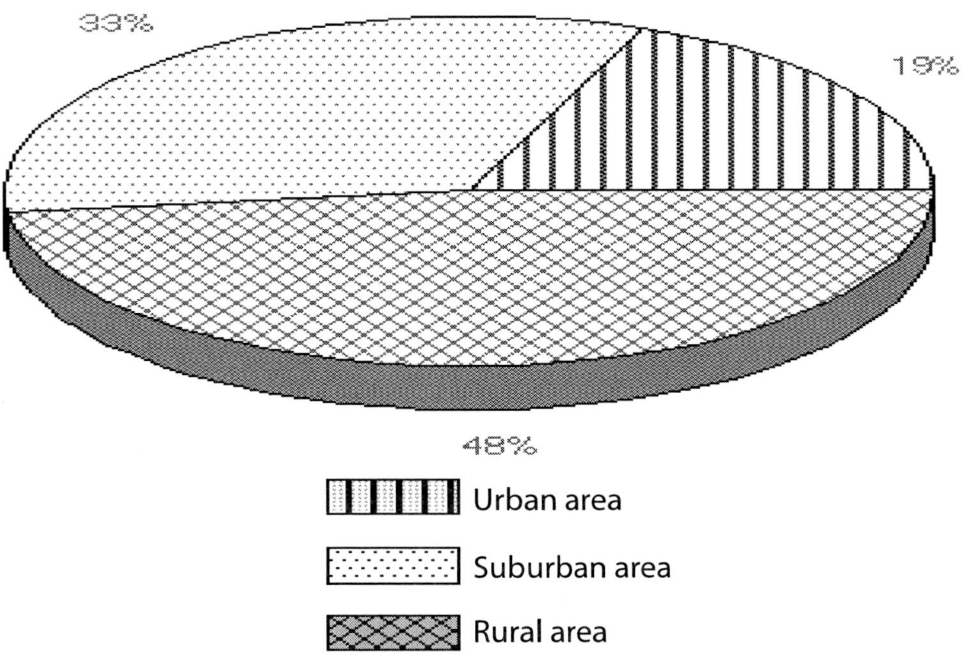

Figure 12.2. Fractions of Manali under urban, suburkan and rural areas.

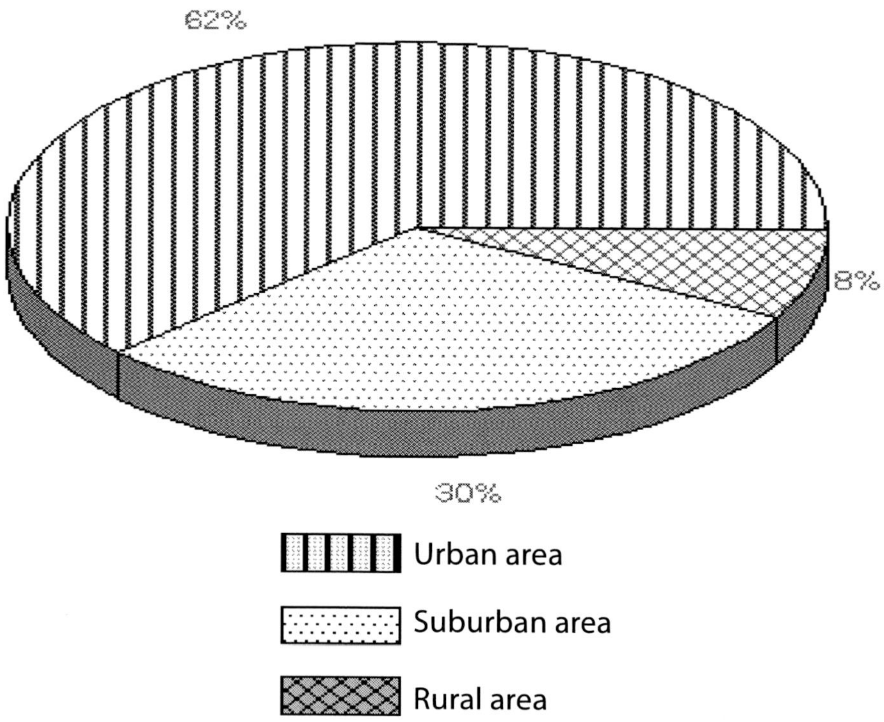

Figure 12.3. Population distribution in urban, suburban and rural areas.

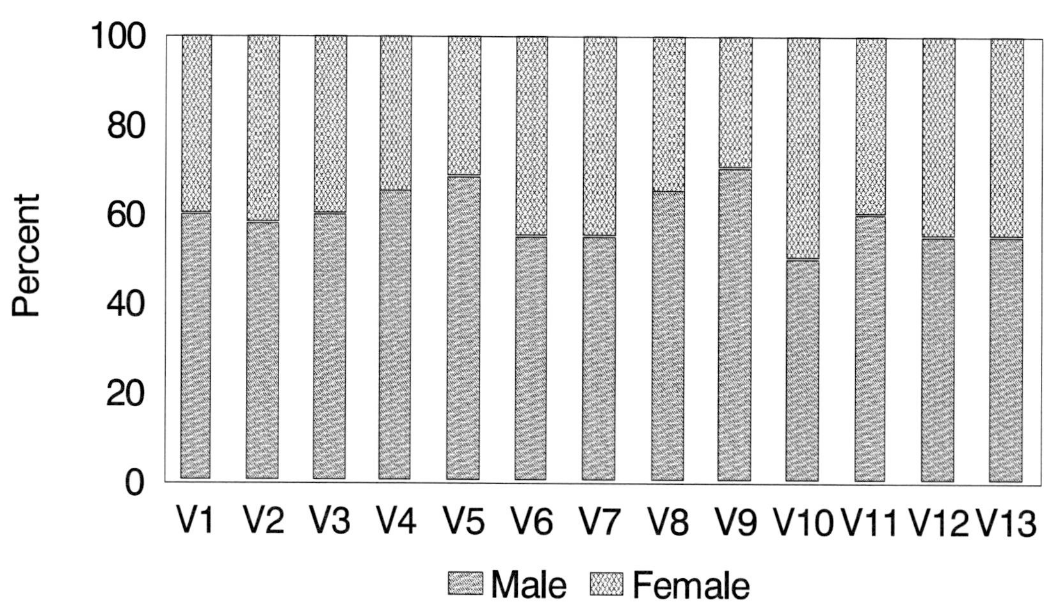

Figure 12.4. Male female ration in the study area.

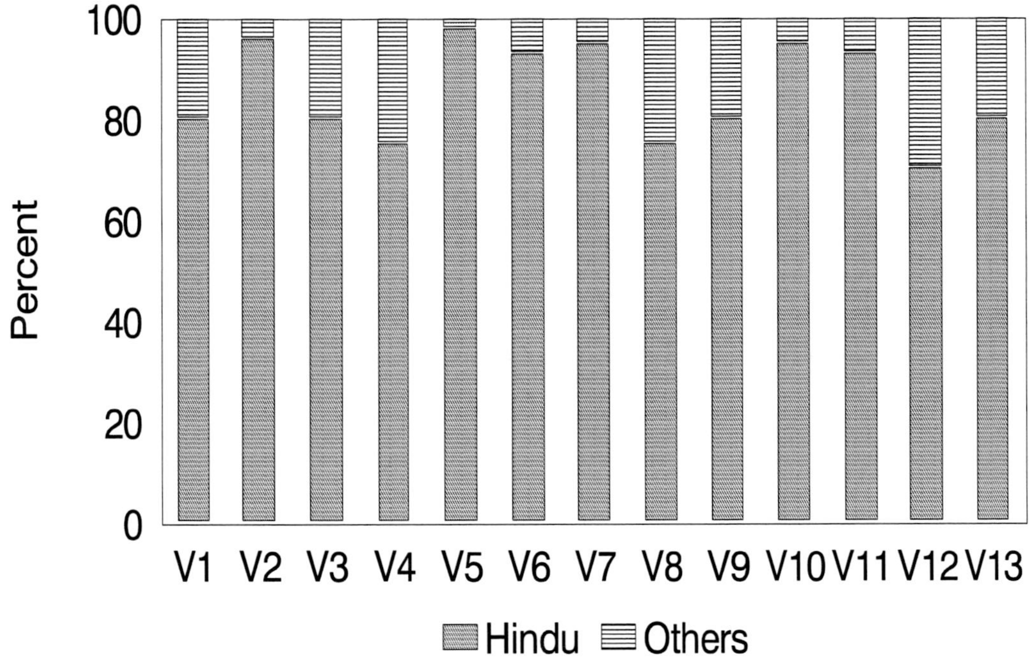

Figure 12.5. Religious details in the study area.

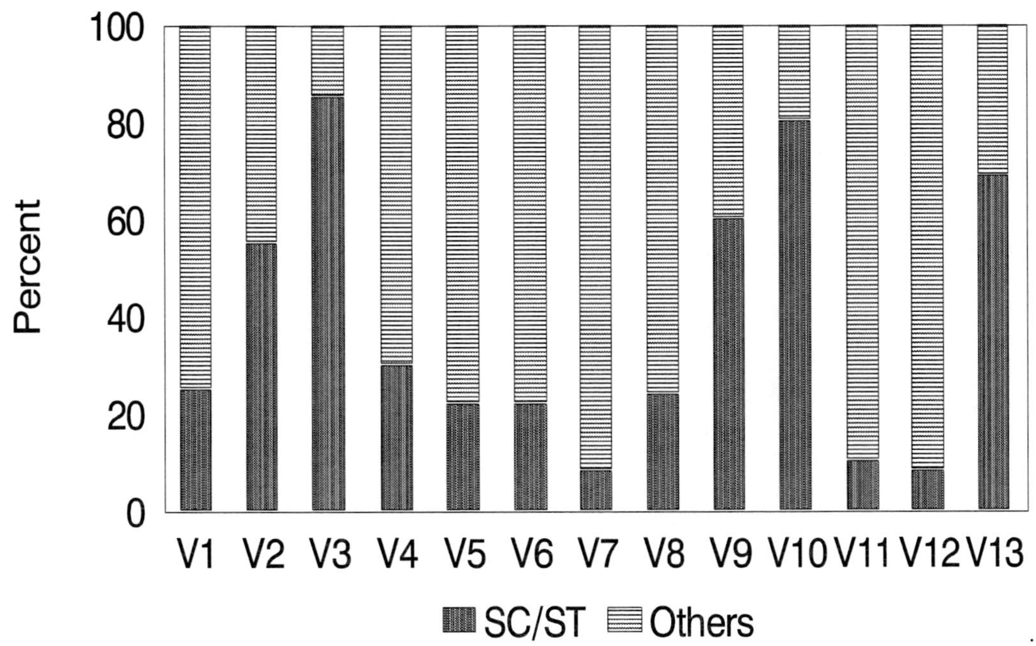

;

Figure 12.6. SC/ST details in the study area.

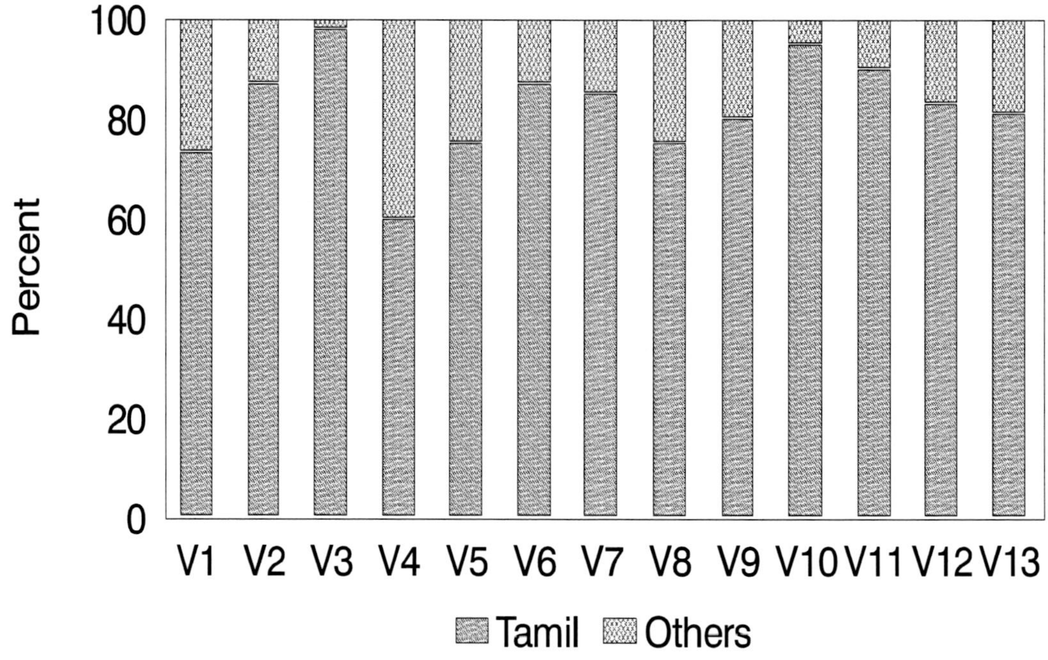

Figure 12.7. Language spoken in the study area.

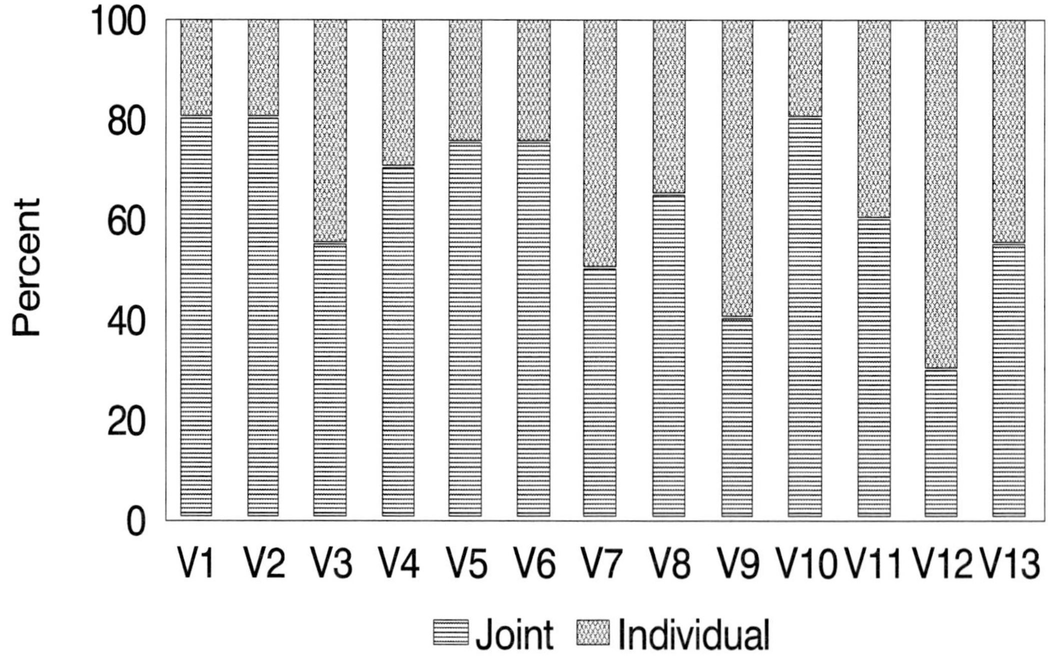

Figure 12.8. Family details in the study area.

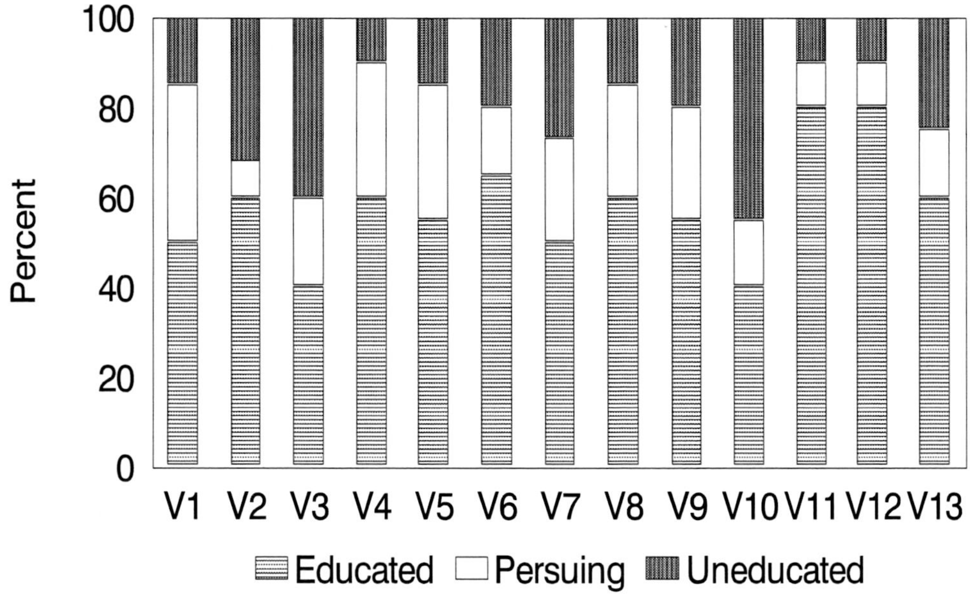

Figure 12.9. Educational status in the study area.

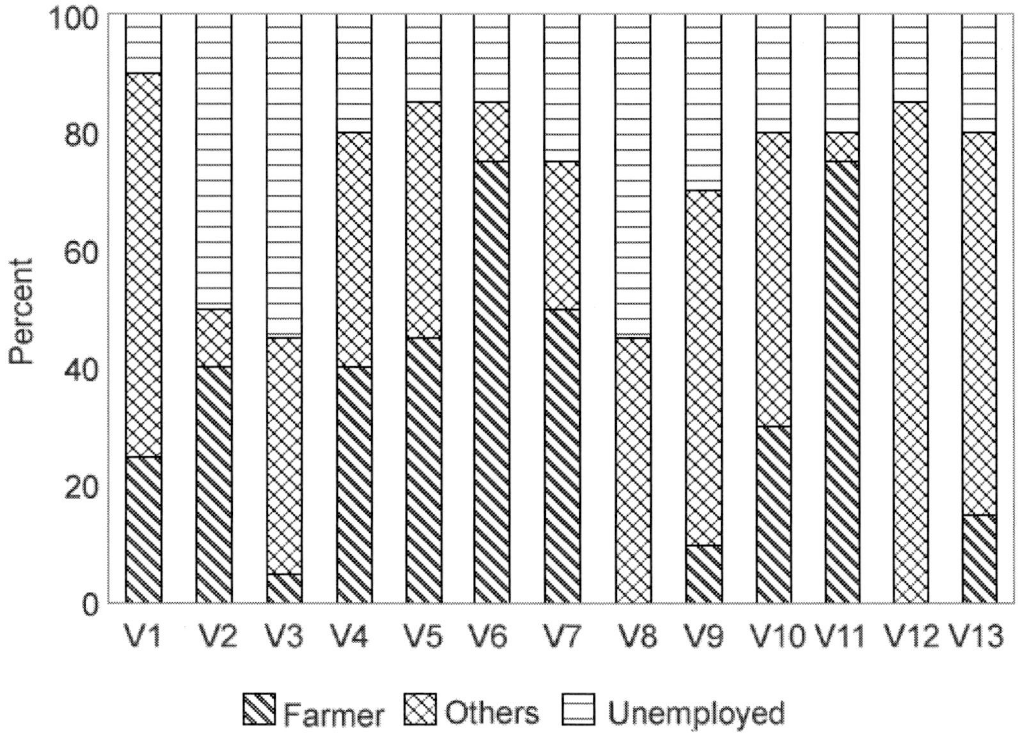

Figure 12.10. Occupational status in the study area (location of villages cf Figure 1).

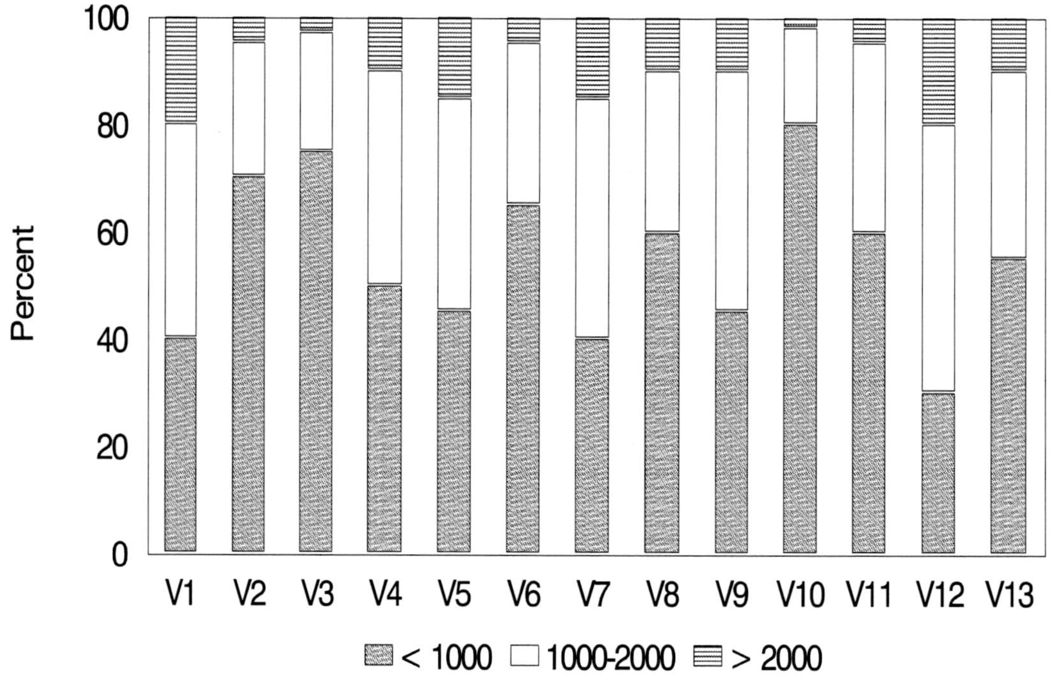

Figure 12.11. Wealth status (Monthly income) in the study area.

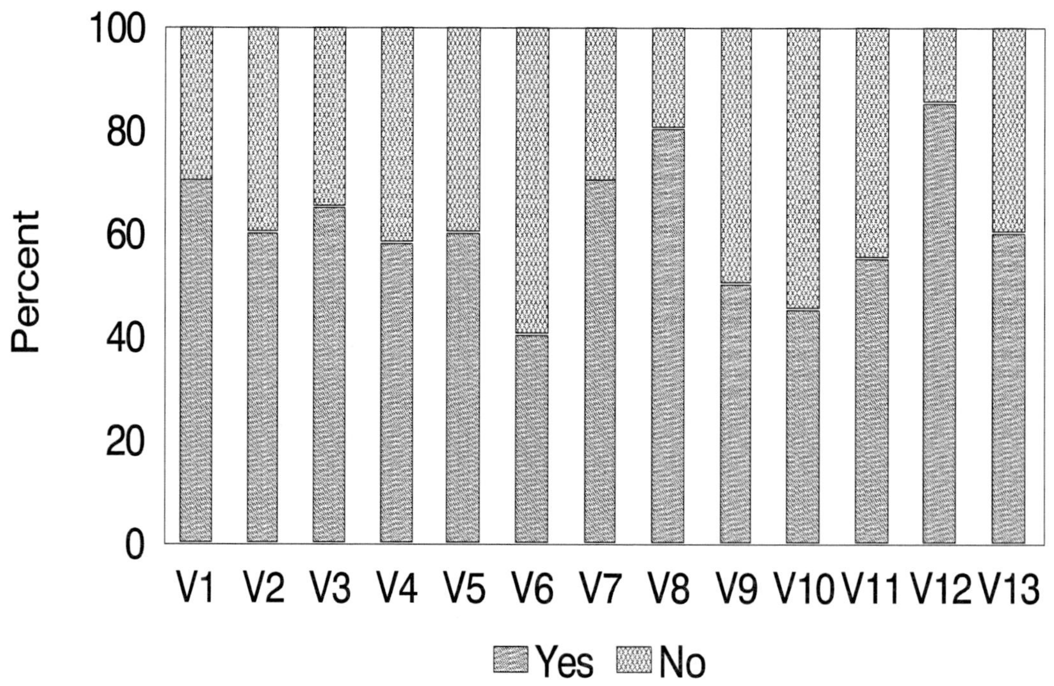

Figure 12.12. Electricity available in the study area.

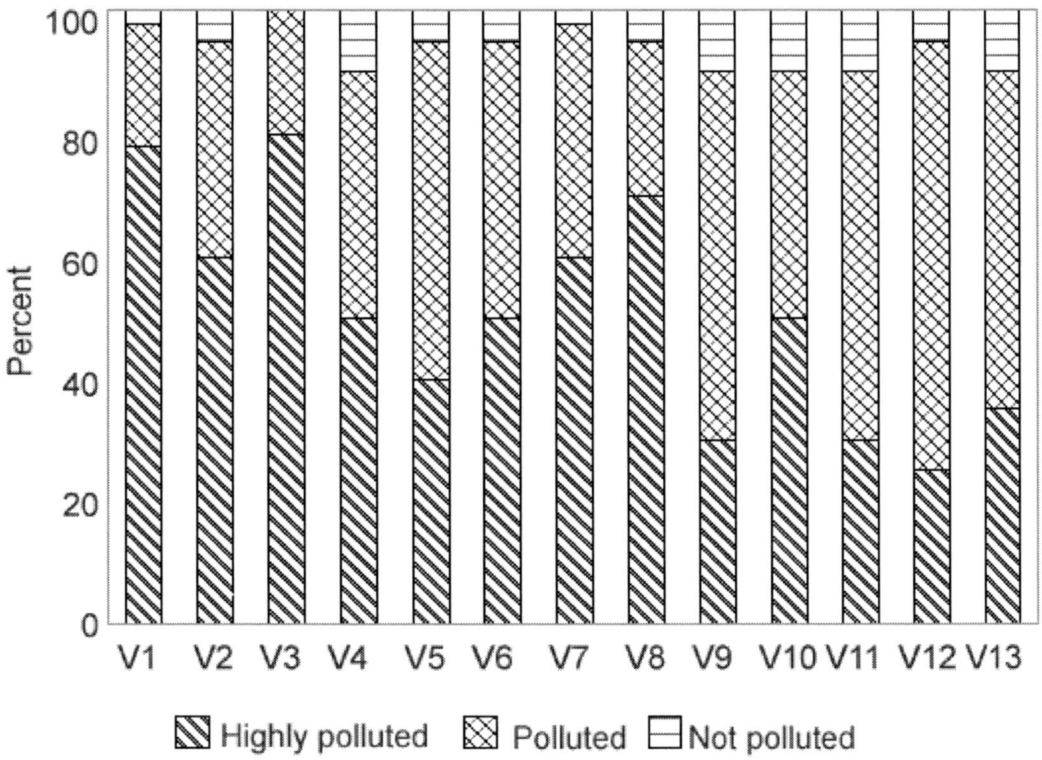

Figure 12.13. Popular opinion about pollution in the study area (location of villages cf Figure 1).

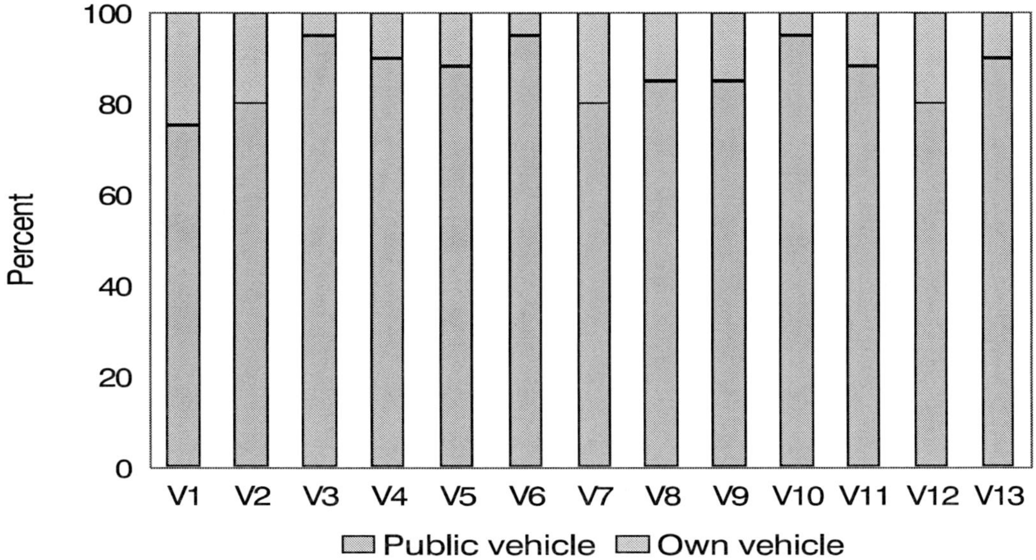

Figure 12.14. Transport facility in the study area.

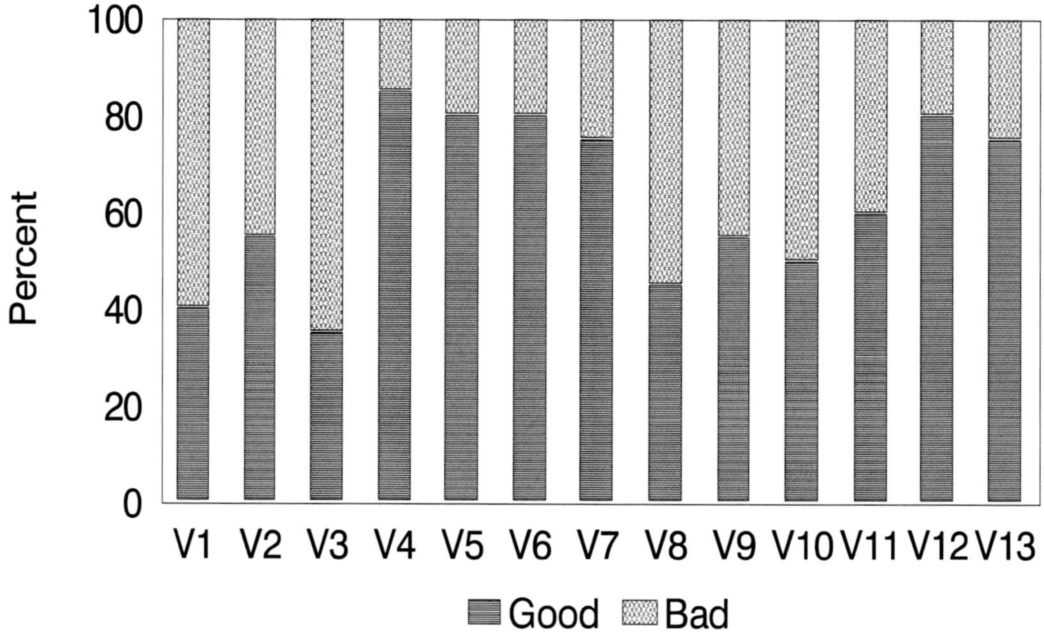

Figure 12.15. Health status in the study area.

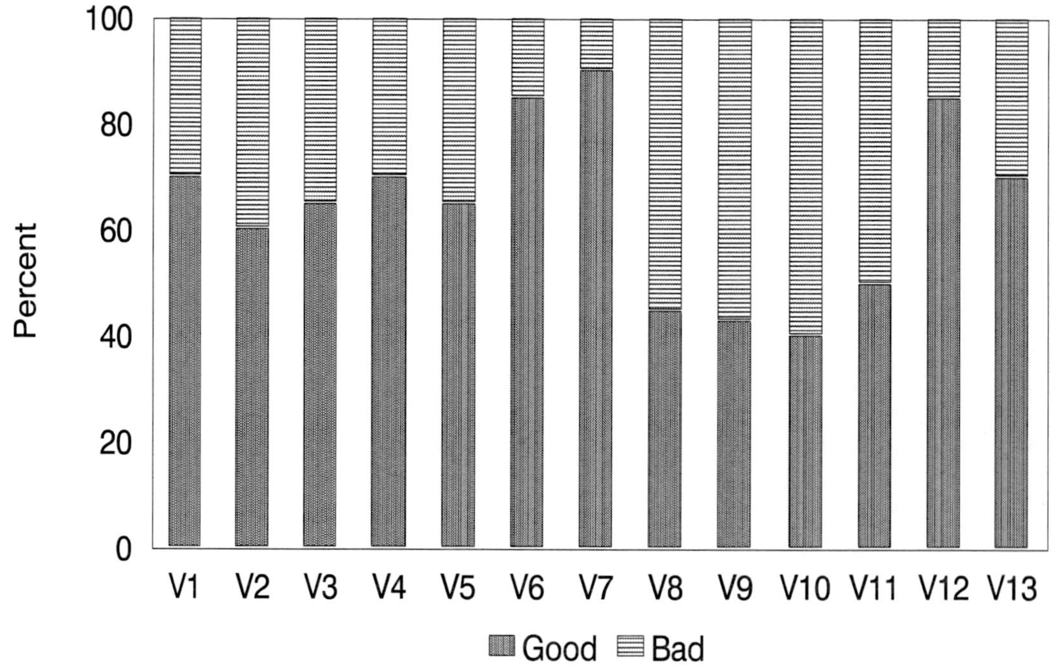

Figure 12.16. Medical facility in the study area.

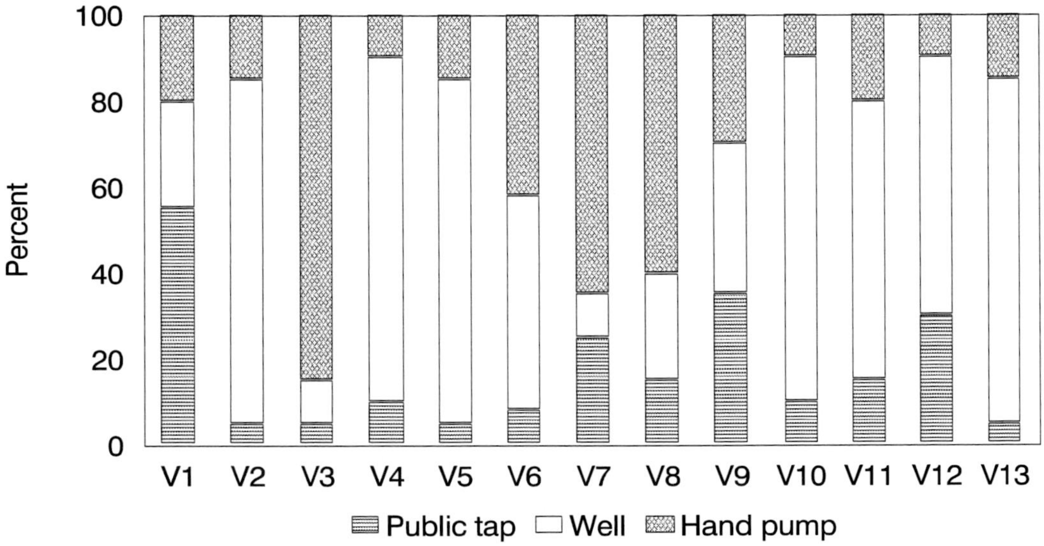

Figure 12.17. Water supply in the study area.

Plates 12.1 and 12.2. Typical stretches of weed-infested barren lands which can be used to raise greenbelts.

(Continued on next page).

(Continued on next page).

Plates 12.3 and 12.4. Gaseous plumes trying to disperse; the *collective* effect of these can be easily felt in the form of assorted malodours that pervade the study area.

PART 2

Chapter 13

Introduction to Part II

13.1. Preamble

As we have detailed in Chapter 14, the history of chemical process industries is replete with accidents of which over 3200 have been major ones. Considering that not all accidents from all countries are allowed to be reported in the international media, the real number is believed to be much larger[6, 7].

With growing industrialization, coupled with ever-increasing demographic pressure all over the world (particularly so in countries situated in the southern hemisphere), the frequency of the accidents and the average damage caused by such accidents has steadily grown[8]. India now has the dubious distinction of having suffered the world's worst–ever industrial accident (the Bhopal disaster, 1984), as also the world's worst accident of the 1990s (the Vishakhapatnam disaster, 1997). We also find that reports of one or the other type of industrial accident, culminating in deaths and destruction, are being received with increasing frequency in the Indian media (for example, please refer Tables 14.9 – 14.16, following chapter).

13.2. The Present Work

The main objective of the present work has been to conduct *regional risk assessment* of the Manali Industrial Complex (MIC) as a companion of the regional EIA of the same area, documented in the preceding part of the book. As such the mendate has been to study the possibilities of occurrence of 'domino effect', or chains of accidents, which can harm more than one industry. However, while conducting risk assessment (RA) on a regional scale – in

---

6 Abdulhamidzadeh, B., Rashtchian, Abbasi, T. and Abbasi, S.A. (2011). Domino effect in process-industry accidents – An inventory of past events and identification of some patterns. Journal of Loss Prevention in Process Industries (Elsevier). 24 (5). 575-593.

7 Tauseef, S.M., Abbasi, T and Abbasi S.A. (2011). Development of a new chemical process-industry accident database to assist in past accident analysis. Journal of Loss prevention in the Process Industries (Elsevier). 24. 426-431.

8 Abbasi, T and Abbasi S.A. (2005). The expertise and the practice of loss prevention in the Indian process industry some pointers for the third world, Process Safety and Environmental Protection. 83, (5B) 413-420.

the process of which we had an opportunity to study some of the risk assessment studies done earlier by others for individual industries – we realised that there is an urgent need for the MIC industries to go for extensive risk assessment including preliminary hazard analysis (PHA), hazard and operability studies (HAZOP), maximum credible accident analysis (MCAA), etc., right up to reworking of disaster prevention/accident management plans. In order to substantiate this suggestion, and to also indirectly illustrate the shortcomings of the existing RAs, we have conducted a few studies of which details are included in the following chapters. We must, at the outset, emphasize that the purpose of this observation is not to belittle the agencies who had done the earlier risk assessment. The science of risk assessment is very new and is expanding fast. The level of know-how at global level has also improved at the same furious pace. By now much more standardized procedures and computer–automated tools are available than before. Also available are experts formally trained in the field of risk assessment. As this centre had the benefit of state-of-the-art tools and expertise, it was perhaps easier to identify the lacunae. Some illustrative examples are given below which pertain to one of the RA reports*:

1) When considering likely accidents in the naphtha storage tank it is assumed that pool fire may occur. But, considering that large quantities of naphtha are involved under *high pressure*, BLEVE followed by fire (fire ball/flash fire) are not only more credible scenarios if a leak were to occur, but also have much greater damage potential. That this is a credible accident is established by the fact that similar accidents have occurred in the past, for example Baton Rouge, (Louisiana, USA); Compana (Argentina), Lake Charles, (Louisiana, USA); Rawley, (England); Kiln, (Germany); Borger, (Texas, USA); Chalmette, (California, USA). These have been listed in the proceeding chapter.
2) A failure in the depropanizer involving butane (which at NTP is in gaseous state) is much less likely to cause pool fire (as envisaged in the said report) but much more likely to generate BLEVE/VCE followed by fireball. The history of past accidents in similar units bears us out. For example accidents which had occurred in similar units at Wilmington, 1955; Port Arthur, 1977; Romeoville, 1984; Juaiman, 1987; Maharastra, 1990; Thane, 1993 etc (which are all listed in Chapter 14, Tables 14.9 to 14.16), had led to fireball and/or explosions, not pool fire.
3) In the other scenarios involving BLEVE (page 68 of the cited report) only radiation effects have been considered whereas substantial damage is also likely to be caused by the shock waves generated in the explosion. Likewise, the possibility of gas explosion in propane and LPG storage vessels has not been considered even as it is one of the most frequently encountered modes of accidents in such vessels, as noted above.
4) The 'worst damage situations' (pages 69&70 of the cited report) have considered only heat radiation effects. Explosions have not been considered at all.
5) Even more worrisome are the omissions in considering likelihood of chain of accidents (domino effect). In Tables 2.1.1 – 2.1.3 of the cited report, for example, possible *initiating* events in various units have been discussed but storage units have

* Risk assessment for Madras Refineries Limited, Volume 1 & II, 1995

been totally left out in the treatment. Considering that much greater quantities of hazardous chemicals are handled in storage units, and considering that the world's worst-ever industrial accident (Bhopal 1984), world's worst accident of 1990s (Vishakhapatnam 1997), and many other major accidents in the past* have involved failures in storage vessels, this is a very serious omission indeed.

6) Then, again, some of the 'worst possible accidents' described in individual units in Volume I of the report have been totally omitted while reporting possible domino effect in the Volume II of the report. Two of the highly credible accident scenarios which may cause secondary and higher order accidents, among several others not considered in the above mentioned report, are:
   - BLEVE followed by fireball in the stabilizer unit. Its impact would encompass an area of 200 m radius, and would trigger chain of accidents. The damage potential of this probable accident encompasses an area of more than ~200 m radius.
   - Instantaneous release of LPG as a vapour cloud; the released chemical on meeting an ignition source would become a fireball. The mishap would have an impact radius of ~750m, generating domino effect.

The cascading effect of both of the above mentioned events may eventually cover the entire MIC.

13.3. Intra-Industry Communication Gaps, and 'No Cause to Worry' Approach

Another reason we have undertaken fresh risk assessment exercises using state-of-the-art tools is the evidence one sees of intra-industry communication gaps, and of 'no cause to worry' attitude. For example representatives of one of the industries told us, "your scenario presumes that PCH is stored under high pressure in our unit but actually the pressure is very less, just about atmospheric, and can't cause BLEVE".

But the industry's own report, sent to us earlier, states clearly that the said chemical in the said unit is stored under a pressure of 3 atmosphere, which is not insignificant by any stretch of imagination. The only plausible explanation for the contradiction between the industry's criticism and their own report, is intra-industry communication gap.

Another example of this type of communication gap, which can have disastrous consequences, is found in the comment on one of the scenarios by the concerned industry saying, 'BLEVE of naphtha storage, in general, is incredible'. But the industry's own earlier report (which they have referred to again and again as an adequate exercise in risk assessment) has cited several past accidents of BLEVE / VCE in naphtha units and in units handling chemicals even Less volatile than naphtha – as many as 10 such past accidents have

* F.P. Lees (1996). Loss Prevention in Process Industries, Butterworts, London, Volme I-III; F.I.Khan, and S.A. Abbasi (1998). Risk Assessment in Chemical Process Industries: Advanced Techniques, Discovery Publishing House, New Delhi, 456 pages

been cited in the industry's previous report (Tables 16.2.1, and Table 16.4, of their report). The representatives of the same industry can't now question our scenario envisaging BLEVE in their naphtha storage units unless there is a major communication gap within the industry.

Risk assessment and safety professionals also always caution industries against overconfidence and 'there is nothing to worry' attitude. One shouldn't be paranoid but smug overconfidence is perhaps the single biggest cause of most of the accidents. It was this kind of over confidence which was behind the most macabre of tragedies associated with process industries — the Bhopal disaster.

In the course of interactions for this book, several industries, emphasized to us that safety provisions have been installed in hazardous units, thereby implying that accidents in these units are no longer within credible realm of probability. This approach is tantamount to saying, "all cars have installed breaks, seat belts, rear-view mirrors, and horns, then why should there be speed-breaks, traffic lights, and accident insurance"? In all hazardous industrial units, appropriate control, forewarning, and accident cushioning systems are installed across the world. This has been done with ever greater care in recent decades. Yet major accidents continue to occur. The most glaring (and catastrophic) Indian examples are the HPCL Vishakhapatnam disaster of 1997 and the Bhopal gas tragedy of 1984; safety systems *were* available but failed to prevent the catastrophe.

13.4. METHODOLOGIES

Risk assessment in chemical process industries is currently one of the fastest growing branches of chemical engineering. As such it is crucial that the concerned professionals keep up with the state-of-the-art. A review of techniques and methodologies for risk assessment in chemical process *industries*, done be the authors, is available in primary literature*.

For the present report we have used the most advanced techniques and tools, including:

I. Most recent indices for fires and explosions, and for toxic release.
II. HAZOP (hazard and operability) studies in their most advanced form.
III. MCAA (maximum credible accident analysis) methodologies using state-of-the-art computer-automated tools including MAXCRED, HAZDIG, MOSEC, and TORAP.
IV. Most comprehensive models and software package for domino effect analysis.

Details of the tools developed or refined by the authors have been published in mainstream referred journals. An illustrative list is presented in Table 1.

* F.I. Khan and S.A. Abbasi (1998), Journal of Loss Prevention in Process Industries, 11 261- 277.

Table 13.1. Techniques and tools for accident forecasting and safety implementation introduced by researchers from India

Authors	Theme	Source
Khan and Abbasi	OptHAZOP	*JLPPI*[a] 10: 191–204 (1997)
Khan and Abbasi	Mathematical model for HAZOP study time estimation	*JLPPI*[a] 10: 249–257 (1997)
Khan and Abbasi	MAXCRED-II	*IJCT*[h] 4: 167–179 (1997)
Khan and Abbasi	TOPHAZOP	*JLPPI*[a] 10: 333–343 (1997)
Khan and Abbasi	DOMIEFECT	*EMS*[d] 13: 163–177 (1998)
Khan and Abbasi	Techniques and methodologies for risk analysis in chemical process industries	*JLPPI*[a] 11: 261–277 (1998)
Khan and Abbasi	Inherently safer design based on rapid risk analysis	*JLPPI*[a] 11: 361–372 (1998)
Khan and Abbasi	Multivariate hazard identification andranking system	*PSP*[c] 17: 157–170 (1998)
Khan and Abbasi	Models for domino effect analysis in process industries	*PSP*[c] 17: 107–123 (1998)
Khan and Abbasi	Cushioning the impact of toxic release from runaway industrial accidents with greenbelts	*JLPPI*[a] 13: 109–124 (1999)
Khan and Abbasi	MAXCRED-III	*JLPPI*[a] 12: 455–469 (1999)
Khan and Abbasi	TORAP	*JLPPI*[a] 12: 299–313 (1999)
Khan and Abbasi	MAXCRED	*EMS*[d] 14: 11–25 (1999)
Khan and Abbasi	Modelling and control of the dispersion of hazardous heavy gases	*JLPPI*[a] 12: 235–244 (1999)
Khan and Abbasi	HAZDIG	*JLPPI*[a] 12: 167–181 (1999)
Khan and Abbasi	PROFAT	*PSP*[c] 18: 42–49 (1999)
Khan and Abbasi	Modelling and simulation of heavy gas dispersion on the basis of modifications in plume path theory	*JHM*[b] A80: 15–30 (2000)
Khan and Abbasi	PROFAT II	*JHM*[b] A75: 1–27 (2000)
Khan and Abbasi	EXPERTOP	*EMS*[d] 15: 67–77 (2000)
Khan and Abbasi	Avoid the domino effect via proper risk assessment	*CEP*[g] 90: 63–72 (2000)
Khan and Abbasi	A new methodology for safety management: SCAP	*JHM*[b] A87: 23–56 (2001)
Khan and Abbasi	SWEHI a user-friendly tool for swift yet comprehensive hazard identification and safety evaluation	*TICE*[f] 79(B): 65–80 (2001)
Khan and Abbasi	Risk analysis of a petrochemical industry using ORA procedure	*PSP*[c] 20: 95–110 (2001)
Khan and Abbasi	Effective design of greenbelts using mathematical models	*JHM*[b] B81: 33–65 (2001)
Khan and Abbasi	A criteria for developing credible accident scenarios for risk assessment	*JLPPI*[a] 15: 467–475 (2002)
Latha et al.	Strategies for quantification of thermally initiated cascade effects	*JLPPI*[a] 5: 18–27 (1992)
Marimuthan	Sodium fire code for pool fire characterization	*JNST*[e] 33: 787–791 (1996)
Sarkar and Abbasi	Enhancing the accuracy of forecasting impact of accidents in CPI by the application of cellular automata technique	*TICE*[f], in press (2005)
Abbasi and Abbasi	FORCIBLE	*JLPPI*[a], communicated (2005)

[a] Journal of Loss Prevention in Process Industries;
[b] Journal of Hazardous Materials;
[c] Process Safety Progress;
[d] Environmental Modeling and Software;
[e] Journal of Nuclear Science and Technology;
[f] Trans IChemE;
[g] Chemical Engineering Progress;
[h] Indian Journal of Chemical Technology.

Chapter 14

Accidents in Chemical Process Industries: Their Assessment and Prevention – An Overview

14.1. Introduction

In December 1984 an accident occurred in India which was not accompanied by any fire, explosion, or even noise. Yet the accident killed or maimed an estimated 200,000 persons and sent such strong shock waves across the world that the resultant vibrations have not died down even till this date (Abbasi and Abbasi 2005; Kletz 1991a).

The accident occurred at 5 minutes past midnight, at 00.05 hrs on December 3, 1984. Almost surreptitiously some 41 metric tons of the lethal methyl isocyanate (MIC) gas leaked out from the factory of the US-based multinational Union Carbide Corporation. Within hours it had not only killed or permanently injured countless people but had also toxifieda huge chunk of land and had left a permanent scar on the face of one of India's most colorful, scenic, and vibrant cities. It is estimated that 50,000 to 500,000 people still suffer from long term effects of the exposure (Willey 1999, Khan and Abbasi 1999; Abbasi and Abbasi, 2005).

The *Bhopal gas tragedy* – as this event has come to be called – is the worst-ever accident in chemical process industry, but it is not something unique. Catastrophic accidents causing massive losses of life and property keep occurring in process industry even as efforts to prevent and control accidents are intensified all over the world, especially in technologically advanced countries. The reason is that new products and processes which involve newer chemicals as also often hazardous process conditions keep adding newer possibilities of accidents even as pre-existing processes are made safer. Moreover, due to increasing population density, especially in the developing countries, larger and larger number of people keep coming under the zone of high risk.

It is important to highlight that accidents in process industry do not cause the kind of loss of life that natural disasters (like earthquake or floods) cause. Nor is it much compared to losses due to individual's mistakes or crimes such as traffic accidents and murders/suicides. This is illustrated from a USA-based study (USBL 1999; Table 14.1). In populous countries like India the number of fatalities due to accidents in chemical process industry may be higher than 3%, yet it is likely to be much lower than fatalities occurring, say, in traffic accidents.

But economic losses are much higher per fatality as is illustrated in another US-based study (Crowl 2003; Figure 14.1).

Thirdly, most accidents in process industry accompany fires and/or toxic releases which cause harm to the environment. Moreover the *impact* of process industry accidents on public confidence, hence industrial growth and nation's economy, is enormously greater in proportion to the losses of life and property. This is because, in public perception, chemical process industries are not seen as much as the provider of so many things essential to good quality of life but as business ventures of rich individuals or protectorates of influential politicians. Any major accident is seen as public harm caused due to the apathy of private interests!

Thus, on one hand, risk of accidents continues to grow, and, on the other hand, the necessity to prevent the accidents is felt with increasing intensity. These concerns, together, have given rise to the engineering science of process safety. This branch of chemical-environmental engineering has acquired high priority status and is beginning to grow rapidly across the world.

In this chapter an overview of accidents in process industry, their causes, their consequences, and the techniques and methodologies for assessing, preventing, and reducing the risk of accidents is presented.

The chapter aims to set the context in which the regional risk assessment, described in the subsequent seven chapters, has been carried out. It also aims to introduce the techniques and methodologies that were used by the authors.

14.2. Definition of an Accident

According to Suchman (1961) an event can be classified as an accident if it is unexpected, unavoidable and unintended. He has proposed three characteristics with which to classify an event as an accident:

- Degree of expectedness
- Degree of avoidability
- Degree of intention

Secondary characteristics are:

- Degree of warning
- Duration of occurrence
- Degree of negligence
- Degree of misjudgement

An event is an accident if it gives little warning, happens quickly, or if there is a large element of negligence and misjudgement leading to it.

Suchman has added that as knowledge increases an event is more likely to be described in terms of its causal factors and less likely as an accident.

Conceptual models to describe the accident phenomena: Houston (1971), has present a conceptual model on the basis of the practices of lawyers and insurers which focuses attention on the `proximate cause'. It is recognised that many factors contribute to an accident, but for practical, and particularly for legal, purposes a principle cause is identified. This approach has a number of defects: there is no objective criterion for distinguishing the principle cause; the relationships between causes are not explained; and there is no way of knowing if the cause list is complete.

Another type of conceptual model is the fault tree model. A simple fault tree model of an accident is presented in Figure 14.2. The initiating event which constitutes a potential accident occurs only if some enabling event occurs or has already occurred. This part of the tree is termed as `demand' tree since it puts a demand on the protective features. The potential accident is realised only if prevention by protective equipment and human action fails. An accident occurs which develops into a more severe accident only if mitigation fails. A similar model based on fault tree has been proposed by Well et al. (1992).

Another approach is that taken by Kletz (1988), who has developed a model oriented to accident investigation. The model is based essentially on the sequence of decisions and actions which lead up to an accident, and shows against each step the recommendations arising from the investigation (Figure 14.3).

A model which emphasises the broader socio-technical background to accidents has been developed by Geyer and Bellamy (1991) as shown in Figure 14.4.

14.3. Wisdom of Hindsight: Past Accident Analysis

To understand the mechanisms of accidents and to develop accident prevention and control strategies, it is essential to know about and learn from past accidents.

This aspect is of particular significance in the area of loss prevention and process safety because it is very expensive and difficult to acquire information on the mechanisms and control of accidents by simulating the accidents (Abdolhamidzadeh et al., 2011; Tauseef et al., 2011). For example full-scale explosions are so destructive and costly that no attempt has been made as yet to experiment with such an explosion for the sake of obtaining precise information on its nature or consequence. Instead, scaled down versions are performed. They do provide very useful pointers but do not fully reflect what happens in the real-life situation. Moreover, a major accident occurring in an actual process plant almost always causes secondary accidents. It is not possible to simulate the vastly different situations that exist on the ground in different process plants even though they all may be dealing with similar chemicals and units. Post mortem of past accidents is, therefore, an essential and irreplaceable means to learn how accidents occur, how their recurrence can be avoided, and how the damage caused by them can be minimized if they do occur.

But ‘past accident analysis’ is easier said than done. As soon as an accident occurs the concerned industry tries to absolve itself of the blame, so do others likely to be implicated. There is also unintended confusion when the accounts of eyewitnesses differ. As a result conflicting versions of what caused an accident and how it unfolded gain currency. There are also disputes over extent of damage. Even as inquiry commissions try their best to reach the truth and process safety experts do periodic post-mortems of major accidents, several aspects of most past accidents still remain shrouded in mystery. For example what exactly happened during the explosions at Felixborough – one of the most extensively studies of all process industry accident which took place in 1974 – is still a subject of doubt and analysis (Venart 2007). Likewise there never has been a complete consensus on whether the Bhopal gas tragedy was a result of laxity by the Union Carbide Corporation (UCC) management or an act of sabotage by someone out to spite the UCC (Abbasi and Abbasi, 2005). Most other accident histories lie under similar shrouds of uncertainty. There are also situations where there is no dispute over the sequence of events but there is uncertainty on the time that elapsed between different events and extents to which each stage influenced the next. The serial fires and explosions that destroyed the refinery at Mexico in 1984 come under this category. These aspects have been discussed by Badoux (1983), Marshall (1987), Kletz (1989), Haastrup and Brockholf (1990), Less (1996; 2005; Abbasi and Abbasi 2005). Unfortunately the tendency of not only the industries but also, in some cases, of the governments to cover up the truth seriously hampers past accident analysis, and, consequently, avoidable recurrence of some accidents. Among these accidents many are due to the repetition of same/similar faults (Kletz-1991a, 1991b). Not only for fixed installation accidents but even for accidents occurring during transportation of chemicals, there is always someone or the other interested in suppressing the facts.

Inspite of these constraints, past accident analysis is an invaluable tool for process professionals.

With the rise in the pace of economic growth, hence increase in the number and size of process industries, in the BRIC countries (Brazil, Russia, India, and China), there is increasing interest in the process safety scenario in these countries. The situation was reviewed recently (Abbasi and Abbasi 2005). The pattern of industrialization in India reflects the trends in other rising economies.

In India there are as many as 29, 716 units which generate hazardous waste as defined by the Hazardous Waste (Management & Handling) Rules, formulated initially in 1989 and updated in 2008 by the Government of India (Table 14.2). Of these, 2744 are large-scale 'highly polluting' industries (Table 14.3). Most of these industries are situated in the midst of densely populated areas. Even when an industry is initially located in an unpopulated location, commercial activity and infrastructure associated with the industrial activity soon attract settlements of people and within a few years the industry is surrounded with thickly populated neighbourhoods (Khan and Abbasi, 1998). This aspect enhances the risk of accidents. For example the death toll caused by the toxic gas leak at Bhopal in 1984 was much higher than it would have been due to the thickly populated slum that had come up all around the factory boundaries (Abbasi and Abbasi 2005).

Tables 14.2 and 14.3 indicate that the Indian states of Maharashtra and Gujarat together account for nearly a third of all major polluting and hazardous industries in India. No wonder, then, that the maximum number of accidents in the past in India have occurred in these two regions. The government has promulgated strict rules for handling and managing hazardous chemicals (Abbasi and Abbasi 2005). Among other things it is mandatory for each industry to report all the accidents that occur in its premises, including very minor ones. But implementation of the rules is never easy and a great deal of laxity goes undetected (Abbasi and Abbasi 2005).

14.4. Major Process Hazards

The major hazards with which the chemical industry is concerned is fire, explosion, and toxic release. Of these three, fire is the most common but explosion is more significant in terms of its damage potential leading to fatalities and damage to property. Toxic release has perhaps the greatest potential to kill a large number of people and reduce an area toxified for several months to several years. Large toxic releases are rare but, as the Bhopal gas tragedy indicates, may have a very high death toll.

Often no distinction is made between fire and explosion losses. The latter are normally included in the overall fire statistics. In fact it is explosions which cause the most serious losses (Doyle-1969, 1981, Norstrom-1982, Davenport-1988, Carson and Maurnfold-1988). As much as two-thirds of the losses in accidents occurring in chemical process industries is attributable to explosions (Health and Safety executive-1988; Lees-1996). Over three-quarters of the explosions involve combustion of explosive materials. The most frequent cause of fires is release and overflow of flammable gases and/or liquids closely followed by overheating/hot surfaces (Table 14.4). Whereas fires contribute about % of the total losses, explosions contribute about 66% (Crowl 2003). Failure of proper reaction controls seems to be the most frequent cause leading to accidents. It account for about 31% to the total number of accidents (Table 14.5). The processing area is the most susceptible location of the accident, and equipment rupture the most common contributing factor.

Bellamy et al (1989) have computed the frequency of accidental leaks as a function of location, operation and type of material. Their findings are summarized in Table 14.6. It may also be seen that about a third of all accidental releases end up in fires or explosions. Of the latter, about 25% events have explosions followed by fire.

It follows that the problem of avoiding major hazards is essentially that of avoiding loss of containment. This includes not only preventing an escape of materials from leaks etc., but also avoidance of an explosion inside the plant vessels and pipe work. Some factors which determine the scale of the hazard are:

1) the inventory
2) the energy factor
3) the time factor
4) the intensity-distance relations
5) the exposure factor, and
6) the intensity-damage and intensity-injury relationships

Data related to major injuries (including fatalities) in chemical and allied industries has been plotted in Figure 14.5. It is evident from the data that in chemical industries the rate of fatal injuries are less compared to mineral oil processing industries (refineries, etc.). But during the last few years the rates seem to have come close to each other for the two types of industries.

14.5. General Causes of Accident

Exactly where lies the trigger of an accident? As explained in Section 2.0 this question is not easy to answer. After an accident occurs, different analysts and pressure groups put different theories of the possible causes. There are almost as many different diagnoses as there are investigators. Unfortunately, there is no acceptance, or consistent cause-effect diagnostic system so that it is often difficult to reconcile different analyses. Till this date the UCC management has refused to accept the general conclusion that the Bhopal disaster was due to their negligence. In fact as soon as a major accident occurs, the concerned industry normally tries its best to disown the responsibility and underplay the public loss. The governmental agencies aren't much helpful either (Abbasi and Abbasi 2005). These aspects have been illustrated in Section 4.0 of the chapter *Environmentalism*. The legal battles for compensation are long and bitter and the innocent victims are generally the biggest losers.

14.6. Reporting of Incidents and Databases

The extent and accuracy of the reporting of incidents and injuries being uncertain, it is particularly difficult to arrive at reliable pointers from the statistical analysis of the data.

For example past incidents in the USA the UK, and some of the EU (European Union) countries have generally been reported in detail and analysed critically but comparable incidents in the erstwhile USSR, China and Balkan states have received much less publicity and assessment.

The problem has been discussed by Badoux (1983). Figure 14.6 shows schematically the probable extent of under-reporting, curve A representing the actual reporting situation and curve B the ideal one (Lees-2005).

There are a number of databases specifically dealing with case histories. They include

- Major Hazards Incident Data System (MHIDAS) and the corresponding explosives data system EIDAS. These are operated by SRD (safety and reliability directorate, UK atomic energy authority).
- The FACTS incident database.
- The Major Accident Reporting System (MARS).
- The FIRE incident database for chemical warehouse fires.
- The offshore Hydrocarbon Release (HCR) database.

The first part of this chapter presents an overview of industrial accidents with a view to identify the factors which lead to accidents and the lessons for loss prevention to be learnt from these accidents. Such information is very helpful in developing newer know-how for process safety and accident damage control. The second major part of the chapter describes the techniques and methodologies that have been developed towards assessing and managing risk of accidents in CPI.

14.7. Classification of Accidents

Accidents dealing with hazardous chemicals can be broadly categorised in to two major groups: fixed installation accidents and transportation accidents. The fixed installation accidents consider all accidents occurring in industries during on-site-storage or different stages of manufacture on-site. The transportation accidents consider accidents occurring during transportation, loading or unloading of chemicals, and can be further categorised according to the different modes of transportation.

A search of literature covering the time span 1926 through to present reveals reports of over 14000 accidents relating to handling/transportation/processing/storage of chemicals (Nash-1976, Lewis, 1984, 1993, Marshall-1987, Hasstrup and Brockhoff-1990, Chowdhury and Parkinson, 1992, Taylor-1993, Thomas, 1995, Lees,1996; Khan and Abbasi, 1996, 1997, Azhagiri 2007, Abbasi and Abbasi 2008). The actual number may be high as reports of all accidents are not available in primary literature. Moreover we have considered here accidents involving losses of more than 1 million and/or at least one fatality.

Of these accidents, 54% are fixed installation accidents, 41% are transportation accidents and 5% are miscellaneous accidents (Figure 14.7). The transportation accidents can be further classified according to the different modes of transportation. Such classification (Figure 14.8) indicates that 37% of them were caused in rail transport, 29% in road transport, 6% in marine transport, 18% in pipe-line transport , 4% in inland waterway transport, and the remaining during loading and unloading of chemicals. There is a controversy whether pipe-lines should be considered as fixed installations or a means of transportation. We feel that the portion of the pipe-line which falls within the confines of the industry should be treated as fixed installation and the portion outside the industrial periphery as a transport vehicle.

Of the different means of transportation, the one involving railroads has higher damage potential as larger quantities are transported by this means. However, if we consider the damage it may cause to life and property, transport by road is more hazardous as roads often pass through populated areas, especially in developing countries (Khan and Abbasi, 1995).

Pipe-line transportation is comparatively safer, provided that rate of transportation, and conditions of transportation (temperature, phase, and pressure) as well as route of the pipe-line are carefully managed. A summary of the worst transport disasters are presented in Table 14.7.

14.7.1. A Fixed Installation Accidents

The survey reveals that over 2000 significant accidents have occurred during the period 1928-2008. A study of major factors (vessels, chemicals, process conditions) leading to accidents is summarised in Table 14.8. It reveals that chemical process plants are most prone to accidents. Ammonia is the most often involved chemical. Of the total accidents, 25% accidents involved fires and explosions, and 71% involved toxic release. The remaining accidents (4%) featured a combination of fire, explosion and toxic release. In terms of harmful consequences, toxic release covers wider areas than fires/explosions. Also if the toxicity of the released chemical is high, as was the case with dioxin leak at Seveso or the MIC (methyl iso-cynate) leak during the Bhopal disaster, the damage may be very severe.

The destructive impact of explosions generally covers wider area than the region of impact of a fire. Secondly, except in certain cases when ambient conditions conspire to enable very rapid spread of a fire, most fires take time to consolidate. If emergency preparedness measures are in place, this time proves crucial in enabling the control of the fire. On the other hand, when an explosion takes place it does so instantly, giving no time to escape from it.

In a very large number of situations, explosions in chemical process industries are either caused by fire, or lead to a fire. A gist of major catastrophic accidents for a period of 1928 to the present has been presented in Tables 14.9 - 14.11. The information has been presented for 3 time spans: the pre-environmental consciousness era (1928-1969), the next 2 decades, and the period since (1990-2008). It gives the alarming pointer that inspite of ever greater attention paid to process safety in each of the eras, the frequency, the enormity, and the global spread of industrial accidents haven't diminished. This information was collected from various literature sources – Nash, 1976; Gugan,1979; Amesz *et al* 1983; Lees,1980;1994,1996; Marshall, 1977, 87; Kletz,1988,91a; Lees and Ang-1984; Kharbanda and Stallworthy 1988; Hasstrup and Brockhoff 1990; Palmer, 1983, Prugh. 1991; Amendola *et al* 1988, TPL-1992; Khan and Abbasi 1996,1997; Koivisto *et al* 1989, Koivisto and Nielson 1994, Chemical industrial digest,1995; Azhagiri 2007; Loss Prevention Bulletins 1980-2008.

The worst-ever accident in chemical process industries involving toxic release occurred at Bhopal in 1984. The worst ever fire-cum-explosion accident (on shore) occurred at Mexico the same year. The worst-ever off-shore accident had occurred at Piper Alpha in 1988.

14.8. Past Accident: Illustrative Case Studies

14.8.1. Accidents in Indian Refineries and Petrochemical Industries

As stated earlier, it is exceedingly useful to study the past accidents and derive knowledge on how to prevent similar accidents from occurring in future. From among thousands of case studies we present below a few which give an idea of the myriad ways in which accidents occur, of how one accident often leads to another, and the resultant harm.

On 22, June 1974, a 16" elbow of pipe carrying potassium carbonate solution in a fertiliser plant at Tamilnadu, India, ruptured suddenly, splashing the hot solution into the nearby control room. The toughened glass panes shattered; 8 persons died in the control room instantaneously, one died in the hospital and others sustained grievous injuries.

On 8 March, 1984, an explosion in a refinery at Kerala destroyed a fire tender along with the shed wherein it was housed, besides a chemical warehouse, cooling tower and other facilities. Later investigations revealed many shortcomings in the plant layout.

On 12 December, 1987, a crude oil storage tank in a refinery at Maharashtra, India, started boiling over, spilling the contents in the dyke around it. Emergency services and fire brigades, who were alerted, tried to evacuate the contents. After 4 hours of pumping out, the tank caught fire and exploded. Spilling its contents. Eight hours of vigorous fire fighting had to be carried out before the fire could be controlled. There was extensive damage to property. A liberal sizing of the dyke and providing a separate dyke for the large tank would have helped to prevent the spread of fire to other tanks.

An accident took place on 18 April, 1989, in a 14" natural gas pipeline in a gas company in India. The pipeline was carrying compressed natural gas at a pressure of about 295 - 298 psig from the compressor station to various consumers. The spot of the accident was about 730 ft. from the compressor station. A security personnel heard a loud sound at about 0950 hrs and saw a huge cloud of black smoke emanating from the ruptured pipeline which caught fire immediately. The flame rose as high as 150 ft during the initial stage.

The fire damaged the general stores and the office of the materials department. Two employees died and six others received burn injuries. Investigations revealed that the portion of the pipeline which had blown off was extensively corroded compared to other portions of the pipeline. The underground pipeline was close to the materials department where old lead cells were stored. The corrosion could have occurred due to the leakage of spent weak acid which seeped through the ground and corroded the buried pipeline.

Panipat Explosion

An explosion followed by a toxic release of ammonia occurred at the National Fertiliser Limited (NFL) fertiliser plant near Panipat. Till today there is no accurate official report available on what happened and how it happened. However, some reliable sources reported that on that evening, an operator observed a leak in one of the vessels and alerted the supervisor. To rectify the problem the vessel was isolated and repaired. After repair the vessel was brought back to the process without checking whether the isolation (slip plate) had been removed or not. As time passed pressure started building up in the vessel and after a few hours an explosion (BLEVE) occurred spilling the contents of the vessel over the surrounding area. As the plant was situated far from the populated areas and the quantity was not too much, the consequences were not very severe.

Four members of operating team and two shift engineer died, more than 25 peoples were injured and more the 1000 peoples were adversely effected. The consequence had controlled due to proper safety arrangements otherwise the death toll and damage would have been higher. An area 2 sqkm was severe damaged and a loss of around $ 20 million had been estimated to have occurred.

Maharashtra Accident

On 5 November, 1990, an explosion at the offside battery of compressors at a gas cracking plant in Maharashtra, India, killed 35 persons besides causing heavy damages to property and business interruption losses. Among the deficiencies in the layout, identified after the disaster, was the location of a contractor's shed dangerously close to the gas compressors. Less publicised, but perhaps of greater consequence, was the lack of facility to shut down the flow of hydrocarbon at the site itself. The plant personnel had to run to the control room faster than the vapour that followed them to close the feed valve.

On 25 November 1997 an explosion occurred in an chemical industry manufacturing rubber products at Halol in Panchmahal district of Gujrat state. Three persons were killed while eleven others were injured (The Hindu, 1997a). The explosion occurred in one of the reactors.

Visakhapatnam Disaster

On September 14, 1997, several large fires and explosions devastated the terminals and storage tanks at the refinery of HPCL (Hindustan Petroleum Corporation Limited) at Visakhapatnam. More than 55 persons were killed and dozens of others seriously injured (The *Hindu*, 1997).

Two bodies were found on the upper storey of the administrative block which had collapsed while three more were seen in the debris underneath by a team of reporters who ventured in later in the evening. The building, housing, the recreation club and canteen were also destroyed.

One of the eight Horton spheres (globe tanks), located near the main gate of the HPCL refinery. Counting either LPG, crude or kerosene caught fire at 6.40 a.m. and exploded, rocking the entire Visakhapatnam city. The storage tanks were full, with crude imports unloaded at the HPCL berth just a few days ago. The second sphere exploded 15 minutes later and before noon the others caught fire. The blaze spread all over the place. Huge tongues of flame and thick black smoke billowed into the sky and joined the hovering monsoon clouds. There was a sharp shower in the morning and people wearing white shirts saw them turn black with soot. The rain water flooding the road also turned black and murky.

With both the entrances to the refinery blocked by burning tanks, neither the fire tenders nor the officials could enter the premises for several hours. Only when the contents in the tanks were burnt out could they venture in. The death toll could have been higher had the fire started half-an-hour later when the first shift staff would have come in.

Luckily, as the accident occurred on a Sunday, a holiday, when, the administrative personnel, numbering over two hundred, were on we are duty. There were some contract labourers along with the HPCL personnel in the Crude Distilling Unit which was shut down for routine maintenance work. The shock of the initial explosion made people think an earthquake had occurred. They ran helter-skelter leaving their belongings.

14.8.2. Accidents in Chemical/Petrochemical Industries across the World

On the evening of 21 September, 1921 two explosions occurred at the Oppau works of Badische Aniline and Sodafabrik (BASF) in aspan of 3 seconds. The explosions created a mammoth crater of 80 m diameter, destroyed the plant, and 700 of the 1000 houses nearby. The explosion was caused by the detonation of some 4500 t of a 50:50 mixture of ammonium sulfate and ammonium nitrate. It was set off by blasting powder, which was being used to break up storage piles of material which had become caked. Exactly the same procedure had been carried out without any mishap some 16000 times previously!

Even houses in the adjacent city of Ludwigshafen and in the Mannheim area were damaged. Walls were dislocated and windows broken. At these places and at Heidelberg, which is about 14 miles from Oppau, the effect of the explosion was first felt as two very heavy earthquake-like shocks. In Mannheim some seconds later and in Heidelberg 82 seconds after the shocks there came an enormous rush of air which broke windows and doors and caused damage to gas holders, oil tanks, and many river barges. The sound of the explosion and the earth shocks reached as far as Bayreuth, at a distance of 145 miles, and the air pressure wave caused considerable damage in Frankfurt, which is about 53 miles from the scene of the explosion. The explosion killed 430 people, including 50 people in the village[2].

On 3 July 1987 an explosion occurred inside an ethylene oxide purification column at a chemicals factory at Antwerp, Belgium (Lees, 1996). The explosion was due to the decomposition of ethylene oxide. It was accompanied by a fireball, which started a number of secondary fires. These together with blasts and missiles, caused extensive damage. Fourteen people were injured.

Faulty operations at the Tomsk-7 fuels reprocessing facility in Russia isbelieved to have resulted in the runaway reaction of a solution of 500 litres of tri-butyl phosphate (TBP) saturated with strong nitric acid, resulting in the explosive failure of the storage vessel and subsequently blowing out a wall of the reprocessing building. TBP is an important organic solvent used in acidic extraction steps in separation processes at fuels reprocessing facilities. Solutions of TBP, hydrocarbon diluent, and HNO_3 (known as "red oil" because of the colour of nitrated hydrocarbons) undergo exothermic reactions that can thermally "run away" if heated to temperature where the heat of reaction exceeds heat losses.

The accident on (May 14, 1997) at Hanford site was the result of a spontaneous (autocatalytic) chemical reaction of the solution stored in tank (Tank A-109) located in the plutonium Reclamation Facility. This 1500 litre tank initially contained a solution of 0.35 Molar (M) hydroxylamine ($OHNH_2HNO_3$) and 0.25 M nitric acid, called CCX solution. The unused solution in the tank had been slowly evaporating. The loss of water concentrated the solution until conditions were reached that caused a spontaneous chemical decomposition reaction. The reaction created a rapid release of gases, which built up pressure inside the tank. The pressure below the lid of the tank and severely damaged the room. No casualties have been reported as no one was near at the time of the accident.

The Flixborough Disaster

The Flixborough Plant of Nypro Limited, UK, was built for the production of caprolactum which is the basic raw material for the production of Nylon 6. Cyclohexanol necessary for the production of caprolactum was produced by the oxidation of cyclohexane. The last named chemical, which in many of its properties is comparable with petrol, had to be stored. More importantly, large quantities of cyclohexane had to be circulated through the reactors under a working pressure of about 8.8 Kg/cm^2 and a temperature of 155 ^oC. The reaction is exothermic; any escape of cyclohexane from the plant was therefore dangerous. The cyclohexane plant at Flixborough consist of a stream of six reactors in series in which cyclohexane was oxidised to cyclohexanone and cyclohexanol by air injection in the presence of a catalyst.

On the evening of 27 March 1974, it was discovered that reactor number 5 was leaking cyclohexane. The following morning an inspection revealed that the leak had extended by some 6 ft. This was a serious state of affairs and a meeting was called to decide the course of action. A decision was taken to remove reactor 5 and to install a bypass assembly to connect reactor 4 directly to reactor 6 so that the plant operation could continue.

The openings to be connected on these reactors were of 28" diameter, but the largest pipe which was available on site and which might be suitable for the by-pass was of 20" diameter. The two flanges were at different heights so that the connection had to take the form of a dogleg of three lengths. Calculations were done to check that a) the pipe had large enough cross-sectional area for the required flow b) that it was capable of withstanding the pressure as a straight pipe.

But no calculations were done which took into account the forces arising from the dog-leg shape of the pipe; no drawing of the by-pass pipe was made other than in chalk on the workshop floor; and no pressure testing was carried out either on the pipe or on the complete assembly before it was fitted. A pressure test was performed on the plant, after the installation of the bypass, but the equipment was tested to a pressure of 9 kgF/cm^2 . Further, the test was pneumatic not hydraulic (Lees, 1996).

The plant was restarted. Initially the bypass assembly gave no trouble. On May 29, 1974 the bottom valve on one of the vessels was found to be leaking. The plant was again shut down for repairs, and restarted, on June 1. A sudden rise in pressure up-to 8.5 kgF/cm^2 occurred early in the morning when the temperature in Reactor-1 was only 110 °C and lesser in other reactors. Later that morning, the pressure reached 9.1-9.2 KgF/cm^2.

During the late afternoon an event occurred which resulted in the escape of large quantities of cyclohexane. This event was the rupture of the dog-leg shaped bypass system. It was perhaps aided by a fire on a nearby 8" pipe. The fugitive cyclohexane soon caught a spark, and there was a massive unconfined vapour cloud explosion. The blast and the fire destroyed the cyclohexane plant as well as several other plants in its vicinity (Figure 14.9).

Of those working on the site at the time, 28 were killed and 36 others suffered injuries. Outside the plant, injuries and damage were widespread but no-one was killed. Of the 28 people who died 18 were in the control room. Some of the bodies had suffered severe damage from flying glass. The main office of the factory was demolished by the blast of the explosion. Mercifully the accident had occurred on a Saturday afternoon when the offices were not occupied. If they had been, the death toll would have been much higher.

Property damage extended over a wide area, and a preliminary survey showed that 184 houses and 167 shops and factories had suffered to a greater or a lesser degree.

The Worms Explosion

On November 21, 1988 an explosion occurred in an liquid storage vessel of Proctor Gamble, Worms, Germany. The explosion was supposed to be the worst ever among explosions in cryogenic storage of liquefied gas, carbon di-oxide. The storage tank was horizontal-high-pressure vessel having nominal capacity of 30 metric tons of carbon oxide and was well connected with relief and safety systems.

The main reasons for the explosion were identified as; a) over heating causing over pressure and failure of the relief valve, b) brittle failure of the ank at or near the normal operational value c) a combination of the above. Detailed investigations have been carried out by Federal Government as well as Proctor Gamble to find the real causes of failure. Finally it was found that the tank suffered brittle failure from two position. Due to high overpressure, the liquid carbon di-oxide escaping from the relief valve attained critical point and sealed the relief valve by forming dry ice. There was no chance for the gas to escape, leading to an explosion.

The explosion intensity was so high that it destroyed two of its neighboring units. An over pressure of more than 1atm was reported over a radius of 1000 meter. The shock wave velocity also exceed 500 m/s. Fragments of the vessel weighing over 100kg were found more than 500 meter from the accident site. Luckily the location of the hazardous units was away from the path of the Missiles, otherwise the damages might have been a lot more severe.

Explosion caused three fatalities at the site, and injured ten others. The monetary damage was estimated to be $ 20 million.

Pepcon Explosion

On May 4, 1988, a massive explosion destroyed the Pacific Engineering and Production Company, (PEPCON) plant near Henderson, about 12 miles south of Las Vegas, USA (Figure 14.10).

PEPCON was one of the only two plants in USA that produced ammonium perchlorate (AP), the other was the Kerr-Mc Gee plant, also located in Henderson about 2 miles from the PEPCON plant. PEPCON reportedly produced about one-third of the AP used as an oxidiser and propellant in solid composite rocket fuels for NASA's space shuttle and missiles.

Although a fire started the PEPCON explosion, the cause of the fire was not easy to explain. After the explosion, PEPCON blamed the fire on a leaking underground pipeline of the Southwest Gas Company that traversed PEPCON'S property. But the natural gas pipeline had been installed about 10 years before PEPCON's plant, and although ruptured, it only contributed to the fire and heat required to detonate the second and the largest explosion.

The fire was also attributed to a welder's torch but one of the reports absolved the welder of the blame. Some blamed the batch dryer's fibre glass insulation which had a history of AP spills into the combustible insulation .

The following were the tell-tale conditions in and around PEPCON.

a) lack of proper storage;
b) combustible fibre glass insulation and sources of fire;
c) glass panel walls in the batch house;
d) inadequate spacing between adjacent process vessels and product storage tanks;
e) no alarm to warn plant personnel, fire departments or Henderson's other citizens;
f) no dependable fire-fighting arrangement with sprinklers and deluge system;
g) no modern dependable radio system to back-up damaged telephone lines needed to call for help and to co-ordinate response teams and warn the community;
h) lack of an effective emergency response plan at PEPCON, the surrounding industrial complex and the town of Henderson.

The explosion caused about $ 100 million in damages to the surrounding community and completely destroyed a neighboring marshmallow plant. About 350 persons were injured. The plant manager and the controller were killed.

The Phillips Explosion

The explosion at the Phillips petrochemical plant in Pasadena, Texas, on 23 October 1989 is one of the worst industrial accidents of the last 20 years.

The immediate cause was simple: a length of pipe was opened up to clear a choke without bothering to see that the isolation valve (which was operated by compressed air) had not been duly closed. The air hoses which supplied power to the valve were connected up the wrong way round so the valve was open when its actuator was in the closed position. Identical couplings were used for the two connections so it was easy to reverse them. According to company procedure they should have been disconnected during maintenance but they were not. The valve could be locked open or closed but this hardly mattered as the lock was missing. The explosion occurred less than 2 minutes after the leak started and two iso-butane tanks exploded 15 minutes later. The explosive force was equivalent to 2.4 tons of TNT: twenty three people- all employees - were killed and over 130 were injured. Nearly 40 tons of ethylene gas leaked and exploded.

The Siberian Accident

Perhaps the most macabre accident - next only to Bhopal gas tragedy in its severity - occurred on June 3,1989, near Nizhnevartovsk in Western Siberia. Engineers stationed there noticed a sudden drop in pressure at the pumping end of an LPG pipeline. The pipeline was commissioned in 1985 to carry mixed LPG to feed the industrial city of Ufa. Instead of investigating the trouble, the engineers responded by increasing the pumping rate in order to maintain the required pressure in the pipeline. The actual leakage point was about 890 miles downstream between the towns of Asma and Ufa where the pipeline was installed about half an away to the side of the Trans Siberian Railway. The smell of escaping gas was reported from the valley settlements in the area but no one did anything about it. The escaping liquefied gas formed two large pockets in the low lying areas along the railway line. The gas cloud then drifted for a distance of 5 miles. Some hours later, after the main leakage had started, a train from Nizhnevartovsk destined for the Red Sea resort of Alder was approaching the leakage area when the driver noticed a fog in the area that had a strong smell. The driver of another train approaching from the opposite direction (Alder to Nizhnevartovsk) saw much the same as he approached the West bound train. Both trains were packed, with a total of 1168 people on board, and as they approached the area, the turbulence caused by them mixed up the LPG mist and vapour with the overlying air to form a flammable cloud. One or the other train ignited the cloud. Several explosions took place in quick succession followed by a ball of fire that was about 1 mile wide and which raced down the railroad tracks in both directions. Trees were flattened within a radius of 2 1/2 miles of the epicentre of the explosions and windows were broken up to 8 miles away (Figure 14.11). The accident left 462 dead and 796 hospitalised with 70% to 80% burn injuries.

Sao Paulo Accident

On February 25, 1984, at least 508 people, most of them young children, were killed in Sao Paulo (Brazil) when a 2-ft diameter gasoline pipe ruptured and 700 tons of gasoline spread across a strip of swamp (Figure 14.12). The cause of the pipe rupture was not reported, though it was said to have been caused due to pressure built up over the safe limit. It was also stated that there was no way of monitoring the pressure in the pipe line.

The Basel Disaster

On 1 November, 1986, a ware house at Sandoz near Basel caught fire and burned. The warehouse was containing ten types of pesticides totalling about 1200 metric tons and 12 metric tons of mercuric fungicide. Most of these chemicals are toxic to both human and animals. Around 70 to 80 % of the stored chemicals had been drained out in different forms due to the fire. Although an alarm was sounded, the citizens of Basel received no relevant information and were in a state of disquiet for several hours. The vitality of the Rhine river flowing nearby was seriously endangered. The river was red for several miles and an innumerable amount of fish perished (Figure 14.13). Not only fish but all aquatic life had been destroyed for several miles. The vegetation also was adversely effected.

About 50,000 persons got effected, approximately 5 km^2 area of river, ground water and 2 Km^2 soil area was contaminated. The damage was estimated at $SFR 100 million.

14.8.3. Accidents Involving Major Toxic Releases

The Seveso disaster: On the morning of Saturday, 10 July 1976, a safety valve vented on a reactor at the Icmesa Chemical Company at Seveso, a town of about 17,000 inhabitants some 15 miles from Milan (Italy). A white cloud drifted over a part of the town; heavy rainfall brought the cloud to earth. The release occurred from a reactor producing trichlorophenol, which is used to make the bactericide hexachlorophenol and the herbicide 2,4,5 trichloro phenoxy acetic acid. The reactor also contained the chemical generally referred to as TCDD - (2,3,7,8, - tetrachloro dibenzo paradioxin). This substance was not an intended reaction product but an undesired by-product. An estimated 2 kg of TCDD were released although this estimate is necessarily approximate.

In normal operations the amount of TCDD made in the reactor was small, but on this occasion the reactor had got out of control. The contents had got overheated and the safety valve had vented. The higher temperature in the reactor favoured the production of abnormal quantity of TCDD.

In the immediate area of the release the vegetation was contaminated and animals began to die. On the fourth day a child fell ill and on the 5th day civil authorities declared a state of emergency in Seveso. An area of some 2 square miles was declared as contaminated and people were asked to avoid contact with the vegetation or eating anything from this area. The contaminated area was later sought to be closed completely. On July 27 the first evacuation of some 250 people took place. By the end of July, 250 cases of skin infection had been diagnosed. Some 100 people had been told to evacuate their homes and some 2000 people had been given blood tests. In early August it was found that the area contaminated was about 5 times larger than was originally thought (Lees, 1996).

There have been accidents involving TCDD release prior to the Seveso disaster. At Ludwigshafen, fifty-five persons were exposed when there was accidental TCDD release in 1953, and many developed severe symptoms of TCDD poisoning. Various measures were taken to decontaminate the plant building, including the use of detergents, the burning off of the surfaces, the removal of insulating material and so on, but these were not effective and eventually the whole building had to be destroyed. In another accident at Duphar in 1963, a leak of 0.03-0.2 kg of TCDD occurred. Some 50 persons were involved in cleaning up the leakage, of whom four subsequently died, and about a dozen suffered occasional skin troubles. The plant was sealed for ten years and then dismantled from the inside brick by brick, the rubble was embedded in concrete, and the concrete blocks were sunk in the Atlantic. Five years later yet another accident involving TCDD release occurred at Bolsover. It involved a runaway reaction in a trichlorophenol reactor, similar to one that later occurred at Seveso. The reaction reached 250 °C, the reactor exploded and the supervising chemist was killed. The plant was closed down, and then reopened after two weeks when it appeared that workers exposed had suffered no ill effects. But within seven months, 79 persons complained of TCDD symptoms. The plant was dismantled and buried in a deep hole. But the story did not end there; three years later contractors on the site developed TCDD symptoms. The only apparent possible source of contamination was a metal vessel which had been thoroughly cleaned and subjected to sensitive testing.

The lesson that emerged from Seveso was that pressure relief valves on plants handling highly toxic substances should not discharge to atmosphere but to a closed system. But several hazardous units didn't bother to implement the lesson learnt from Seveso. The pesticide factory of the Union Carbide Corporation, situated in Bhopal, was one such unit. Its pressure relief volves vented tones upon tones of lethal methyl iso cyanide gas into the Bhopal air, leading to a horrible disaster.

The Bhopal Disaster

The worst-ever disaster in the history of the chemical industry occurred in Bhopal, India, on December 3, 1984. A leak of methyl isocyanate from a chemical plant, where it was used as an intermediate in the manufacture of a pesticide, spread beyond the plant boundary and caused death by poisoning of over 2,500 people -injuring about 100 times as many .

Methyl isocyanate boils at about 40 °C at atmospheric pressure. According to press reports, the contents of the storage tank got overheated and boiled, causing the relief valves to lift. The discharge of vapour -- about 25 tons -- was too great for the capacity of the scrubbing system. The escaping vapour spread beyond the plant boundary where a shanty town had sprung up. The cause of the overheating was contamination of the methyl isocyanate, by water or other materials, and several possible mechanisms were suggested. According to some reports cyanide was produced. Had Union Carbide conducted risk analysis, (specifically maximum credible accident analysis) during the design of the MIC system or even later, it would have learnt that in the event of a MIC leak the scrubbing system would be inadequate. This would have enabled the industry to install better emergency handling systems, thereby saving thousands of lives (Abbasi et al, 1997, Abbasi and Abbasi 2006).

14.9. Fatality-Frequency (FN) Analysis of Accidents

FN curves, also known as *social risk plots*, represent the probability of fatalities as a function of the number of fatalities. Maximum fatalities have been observed in fixed installation accidents (47%) followed by transportation accidents (34%). However, the FN curve for transportation accidents (Figure 14.14) is more flat than the curve pertaining to fixed installation accidents indicating that the probability of fatality is more in transportation accidents compared to fixed installation accidents; in other words, fatalities per accident is more during transportation than in a fixed installation. This apparently is due to the *additional* risks a hazardous unit faces of causing an accident when it is in transit compared to when it is fixed. Further if an industry where a fixed installation accident takes place has an efficient emergency preparedness program in place, the likely damage may be contained. Such damage control is rarely possible if a vessel fails due to a transportation accident, or during transit.

The FN curves also reveal that fires and explosions cause more number of fatalities per accident in comparison to toxic release (Figure 14.15). The possible reason may be that only deaths occurring immediately after the toxic release are reported in literature, long-term chronic impacts—which could be very significant—do not normally come to light. Of the total fatalities due to fixed installation accidents, ~49% have been attributed to fires and explosions, 38% to toxic release and 13% to combinations of these effects. The average fatality per accident in a fixed installation is 2.32; it is 3.27 for fire and explosion, and 2.49 for toxic release. The FN Curves for fires and explosions includes fire, vapour cloud explosion (VCE), confined vapour cloud explosion (CVCE) and boiling liquid expanding vapour cloud explosion (BLEVE). The number of explosions (either type) have been plotted as function of a five years moving average (Figure 14.16). It is evident from the figure that during 1975-1979 a large number of explosions were reported, while subsequently there was a sharp decrease. To study the consequences of each accidental event, FN curves for various accidental events are presented in Figure 14.17. It is observed that the curve for VCE is most flat while the curve for fire is the steepest indicating that a VCE has the highest risk of fatalities while fire has the least.

An illustrative table of fatality rates due to accidents in different countries is presented in Table 14.12. The Netherlands has the lowest fatality rate per accident while Austria and Belgium have the highest.

14.10. Summary of the Prceding Sections

The information presented in this chapter thus far leads to the following important conclusions.

I. most of the accidents take place due to malfunctioning of one or other component of equipment and/or minor negligence of personnel during operation or maintenance.

II. Whereas the number of accidents per year has declined through the 1980s, the *extent* of damage per accident has increased substantially. This is particularly in developing countries such as India.

III. The damage potential of an accident depends upon the chemical in use, causative factors, operating conditions, and site characteristics.

IV. The damage potential in terms of areas getting affected is maximum for toxic release and depends upon the type of chemical, meteorological conditions and site characteristics.

V. The impacts of fires and explosions extend too much smaller areas but the devastation caused is more immediate and severe than in most cases of toxic release (except in cases such as the MIC leak at Bhopal in 19898). Fire and explosions can also cause *domino effect* (chain of accidents).

VI. The number of fatalities per accident is the highest for accidents involving explosions.

VII. Pipe line transport of chemicals is comparatively safer provided that the line is carefully maintained and its route does not pass through populated areas.

VIII. With an increase in density of industries in a complex, the probability of accidents as well as the probability of domino effect increase sharply.

The material presented thus far highlights the need of accident forecasting, consequence assessment, and development of up-to-date emergency preparedness and disaster management plans in chemical process industries. The rest of the chapter presents techniques and methodologies to accomplish these objectives.

14.11. Proscess Safety: Assessment of Risk of Accidents and Accident Prevention

The science of process safety and loss prevention, which has emerged in recent years with ever-increasing importance being attached to it, deals with the following key aspects of accidents in chemical process industries:

I. Development of techniques and tools to forecast accidents

II. Development of techniques and tools to analyse consequences of likely accidents. Such consequence analysis fulfills two objectives: a) it helps in siting of industries and management of sites so as to minimize the damage if accident do occur; b) it provides feedback for other exercises in accident forecasting and disaster management.

III. Development of managerial strategies for 'emergency preparedness' and 'damage minimization'.

14.11.1. Risk Assessment

The terms hazard and risk are sometimes used interchangeably by lay persons. In reality hazard relates to the source of harm, while risk is the probability and the extent of the harm likely to be experienced (Lees, 1996; Greenberg & Cramer, 1991; Khan & Abbasi, 1995a; Abbasi & Venilla, 1994; Abbasi *et al.,* 1998; Khan & Abbasi, 1998a). For example a large stone lying in the middle of a road is *a hazard*. But the *risk* posed by it would depend on factors like a) density of traffic (greater the density more will be the probability of an accident being caused by that stone as also the likely harm), b) presence of other hazards (if there are other stones lying nearby or if there are pot-holes close to where the stone is lying, the risk posed due to that stone will go up) c) quality of vehicles passing by (better the vehicles equipped with safety features, lower will be the risk); d) weather conditions e) fitness of the driver; etc etc.

Another example can be of a large cylinder filled with liquid chlorine. Such a cylinder kept in a desert will pose little risk. If it is shifted to a populated area, the risk posed by it will increase. If the same cylinder is kept on a busy highway, the risk posed would be enormous.

In summary ‘risk’ is a combination of the probability of a hazard causing harm (to human beings, property, society, environment etc) as well as the extent of the likely harm. On the other hand ‘hazard’ is a source of harm; it can be a physical or chemical object or condition, and the risk a hazard poses is dependent on the situation in which the hazard is present. If risk is the combination of three attributes— What can go wrong, how bad could it be, and how often might it happen—than hazard refers to the first of these attributes (CCPS 2007).

14.12. Techniques and Methodologies for Risk Assessment

Several techniques and methodologies have been proposed for risk and safety study (Less 1996, 2005; CCPS 2007; Khan & Abbasi, 1998; Abbasi & Abbasi 2006, 2007, 2008). A brief review of the important ones is presented below.

14.12.1. Checklist

Checklist (Balemans, 1974; Rose et al., 1978; Hessian & Rubin, 1991; Khan & Abbasi 1998; Oyeleye & Kramer, 1988) represents the simplest method used for hazard identification. A checklist is a list of questions about plant organization, operation, maintenance and other areas of concern to verify that various requirements have been fulfilled and nothing is neglected or overlooked. Checklist is primarily based on the preparers' prior experience, but it can also be based on codes and standards (Hessian & Rubin, 1991; Oyeleye & Kramer, 1988). The checklist has to be maintained during the life of the project and should be updated after each modification and after every major outage when equipment is replaced or modified substantially.

Although checklist development requires trained and experienced personnel, even relatively untrained personnel can use them effectively. The main limitations of this methodology are:

- it takes a long time to develop a checklist but it yields only qualitative results, with no insights into the system. It merely provides the status of each item in terms of 'Yes' or 'No'.
- a checklist can focus only on a single item at a time, so it can't identify hazards as a result of interaction among different units or components (equipment).
- it is only as good as the ability and prior experience of the person preparing it. There is always a significant probability of some critical item being neglected.
- it is unable to identify hazard due to the type of unit operation (reaction, heat transfer, storage etc.), severity of operating conditions (temperature, pressure), and any mis-operation (leak or excess heat generation etc.).

Due to the above-mentioned drawbacks this technique is not recommended for detailed risk analysis. However, it continues to be used (Eley, 1992; Ozog & Stickles, 1991).

14.12.2. HAZOP

HAZOP (lCI, 1974; Lawley, 1974;; Kletz, 1983, 1985; Freeman, 1991; Venkatasubramanian & Vaidyanathan, 1994; Medermid *et al.*, 1998; Mahnken, 2001; Yang and Liu, 2005; Jensen, 2006; Verna and Stevens 2008) is a simple yet structured methodology for hazard identification and assessment. It had been developed at Imperial Chemical Industries (lCI) in 1974 and later went through several modifications (ICI, 1974; Kletz, 1985; Andow et al., 1980; Knowlton, 1982, 1989; McKelvey, 1988; Montague, 1990.) The basic principle of a HAZOP study is that normal and standard conditions are safe, and hazards occur only when there is a deviation from normal conditions. It is a procedure that expects its user to make intelligent guesses in the identification of hazard and operability problems.

In a typical HAZOP study, design and operation documents (PI&Ds, PFD, material flow diagrams, and operating manuals) are examined systematically by a group of experts. Abnormal causes and adverse consequences for all possible deviations from normal operation that could arise are identified for each unit of the plant. HAZOP is considered by a multi-disciplinary team of experts who have extensive knowledge of design, operation and maintenance of the process plant. To cover all the possible malfunctions in the plant the imagination of the HAZOP team members is guided systematically with a set of guide words for generating the process variable deviations. A list of guide words and their definitions is given in Table 14.13. The salient features of HAZOP study are:

- it gives an idea of priorities for detailed risk analysis,
- it provides first information of the potential hazards, their causes, and consequences,
- it indicates some ways to mitigate the hazards,
- it can be performed at the design stage as well as the operational stage,
- it provides a basis for subsequent steps in the total risk management program.

A number of applications of HAZOP in the chemical process industries (CPI) have been reported in literature; for example by Freeman *et al.* (1992), Sweeny (1993), Pully (1993), Kolodji (1993), Shafagi and Cook (1988), Mulvihill (1988), Parmer & Lees, 1987, Piccinini and Levy (1984) *etc.* Tables 14.14 and 14.15 illustrate two of the typical HAZOP tables.

In its original, and thus far the most widely used form, HAZOP has some limitations; these limitations are of two kinds. The first kind arises from the assumptions underlying the method and is a limitation (perhaps intended) of scope. The method assumes that the design has been carried out in accordance with the appropriate codes. For example, it is presupposed that the design is appropriate for the requirements of normal operating conditions. If it isn't, HAZOP has no capability to identify the design flaw, less so to correct it. A HAZOP only tries to identify deviations from the supposedly 'ideal' situations.

The other kind of limitation is one which is neither intended, nor desirable, but is inherent in the method. For example HAZOP is not inherently well-suited to deal with spatial features associated with plant layout and their resultant effects. Furthermore HAZOP needs large inputs of time and expert manpower. As the efficiency and accuracy of the study is fully dependent on the experience and sincerity of the expert team members, any limitaions in manpower selection of performance can seriously harm the success of any HAZOP.

Khan and Abbasi ( ) built upon the earlier efforts to overcome some of the shortcomings of HAZOP. They have developed mathematical models to estimate the time required for a HAZOP, and suggested ways to minimize the time and effort conventional HAZOP takes (Khan and Abbasi)….. They have also developed expert systems to automate HAZOP ( ). According to them the duration of the study can be reduced drastically using automated systems to study the commonly occurring equipment. This may reduce the workload of team members and increase the efficiency and reliability of the study.

Inspite of its limitations, HAZOP remains the most favoured technique for hazard identification and assessment. Efforts to automate HAZOP also continue to be made

(Venkatasubramanian and Vaidyanathon 1994; Chung and McCoy, 2005; McCoy *et al* 2005; Eizenberg *et al* 2006; Zhoa *et al*; Palmer et al 2008).

14.12.3. Fault Tree Analysis (FTA)

Fault tree analysis or FTA (Lapp & Powers, 1976, 1979; Parmer & Lees, 1987; Haupmnanns, 1988; Khan and Abbasi.....) is an analytical tool that uses deductive reasoning to determine the occurrence of an undesired event. FTA, along with component failure data and human reliability data, can enable determination of the frequency of occurrence of an accidental event.

FTA was developed in 1960s by Bell Laboratories during the Polaris missile project. It was initially applied in the aerospace industry. Later its use was extended to nuclear and chemical industries (Lees,1996, 2005; Greenberg & Cramer, 1991; Lapp & Powers, 1979; CCPS, 1989; Rauzy, 1993; Cummings *et al.*, 1983; Hauptmanns & Yllera, 1983; Ulerich, 1988; Guymer *et al.*, 1987, Khan and Abbasi 1998,...), FTA yields both qualitative as well as quantitative information.

FTA has the following advantages.

I. it directs the analyst to ferret out failures deductively;
II. it points out the aspects of the system which is relevant to an understanding of the mechanism of likely failure;
III. it provides a graphical aid enabling those responsible for system management to visualize the hazard; such persons are otherwise not associated with system design changes;
IV. it provides avenues for system reliabilit analysis (qualitative, quantitative);
V. it allows the analyst to concentrate on one particular system failure at a time;
VI. It provides the analyst with genuine insights into system behaviour.

Yllera (1988) and Lai *et al* (1986) have drawn attention to the difficulties associated with FTA. According to them FTA is a sophisticated form of reliability assessment and requires considerable time and effort by skilled analysts. Although it is the best tool available for a comprehensive analysis, it is not foolproof and, in particular, it does not of itself assure detection of all failures, especially common cause failures. The accuracy of prediction is limited and depends upon the reliability and failure data of components of the fault tree.

In many real-world applications, it may be difficult to assign exact values to the probabilities of occurrence of the fundamental events. This problem is likely to arise in dynamically changing environments or in systems in which accidents occur so frequently that reasonable failure data are not available. In the absence of genuine probability data, estimates of failure probabilities are customarily supplied by personnel familiar with the operation of the system. Usually they prefer to express their knowledge in general terms and find it extremely difficult to specify the exact numerical values that are required in conventional fault tree analysis.

To cope with this problem associated with the assignment of exact numerical values to failure probabilities, modifications have been suggested to dilute FTA's dependency on reliability data and cut short the time of analysis using fuzzy mathematics (Rauzy 1993; Khan and Abbasi…; Jia and Shi, 2001; Post *et al* 2002; Chang *et al* 2002; Yang *et al* 2004; Wang and Jiang, 2005;He *et al* 2005, 2007).. Algorithms have also been proposed for computer-aided fault tree design and analysis, which seem to be useful. Lapp (1991); Bossche (1991); Khan and Abbasi…; Shastri *et al* 2004; Chen *et al* 2004; Chen and Chen 2006; Zhao *et al* 2006; Ferdous *et al* 2007; Chen and Yea 2007).

14.12.4. Failure Mode Effect Analysis (FMEA)

FMEA (Lees, 1996, 2005; Greenberg & Cramer, 1991; Khan & Abbasi, 1995a; Schwab and Urban 2003; Rabuse 2008) is an examination of individual components such as pumps, vessels, valves, etc, to identify the likely failures which could have undesired effects on system operation. FMEA involves following steps:

I. identification of each failure mode, of the sequence of events associated with it, its causes and effects;
II. classification of each failure mode by relevant characteristics, including deductability, diagnosability, testability, item replaceability, and compensating and operating provisions

Typical information required for an FMEA includes:

I. system structure;
II. system intimation, operation, control and maintenance;
III. system environment;
IV. system modeling;
V. system software;
VI. system boundary;
VII. system functional structure;
VIII. system functional structure representation;
IX. block diagrams; and
X. failure significance and compensating provisions.

FMEA is a qualitative inductive method and is easy to apply (Fechter and Barba, 2004). FMEA is assisted by the preparation of a list of the expected failure modes in the light of (I) the use of the system, (2) the elements involved, (3) the mode of operation, (4) the operation specification, (5) the time constraints and (6) the environment.

FMEA is effective in analyzing elements which can cause failure of the whole, or of a large part, of a system. It works best where the failure logic is essentially a serial one. It is much less suitable where complex logic is required to describe system failure (Klaassen & Van Pepper, 1989; Lees, 2005).

In essence FMEA is an inductive method. FTA serves as a complementary deductive accessory of FMEA and is needed where analysis of complex failure logic is required. FMEA is good for generating the failure data and information at component level (Cui *et al* 2007; Gipp, 2008). It has been recommended for use as a hazard identification technique mainly for systems dealing with low/moderately hazardous operations and the ones which cannot support the expensive and time-consuming HAZOP study (AIChE, 1985).

It has been stated that FMEA can be a laborious and inefficient process unless judiciously applied. FMEA is unable to deal with the interaction among different components and needs a highly expert team with sufficient experience and time to carry out the study. More seriously, FMEA is restricted up to component level, while actual hazard may start at sub-component level (failure of transmission line, failure of temperature transducer, failure of controller etc.). In recent years FEMA is being applied with fuzzy logic (Cui *et al* 2007), and the use of FEMA software is on the increase (De Kemos, 2004).

14.12.5. What-if Analysis

The What If method (CCPS, 1989; Khan and Abbasi 1998; Lees 2005; Sivapirakasam *et al* 2008) involves asking a series of questions beginning with what if as a means of identifying hazards. Apart from checklists, What if analysis is possibly the oldest method of hazard identification and is still popular (Sivapirakasam *et al* 2008).What if analysis is performed with questions such as:

- What if the pipe leaks?
- What if the flow controller fails?

The questions need not necessarily start with 'what if'; other phrases may also be used.

The method essentially involves a review of the entire design by a team using questions of this type, often using a checklist.

The advantages of this technique are:

I. no specialized technique or computational tool is required,
II. once the questions have been developed they can be used throughout the life of the project with 0nly slight modifications,
III. provides a simple tabular summary.

The major disadvantages are:

I. it requires a team of experts to perform the study; it thus has disadvantages (in terms of expertise available and costs) similar to HAZOP;
II. the heavy reliance on the experience and intuition of the study team both to develop questions imaginatively and to get the answer implies that any limitations in this aspect of the study can render the study totally useless (worse still, it can be misleading);
III. it is not as systematic as HAZOP, and FMEA;
IV. It gives only qualitative results with no numerical prioritization.

Due to these disadvantages 'what if' analysis is considered inferior to HAZOP and FTA(CCPS 1989). AIChE (1985) and IChemE (1985) have recommended this technique only when the other two—HAZOP and FMEA —are not applicable or the cost of study is the main consideration.

14.12.6. Hazard Indices

A number of indices have been developed to provide measures of hazards in different contexts. These include the Dow Index, the Mond Index and the IFAL Index.

14.12.6.1. Dow Index

It is by far the most widely used of hazard indices. It was developed by Dow Chemical Company for fire and explosion hazards.

The Dow Guide, describing the Dow Index, was originally published in 1964 and has gone through seven editions (Dow Chemical Company, 1964, 1994; Scheffler, 1994). In the first three editions the methods of determining the index values were developed and refined. In the fourth edition a simplified version of the index was described and two new features were introduced: the maximum probable property damage (MPPD) and a toxicity index. The fifth edition escribed a new framework for making the risk evaluation. It also included improvements in the method of calculating the indcx and several other new features-loss control credits and maximum probable days outage. In the sixth edition, a risk analysis package, including business interruption and a toxicity penalty to reflect emergency responses, was introduced. The seventh edition updates the sixth edition with respect to codes and good practice, but includes no major conceptual changes.

The overall structure of the methodology is shown in (Figure 14.18). The procedure it to calculate the fire and explosion index (F&EI) and to use this to determine fire protection measures and, in combination with a damage factor, to derive the base MPPD. This is then used, in combination with the loss control credits, to determine the actual MPPD, the maximum probable days outage (MPDO) and the business interruption (BI) loss (AIChE, 1994).

14.12.6.2. Mond Index

The Mond fire, explosion and toxicity index is an extension of the Dow Index. This index was developed at the Mond Division of ICI. The original version was described by Lewis (1979). Other accounts have been given by Tyler (1982, 1994).

The Mond method involves making an initial assessment of hazard in a manner similar to that used in the Dow Index, but taking into account additional hazard considerations. The potential hazard is expressed in terms of the initial value of a set of indices for fire, explosion and toxicity. These include:

I. Fire load index,
II. Unit toxicity index,
III. Major toxicity incident index,
IV. Explosion index,
V. Aerial explosion index,
VI. Overall index, and
VII. Overall risk rating.

14.12.6.3. IFAL Index

The instantaneous fractional annual loss (IFAL) Index was developed by the Insurance Technical Bureau (1981), UK, primarily for insurance assessment purposes (Singh & Munday, 1979; Whitehouse, 1985). Procedure for the calculation of the index is described in the IFAL Factor Workbook (Insurance Technical Bureau, 1981). It involves considering the plant as a set of blocks and exarrurnng each major item of process equipment in tum to assess its contribution to the index. The main hazards considered in the index are:

I. pool fires,
II. vapor fires,
III. unconfined vapor cloud explosions,
IV. confined vapor cloud explosions,
V. internal explosions.

In contrast to the Dow and Mond Indices, the IFAL Index is rather complex for manual calculation and needs a computer.

12.6.4. Other Indice

Attempts have been made to develop indice which are sensitive to larger number of risk-enhancing factors than the three oft-used indice described above. These include AHI, HIRA, and SwEHI described later in section 14.0.

14.13. Proposed Schemes of Risk Assessment Based on a Combination of Various Techniques

14.13.1. The WHO Scheme

The World Health Organization (WHO) and International Labor Office (ILO) have jointly proposed a scheme for conducting hazard assessment (WHO, 1984). The scheme consists of three-step procedure (Figure 14.19). As most of the constituent techniques are quantitative in nature this procedure is not as amenable to quantification as some of the other procedures described below.

14.13.2. The ISGRA Scheme

This scheme, formulated by the International Study Group on Risk Analysis (ISGRA, 1985), comprises three steps, (l) hazard identification, (2) consequence analysis, and (3) quantification of risk. The hazard identification step identifies and assesses hazards based on the chemical properties, capacity, and deviation in operating parameters. HAZOP, FMEA, and FTA/ETA have been recommended for this step. The consequence analysis step is to estimate the damage potential using standard mathematical expressions. The last step—quantification of risk—is based on the frequency of occurrence of an accident and its damage consequences. The frequency of occurrence is estimated based on the past history of similar accidents.

Unless the user of this scheme is very well-versed with the techniques and tools of risk assessment, he/she may falter and bypass some of the causes of hazards and frequency of their occurrence. As these are crucial inputs for any risk assessment study, it may lead to wrong conclusions.

14.13.3. Maximum Credible Accident Analysis (MCAA)

MCAA (AIChE, 1985; API, 1992; Khan & Abbasi, 1998, 2002; Less 2005) is an approach for forecasting the damage likely to be caused if an accident takes place in a chemical plant. MCAA comprises of the following main steps:

1) Study of the plant to identify hazardous materials, thus saving the effort and duration which would go to waste if non/less hazardous units were to be studied. This provision is not available in QRA, and to estimate the same parameters using Dow's Index and/or Mond's Index requires extra information and calculations.
2) Development of credible accident scenarios.
3) Assessment of damages likely to be caused in each scenario using mathematical models.
4) Delineation of the maximum credible accident scenario.

The first step identifies the hazards in any process industry on the basis of properties and capacities of the chemicals and by employing different indices such as the System of Hazard Identification (SYHI,1993), Extremely Hazardous Substance (EHS) (Environment Protection Act, 1987), and National Fire Protection Agency index (NFPA, 1991). On the basis of the storage or handling situations in the industry, different accident scenarios are generated, representing plausible accidental events. The next step—the consequence assessment step—estimates the consequences of each accident scenario in terms of likely extent of damage. Finally, based on the probability of occurrence and damage potential, the worst disaster scenario is identified.

MCAA has been used extensively in risk assessment and forms the basis of QRA (quantitative risk assessment) schemes proposed and applied, among others, by Arendt (1990a); Van Sciver (1990); and Khan & Abbasi (1996, 1997, 1998). Khan and Abbasi (2002) have developed a criteria by which *credible* accident scenarios can be identified from among a large number of possibilities. They have proposed a 4-step method to identify maximum credible accident sceneries followed by a procedure for estimating the damage potential of each credible scenario.

MCAA forms one of the key steps in any elaborate risk assessment exercise but necessitates the use of other techniques for identifying causes of hazards and estimating frequency of likely accidents.

14.13.4. Safety Analysis

Safety analysis is defined as a systematic examination of structure and function of a system aiming to identify accident contributors, modelling sequence of potential accidents, estimation of risk, and fixing risk-reducing measures. Safety analysis can be extended to risk analysis.

The procedure starts with identification of hazards using HAZOP and FMEA. This is followed by identification of different accidents and their causes. FTA (fault tree analysis), ETA (event tree analysis) and CCA (cause-consequence analysis) have been recommended for this step. This logical model is later analysed for further results (frequency and loss in terms of economic and fatal). The procedure can be extended to use in risk analysis by incorporating the consequence analysis step.

14.13.5. Quantitative Risk Analysis

Quantitative Risk Analysis (QRA) has been in existence for many years. Before its use in the chemical process industries (CPI), it was used extensively in the nuclear industry. Unfortunately, the application of QRA in the CPI is much more difficult than in the nuclear industry. This is because of the greater diversity of processes, hazardous materials, equipment types and control schemes in the CPl. This diversity requires continuous addition of new capabilities in QRA (CCPS, 1989; Arendt, 1990; Van Sciver, 1990; ICI, 1982; CMA, 1985; CCPS, 1994; Arendt, 1990; Khan and Abbasi 1998; Lees 2005). A typical QRA comprises four steps:

I. hazard identification;
II. frequency estimation;
III. consequence analysis, and
IV. measure of risk.

The first step seeks an answer to the question: what can go wrong? This is the most important step because hazards that are not identified will not be quantified, leading to an underestimated risk (Van Sciver, 1990). The techniques used for hazard identification include HAZOP studies, FMEA, 'What If' analysis, and checklists. After the hazards are identified, the scope of a QRA is defined.

The second step involves another key question: 'how likely is each accident'? Answering this question involves quantification of the probability of each accident scenario. FTA may be used for a this purpose.

The third step of consequence analysis aims to quantify the negative impacts of the likely events. The consequences are normally measured in terms of the number of fatalities, although they could also be measured in terms of number of fatalities although they could also be measured in terms of number of injuries or value of the property lost. The analysis of consequences in the CPI is very complex due to the great variety of materials, chemical reactions, and technologies involved. Consequence analysis is the aspect of QRA that is growing most rapidly.

The last step of a QRA is to calculate the actual risk. This is done by estimating the areas that are at risk, and the extent of that risk.

Inspite of the lengthy procedure which needs a lot of time for implementation, high cost of implementation (due to the need of highly expert professionals of various disciplines for a long duration), and need for sophisticated tools and data, it is the most favoured and presently most frequently used scheme for the risk analysis of chemical process industries (Lees, 1996, 2005; Roy *et al* 2003; Cozzani and Salzano 2004 a,b; Abdolhamidzadeh *et al*, 2006; Van der Voort *et al*, 2007; Heuselwood and Phillips, 2008).

Improvements in terms of reducing the duration of the implementation of various steps by screening the nonhazardous units, cutting short the time of each step (use of an already developed information base) ,etc, would bring down the cost of a QRA study drastically and thus make the study optimal in all respects (cost, duration, and reliability of results).

14.13.6. Probabilistic Safety Analysis

Guymer et al. (1987), Popazoglou *et al.* (1992), Kafka (1991, 1993), and Khan and Abbasi (1998) have proposed a combination of different techniques for probabilistic safety analysis (PSA) in chemical process industries. PSA provides a framework for a systematic analysis of hazards and quantification of the corresponding risks. It also provides a basis for supporting safety-related decision-making. The methodology and the procedures followed for the PSA of a typical chemical installation involved in handling a hazardous substance can be outlined in the following seven major steps (Popazoglou et al., 1992) (Figure 14.20).

14.13.6.1. Hazard Identification

The main potential sources of hazardous substance releases are identified and the initiating events that can cause such releases are determined.

14.13.6.2. Accident Sequence Modeling

A logical model for the installation is developed. The model includes each and every initiator of potential accidents and the response of the installation to these initiators.

14.13.6.3. Data Acquisition and Parameter Estimation

Parameters which must be estimated include the frequencies of the initiating events, component unavailability and probabilities of human actions.

14.13.6.4. Accident Sequence Quantification

This step quantifies the accident sequences, and calculates their frequency of occurrence. In particular, the plant model built in the second step is quantified using the parameter values estimated in the third step.

14.13.6.5. Hazardous Substance Release Categories Assessment

Release categories of the hazardous substance are defined in order to streamline the calculation of the consequences of the accidents and the associated frequencies.

14.13.6.6. Consequence Assessment

Undesirable consequences and associated probabilities are calculated for each release category. If the hazardous substance is toxic, immediate health effects can be estimated by calculation of the atmospheric dispersion of the released substance, the assessment of the dose an individual would receive at each point around the site, and by establishing a dose/response model.

14.13.6.7. Integration of Results

Integrating the models and the associated results developed in steps 4, 5 and 6, results in the establishment of a range of possible consequences and the associated uncertainties.

Beckjord et al. (1993) have reported a few applications of PSA in chemical process industries. For the same level of accuracy PSA takes about 50% more time than QRA, which makes the technique very expensive. Moreover, the application of PSA is limited to the operational stage because many of its steps (data acquisition and parameter estimation, and accident sequence quantification) need precise operational data, which are available only during operation.

A board pattern of 9 risk assessment exercise is presented in Figure 14.21. It also illustrates the domains of safety analysis QRA and probabilistic Risk Assessment.

14.14. The Work Done at the Centre for Pollution Control and Energy Technology, Pondicherry University

It emerges from the foregoing review that several of the existing methodologies are useful in conducting one or other aspect of risk analysis. For example, HAZOP is a powerful technique for identifying and assessing hazards qualitatively, while MCAA is widely applicable in consequence analysis. All conventional risk analysis procedures require a combination of these methodologies, As some of them—such as HAZOP—are cumbersome and costly, and some others—such as FMEA and FTA—require extensive reliability data which might not be easy to obtain, the conventional RA procedures become tedious, costly, and prone to serious errors (when precise basic data is required but is not available).

The authors and coworkers, among others, have tried to improve the situation by modifying some of the conventional methodologies and strengthening some others in terms of enhancing their analytical and computational capabilities. The author's efforts have led to the following:

1) HIRA
2) optHAZOP
3) TOPHAZOP
4) EXPERTOP
5) PROFAT
6) HAZDIG
7) MOSEC
8) DOMIFFECT
9) MAXCRED

10) ORA
11) FORCIBLE
12) SWEHI
13) Application of cellular automata

Each of these is briefly described below.

14.14.1. Hazard Identification and Ranking: HIRA

HIRA (Hazard Identification and Ranking Analysis) is an index-based technique proposed by the authors to conduct the very first step of risk analysis (hazard identification and ranking). The objective of this step is to identify the chemicals and unit operations that constitute a potential hazard. HIRA is based on a multi-attribute hazard identification and ranking method detailed elsewhere (Khan & Abbasi, 1997,). It considers hazard potential in a unit as a function of material, capacity, type of unit operation, operating conditions, and surroundings (degree of conjunction, location of other hazardous units etc.). The output of HIRA gives two indices, damage index for fire and explosion hazard, and risk index for toxic release and dispersion hazard. HIRA has provided a platform for development of several process specific as well as general indice (Adu *et al* ; Al-Sharah; Hassim and Edwards: Paralikas and Lygeros ; Shah *et al* and Tixier *et al*).

14.14.2. Qualitative Hazard Assessment: Opthazop and TOPHAZOP

14.14.2.1. OpHAZOP

OPTHAZOP (optimal and effective HAZOP) proposed by Khan & Abbasi, (1997) sets out to conduct hazard study in such a way that the duration of the study is optimum. Most of the hazards are sought to be identified and assessed, with better efficiency, good reliability of results, and in lesser time so that the recommendations made by the study can be followed easily and economically. To fulfill the above objective a systematic procedure along with various recommendations has been developed. It uses an already developed expert knowledge-base (Figure 14.22). This knowledge-base is a large collection of facts, rules and information regarding various components of a process plant. Along with the use of knowledge it also suggests a few recommendations to reduce the time of discussion and produces effective and reliable results (Khan & Abbasi, 1997,). Others have since built-up on optHAZOP (Arunraj and maiti ; ; Zhzo and Cu; Garcia-Serna *et al* ; clarizia *et al ;* Khan *et al ; Yet-Pole).*

14.14.2.2. TOPHAZOP

optHAZOP, described above, consists of several steps, the most crucial one requires use of a knowledge-based software tool which would significantly reduce the requirement of expert man-hours and speed up the work of the study team. TOPHAZOP (Tool for Optimizing HAZOP) has been developed to fulfill this need (Khan & Abbasi, 1997). TOPHAZOP is a user-friendly software which seeks to conduct HAZOP study in a comprehensive, effective, and efficient manner within a short span of time. TOPHAZOP overcomes several major limitations (time, effort, repetitious work, etc.) of the existing HAZOP procedure. The software has an in-built knowledge-base which is extensive and dynamic. It incorporates process units, and works out numerous modes of failure for certain input operational conditions. It drastically minimizes the need of expert time.

The knowledge-base has been developed in two segments: process general knowledge, and process specific knowledge. The process specific knowledge segment handles information specific to a particular process unit in a particular operation, whereas the process general knowledge segments handle general information about the process unit. At present the knowledge-base incorporates information pertaining to 15 different process units including their characteristics and modes of failures. The availability of on-line help and graphical user-interface enhances its user-friendliness so that even an inexperienced professional can utilize the software with relative ease.

14.14.3. Probabilistic Hazard Assessment: PROFAT

As detailed earlier, fault tree analysis involves identification of causes of an accident, frequency of occurrence of an accident, and contribution of each cause to the accident. It is a useful methodology but is besieged with the same types of limitations which we find with other methodologies, such as a) need of large volumes of precise data, and b) requirement of much expert time. Attempts have been made to overcome these limitations by incorporating a combination of analytical method (Hauptmanns, 1988), and Monte Carlo simulation technique (Rauzy, 1993; Hauptmanns & Yllera, 1983) with fuzzy set theory (Tanaka et al., 1983; Khan & Abbasi, 1997). A software PROFAT (Probabilistic Fault Tree Analysis) has been developed on the basis of this recipe (Khan and Abbasi).

14.14.4. Consequence Analysis: MOSEC, HAZDIG and DOMIFFECT

Consequence analysis involves assessment of likely consequences if an accident scenario does materialize. The consequences are quantified in terms of damage radii (the radius of the area in which the damage would readily occur), damage to property (shattering of window panes, caving of buildings) and toxic effects (chronic/acute toxicity, mortality), The assessment of consequence involves a wide variety of mathematical models. For example source models are used to predict the rate of release of hazardous material, the degree of flashing, and the rate of evaporation. Models for explosions and fires are used to predict the characteristics of explosions and fires. The impact intensity models are used to predict the damage zones due to fires, explosion and toxic load. Lastly toxic gas models are used to predict human response to different levels of exposures to toxic chemicals.

14.14.4.1. MOSEC

A software MOSEC (MOdeling and Simulation of fire and Explosion in Chemical process industries) has been developed specifically to estimate the impacts of accidents involving explosion and/or fire (Khan & Abbasi, 1997). MOSEC comprises state-of-the-art models to deal with: (i) pool fire, (ii) flash fire, (iii) fire ball, (iv) jet fire, (v) boiling liquid expanding vapor explosion (BLEVE), (vi) confined vapor cloud explosion (CVCE), (vii) unconfined vapor cloud explosion (UVCE), and (viii) vented explosion. The software has been developed in object-oriented programming environment using C + + as a coding too!. It has been made user-friendly by incorporating such features as graphics, on-line help, ready-to-use output format, etc.

14.14.4.2. HAZDIG

HAZDIG (Khan & Abbasi, 1998) (HAZardous DIspersion of Gases) is a computer software specifically developed to estimate the consequences (damage potentials and risks) due to release of toxic chemicals, accidentally or voluntarily (Khan & Abbasi, 1997). The modular structure of HAZDIG (developed in object oriented environment) enables swift processing of data and computation of result. It is also easy to maintain and upgrade.

HAZDIG incorporates well tried and tested models for estimating atmospheric stability (Van Ulden & Hostlag, 1985) and dispersion (Van Ulden, 1988; Erbink, 1995; Pasquill & Smith, 1983; Erbink, 1993; Khan & Abbasi, 1997). The data needed to run the models is easy to obtain and feed—properties of chemicals, operating conditions, ambient temperature, and a few commonly available meteorological parameters. A database containing various proportionality constants and complex empirical data has been built into the system. It is capable of handling various types of release and dispersion scenarios: two phase release followed by dispersion, momentum release followed by dispersion, dispersion of heavierthan-air gases, etc. The graphics option enables the user to draw any industrial site/layout using freehand drawing or using any already defined drawing tool. The contour drawing option has the facility for drawing various damage/risk contours over the accident site.

14.14.4.3. DOMIFFECT

Most of the risk analysis methodologies deal with accidents in a single industry, more so in one of the units of an industry. But it is always possible that a major accident in one unit-sayan explosion or a fire-can cause a secondary accident in a nearby unit which in tum may trigger a tertiary accident (Khan & Abbasi, 1997k; Pasman et aI., 1992). The probability of such domino or 'cascading' effects occurring is increasing day by day with more new industries coming up in already congested industrial areas (Khan & Abbasi, 1997k).

We have developed a computer automated methodology DOMIFFECT (Khan & Abbasi, 1997) (DOMIno effECT) which enables one to know (a) whether domino effects are likely to occur in a given setting, (b) if they do what would be the likely accident scenarios, and (c) what would be the likely impacts of the different scenarios. Finally, the tool guides us towards strategies needed to prevent domino effects (Khan & Abbasi, 1997). DOMIFFECT is menu driven and nteractive. Capable of the following:

- estimation of all possible hazards from toxic release to explosion;
- handling of interaction among different accidental events (generation of domino or cascading accident scenarios);
- estimation of domino effect probability;
- estimation of domino effect consequences.

14.14.5. Rapid Risk Analysis: MAXCRED

A total risk assessment exercise covering all steps exhaustively from beginning to end is expensive in terms of time as well as monetary and personnel inputs (Greenberg & Cramer, 1991; Khan & Abbasi, 1995b; CCPS, 1989; AIChE, 1985; WHO, 1984; Suokas, 1988; Popazoglou et al., 1992; Pasman et al., 1992). It often becomes necessary to conduct rapid risk assessment (RRA) to draw the same conclusions that a full fledged risk assessment would lead to, albeit with lesser (yet practicable) accuracy and precision (Khan & Abbasi, 1996, 1997h, i; Khan et al., 1998).

We have proposed a software package, and the system of methodologies on which the package is based, for conducting RRA in chemical process industries. The package is named MAXCRED (MAXimum CREDible rapid risk assessment) (Khan & Abbasi, 1996). The package, coded in C++, has the following attributes:

I. it incorporates a larger number of models to handle a larger variety of situations useful in RRA;
II. it includes more precise, accurate, and recent models than handled by existing commercial packages;
III. greater user-friendliness;
IV. ability to forecast whether second or higher order accidents may occur.

14.14.6. Optimal Risk Analysis (ORA)

We have combined the first seven methodologies described above into a framework, named ORA (Optimal Risk Analysis). ORA involves four steps: (i) hazard identification and screening, (ii) hazard assessment (both qualitative and probabilistic), (iii) quantification of hazards or consequence analysis, and (iv) risk estimation. These steps of ORA and he corresponding methodology to be used in each step are presented in terms of ORA algorithm (Figure 14.23).

To compare the performance of ORA with the other commonly used schemes we have conducted a preliminary Delphi. Experts in safety engineering were asked to give weightages on a scale of 0-10 to eight attributes of seven well-known methodologies (Table 14.16). After second-round corrections and averaging the average weightage as obtained is presented in Figure 14.24 and 14.25. Of these Figure 14.21 compares seven of the old methodologies and Figure 14.25 compares QRA with ORA. All-in-all ORA appears to be ahead of the other seven methodologies. These findings would gain firm quantitative footing only after ORA has been extensively used by persons other than the authors. For the present we can say that ORA appears to be a virtuous scheme, with the following features:

I. is swifter,
II. it is less expensive,
III. It is as (or possibly more) accurate and precise.
IV. The features become apparent when we consider the following

1) Use of HIRA in ORA gives directly applicable results: damage radii (radius of the area under the probability of 50% damages due to fire and/or explosion), and the areas with high probability of lethal impacts. This makes it easy to screen the various units in terms of their risk potential.
2) Conducting HAZOP by the computer-automated tools optHAZOP and TOPHAZOP saves about 45% of the time otherwise taken by the conventional HAZOP (Khan & Abbasi, 1997a, b).
3) Use of PROFAT (based on a combination of analytical method and Monte-Carlo simulation) saves not only computational time, and overall duration of the study, but also increases the effectiveness of the results by doing the computations in fuzzy probability space. The Provision for modelling the complex problem into smaller and simpler modules further enhances the ease and speed of computation.
4) Use of HAZDIG, MOSEC and DOMIFFECT (based on state-of-the-art models) enables easy, fast, and reliable consequence assessment.

DOMIFFECT enables study of the possibility and likely impacts of domino effects; without such a study no risk assessment exercise can be considered complete or 'safe'.

REFERENCES

Abbasi, T., Abbasi, S.A., (2006), The boiling liquid expanding vapour explosion (BLEVE): mechanism, onsequence assessment, management, *Journal of Hazard Materials.*

Abbasi, T., & Abbasi, S. A., (2007), Boiling liquid expanding vapour explosions (BLEVEs), echanism, consequence assessment, management, *Journal of Hazardous Materials.* 141, 489– 519.

Abbasi T., Abbasi S.A., (2008), The boiling liquid expanding vapour explosion (BLEVE) is fifty ... and lives on!, *Journal of Loss Prevention in the Process Industries*, 21, (4) 485-487.

Abbasi, S. A., Krishnakumari, P., & Khan, F. I., (1998),*Hot topics, Global warming, acid rain, ozone hole, hazardous waste, industrial disasters, disinfection,* New Delhi, Oxford University Press (in press).

Abbasi, S.A., Krishnakumari, P., and Khan, F.I., (1997), HOT Topics: global warming, ozone hole, diversity, acid rain, industrial hazards, and dis-infection, *Oxford press*, India.

Abbasi, S. k, & Venilla, V. (1994), *Risk assessment,* Encyclopaedia for Environmental Engineering, R. K. Trivadi, Karad: Enviro Media, 239-254.

Abdolhamidzadeh, B., Hashemi, V., Rashtchian, D., (2006) Quantitative risk assessment for accidental release of ethylene oxide from purification column of an ethylene oxide production unit, *CHISA 2006 – 17th International Congress of Chemical and Process Engineering.*

Amendola. A, Contini, S, and Nichele, P. (1988), MARS; the major accident reporting system. *Preventing major Chemical and Related Process Accidents,* 445

Amesz, J., Francocci, G., Primavera, R. and Van der pas, T., (1983), The European abnormal occurrences reporting system, *Euredata 4*, 6.2.

Andow, P. K., Lees, F. P" & Murphy, C. P., (1980), *Int. Chem. Eng. Sym. Ser.,* 58, 225.

API. (1992), *Management or Process Hazards, Recommended Practice 750.* Washington, DC: American Petroleum Institute.

AIChE. (1995), *Guidelines for Hazard Evaluation Procedures,* prepared by Battle Columbus Division. New York.

AIChE. (1994), *Dow's Fire and Explosion Index Classification Guide, A1ChE Technical Manual,* LC 80-29237, New York.

Arendt, I. S. (1990), *Reliability Engineering and System Safety, 29,* 133-149.

Balemans, k W. M., (1974), *Loss Prevention and Safety Promotion, I, 7-15.*

Badoux, R.A.J., (1983), Some experience of a consulting statistician in industrial safety and reliability, *Reliab.Engng,* 1985, 10, 219.

Bellamy, I.J., Geyer, T.A.W. and Astley, J.A., (1989), Evaluation of the Human Contribution to pipework and In-lineEquipment Failures frequencies, *HSE Contract Res. Rep.15/198,* Technica Ltd, London.

Bruce, R.A.P., (1994), The offshore hydroc arbon releases (HCR) database, *Hazards XII,* 107.

Bossche. A. (1991), *Reliability Engineering and System Safety, 32,* 217-241

Beckjord, E. S., Cunningham, M. A" & Murphy, J. k (1993), *Reliability Engineering and System Safety,* 39, 159-170.

Carson, P. A., and Maurnfold, C.J., (1988), Reporting and analysis of industrial incident 1981-86, *Industry and Environment,* **II** (2).

Chemical industrial digest (1995), *Blockdale publication*, June, 1995.

Chen, J., Chen, H.H., (2006), On-line batch process monitoring using MHMT-based MPCA, *Chemical Engineering Science,* 61, pp. 3223-3239.

Chen, G., Liang, J., Qian, J., (2004), Process monitoring and fault detection based on blind signal processing with temporal structure, *Huagong Xuebao/Journal of Chemical Industry and Engineering (China)* 55, pp. 1837-1841.

Chen,J., Yea, Y., (2007), Controlled output variance based diagnosis tree for feed forward/feedback control systems, *Chemical Engineering Science,* 62, pp. 943-956.

Chowdhury, J., and Parkinson, G., (1992), OSHA tightens its hold, Chem.Engng, 99, 37.

Cummings, D. L., Lapp, S. A., & Powers, G. J., (1983), 1. *IEEE Transaction on Reliability. R-,* 32, 140.

CCPS., (1989), *Guidelines for Chemical Process Quantitative Risk Analysis*, Center for Chemical Process Safety, American Institute of Chemical Engineers, New York,.

CCPS. (1994), *Guidelinesfor Evaluating the Characteristicsfor Vapor Cloud Explosions, Flash Fires, and BLEVEs.* New York, AIChE.

Chang, S.Y., Lin, C.R., Chang, C.t., (2002), A fuzzy diagnosis approach using dynamic fault trees, *Chemical Engineering Science* 57, pp. 2971-2985.

Chung, P.W.H., McCoy, S.A., (2005), Computer-aided hazop of batch processes, *Institution of Chemical Engineers Symposium Series* pp. 631-640.

CMA. (1985), *Risk Analysis in the Chemical Industry.* Rockville, MD: Chemical Manufacturers Association. Government nstitute Inc.

Cozzani, V., Salzano, E., (2004), The quantitative assessment of domino effect caused by overpressure: Part I. Probit models, Journal of Hazardous Materials, 107 (3), pp. 67-80,

Cui., W.B., Wu, G.I., Sun, P.I., hang, Y.Q., (2007), Ship safety assessment based on FMEA and fuzzy comprehensive evaluation methods, Harbin Goncheng Daxue Xuebao *Journal of Harbin Engineering Unversity* 28, (3), pp. 263-267+276.

Davenport, J.A., (1988), Hazards and protection of pressure storage and transport of LP-gas, *J. Hazardous Materials,* 20, 3.

Doyle, W.H., (1969), Industrial explosions and insurance, *Loss Prevention,* 3, 11.

Doyle, W.H., (1981), Case histories in duct fire incident, *Loss Prevention,* 14, 181.

Dow Cbemical Company, (1964), *Dow's Process Safetv Guide.* Midland.

Dow Chemical Company, (1994), *Dow's Chemical Exposure Index.* New York: AIChE.

Drogaris, G. (1991), *Major Accidents Reporting System*: Lessons Learned from Accidents Notified, Community Documentation Centre on Industrial Risk.

Drogaris, G., (1993), Major Accident Reporting System, *Elsevier,* Amsterdam.

Eizenberg, S., Snacham, M., skofteland, G., (2006), Combining HAZOP with dynamic simulation – Applications for safety education, *CHISA 2006 17th International Congress of Chemical and Process Engineering.*

Erbink, I. J., (1995), *Turbulent diffusion model from tall stacks.* Ph.D. thesis submitted to VRIJ University, The Netherlands.

Erbink.J. J. (1993), *Workshop on Intercomparison ofAdvanced Practical Short Range Atmospheric Dispersion Modeling,* Manno, Switzerland.

Eley, C., (1992), *Hydrocarbon Process,* 71, (8), 97. *Environment Protection Act-40CFR part* 355. (1987), Washington, DC.

Ferdous, R., Khan, F.I., Veitch, B., Amyotte, P.R.,(2007), Methodology for computer-aided fault tree analysis, *Process Safety and Environmental Protection*, 85, pp. 70-80.

Freeman, R. A., (1991), *Plant/Operation progress, /0(3), 155.*

Freeman, R. A., Lee, R., & MeMamara, T., (1992), *Chemical Engineering Progress,* (August).

Geyer, T.A.W., and Bellamy, L.J., (1991), Pipe -work Failure, Failure Causes and the Management Factor, *Paper C425/012*, Instn Mech. Engrs, London.

Greenberg, H. R., & Cramer, J. J. (1991), *Risk Assessment and Risk Management for Chemical Process Industries.* New York: Van Nostrand Reinhold.

Guymer, P., Kaiser, G. D., & Mckelvey, T. C., (1987), *Chemical Engineering Progress.* (January) 37-45.

Gugan, K., (1979), Unconfined Vapour Cloud Explosion, *Instn Chem Engrs,* Rugby.

Gupta, K.C., (1990), Major accident hazard control system in India, *International conference on Hazard Assessment and Disaster Mitigation in Pterolium and Chemical Process Industries*, December 10-14, Chennai.

Hasstrup, P., and Brochoff, K.C., (1990), Severity of accidents with hazardous materials, *J. Loss Prevention in Process Industries*, 3, 33.

Health And Safety Executive (1988), *COSHH Assessments* HM Stationery Office, London

Houston, D.E.L., (1971), New approaches to the safety problem, *Major Loss Prevention,* 210.

Hauptmanns, U., (1988), *Fault Tree Analysis tor Process Industries: Engineering Risk and Hazard Assessment,* (Kandel and Avani, Eds.). Florida: CRC Press Inc.

Hauptmanns, U., & Yllera, I. (1983), *Chemical Engineer, 90, 91-103.*

He, N., Xie, L., Guo, M., Wang, S., (2005), Fault detection and diagnosis incontinuous dynamic multivariable processes using independent component analysis, *Huagong Xuebao/Journal of Chemical Industry and Engineering (China)* 67, pp. 885-889.

He, Y.F., Liu, Q.Y., Zhu, Y.Y., Xiao, C.Y., (2007), Fuzzy comprehensive estimation for atmospheric-vacjmm distillation unit based on fault tree, *Petrochemical Equipment* 36, pp. 18-21.

Hessian, R. T., & Rubin, J. N., (1991), *Risk Assessment and Risk Man- agemen! For the CPI* (Greenberg and Creamer, Eds.). New York: Van Nostrand.

Henselwood, F., Phillips, K.G., (2008), Managing the intolerable, *AIChE Annual Meeting, Conference Proceedings.*

ICI., (1974), *Hazard and Operability Studies,* Process Safety Report 2. London: Imperial Chemical Industries.

ICI. (1982), *The Chemical Engineer,* 385, 355-367.

IChemE, (1985), *Risk Analysis in the Process Industries,* EFCE Publication Series No. 45. Rugby, UK: IChemE.

Insurance Technical Bureau. (1981), *IFAL Factor Workbook.* London.

ISGRA. (1985), Risk analysis in the process industries-an ISGRA update. *Plant/Operation Progress* (April), 4(2), 63-67.

Johnson, J., (2006), No middleground in senate debate, *Chemical Processing* 70, pp. 36-40.

Jia, Y., Shi, X., (2001), Analysis and simulation on reliability based on fault tree model, *Xiangtan Daxue Ziran Kexue Xuebao* 23, pp. 101-105.

Kafka. F L. (1984), *The 191'14 European Major Hurards Conference,* London.

Kafka, P. (1991), *Probabilistic Safety Assessment: Quantitative Process to Balance Design, Manufacturing and Operation for Sajetv of Plant Structures and Systems.* Principal Division Lecture, Transactions SMiRT II, A Tokyo.

Kafka, P., (1993), Important issues using PSA technology for design of new system and plants. GRS mbH 85748 Garchirg, Germany.

Khan, F. **I.**, & Abbasi, S. A., (1995), Analytical Simulation*: a Procedure to Conduct FTA in Chemical Process Industries,* CPCEIRA 012/95. Pondicherry University, Pondicherry

Khan, F.I., and Abbasi, S. A., (1997), Risk analysis of epichlohydrin manufacturing industry using new computer automated tool MAXCRED, *J Loss Prevention Process Industries,* 10, 91-100.

Khan, F.I., and Abbasi, S.A., (1996), Simulation of accidents in a chemical process industry using software MAXCRED, *Indian J of Chemical Technology,* 3**,** 338-344.

Khan, F.I., and Abbasi, S.A., (1995), Accident during transportation of hazardous chemicals, *Chemical Weekly*, May 15-19, 9.

Khan, F. **I.**, & Abbasi, S. A., (1995), Analytical Simulation*: a Procedure to Conduct FTA in Chemical Process Industries,* CPCEIRA 012/95. Pondicherry University, Pondicherry

Khan, F.I., & Abbasi, S.A. (1996), Simulation of accidents in a chemical process industry using software MAXCRED. *Indian Journal of Chemical Technology*, *3*, 338–344.

Khan, F.I., & Abbasi, S.A., (1997), Risk analysis of epichlohydrin manufacturing industry using new computer automated tool MAXCRED. *Journal of Loss Prevention in Process Industries*, *10,* (2) 91–100.

Khan, F.I., and Abbasi S. A., (1997), OptHAZOP—an effective and optimum approach for HAZOP study *Journal of Loss Prevention in the Process Industries,* Volume 10, Issue 3, May Pages 191-204

Khan, F I., & Abbasi, S. A., (1997), *Hazard Identification and Ranking: a Multi-attribute Technique for Hazard Identification.* Research Report CPCEIRA 22/97. Pondicherry University, Pondicherry.

Khan, F. I., & Abbasi, S. A., (1997), *2nd International Specialty Conference on Environment Progress in the Petroleum and Petrochemical Industries.* (17-19 November) Bahrain.

Khan, F. I., & Abbasi, S. A., (1997), Svmp. *Of Air Quality Management at an Urban, Regional, and Global Scale* (23-26 September), Istanbul.

Khan, F I., & Abbasi, S. A., (1997), Hazard Identification and Ranking*: a Multi-attribute Technique for Hazard Identification.* Research Report CPCEIRA 22/97. Pondicherry University, Pondicherry.

Khan, F I., & Abbasi, S. A., (1997), Risk analysis of a chloralkali industry situated in densely populated area, *Process Safety Progress, 16, (3*), 172-185.

Khan, F.I., & Abbasi, S.A., (1997), Mathematical model for HAZOP study time estimation *Journal of Loss Prevention in the Process Industries, 10, (4), 249-257.*

Khan. F I., & Abbasi, S. A., (1997), Environmental Modeling and Software (in press).

Khan. F I., & Abbasi, S. A., (1997), *MOSEC: MOdeling and Simulation of Fire and Explosion in Chemical Process Industries.* Research report *CPCEIRA* 21/97. Pondicherry University, Pondicherry.

Khan. F I., & Abbasi, S. A., (1997), *Journal of clear production* (In press).

Khan, F.I., & Abbasi, S.A., (1997), Rapid Risk analysis of a typical chemical Industry using. MAXCRED – II Indian *Journal of Chemical Technology.* 4, 167-179.

Khan. F I., & Abbasi, S. A. (1998), *Risk assessment in Chemical Process Industries: advanced techniques,* New Delhi, Discovery Publishing House (in press).

Khan, F.I., Deepa Raniv L, & Abbasi, S.A., (1998), *Korean Journal of Chemical Engineering"* /5(I), 1-12.

Khan,. F.I. Abbasi, S.A., (2002), A criterion for developing credible accident scenarios for risk assessment, *J. Loss Prevent. Process Ind,* 15, 467– 475.

Kharbanda, O. P., and Stallworthy, E. E., (1988), Safety in chemical industry, - lessons from major disasters, *Heinemaun Professional Publishing Ltd.*, London.

Kletz, T. A., (1983), *HAZOP and HAZAN Notes on the identification and assessment l!f harards.* Rugby: The Institution of Chemical Engineers.

Kletz, T. A., (1985), *Chemical Engineer,* 92, 48-56.

Kletz, T.A., (1991a), The Philips explosion - this is where I came in, *Chem Engr*, , 493, 47.

Kletz, T.A., (1991b), Process safety - an engineering achievement. *Proc. Instn Mech. Engrs, 205*, 11.

Kletz, T.A., (1988), Piper Alpha: latest chapter in a long history, *The Chemical Engineer*, 4 July, 277.

Kletz, T.A., (1989), What went wrong, *Gulf publication*, London.

Knowlton, R. E., (1982), *An Introduction to Hazard and Operability Studies.* Vancouver, Canada: Chernetics International Ltd.

Knowlton, R. E., (1989), *Hazard and Operability Studies, The Guide Word Approach.* Vancouver, Canada: Chematics International Co.

Koivisto, R., Vaija. P., and Dohnal,M., (1989), Intelligent interface of accident databases, *Loss prevention and Safety Promotion,* 6, 98.1.

Koivisto, R. and Nielsen, D. (1994), FIRE - a database on chemical warehouse fires, *J. Loss Prevention Process Ind.,* 7, 209.

Kolodji, B. P., (1993), *Process Sajet» Progress,* 12, (2), 127-731.

Klaassen, K. B., & Van Pepper, J. C. L. (1989), *System Reliability Concept and Applications.* New York: Chapman and Hall Inc.

Lai, F S., Shenoi, S., & Fan, L. T., (1986), *Engineering Risk and Hazard Assessment.* (Kandel and Avni Eds.) Florida: CRC Press.

Lapp, S. A., & Powers, G. I., (1979), *Journal of IEEE Transaction on Reliability, R-,* 28, 12.
Lapp, S. A., & Powers, G. J., (1976), *Chemical Engineering Progress,* 72, (4), 89.
Lapp, S. A., (1991), *Chem'Tech.,* 7(x)-704.
Lawley, H. G., (1974), *Chemical Engineering Progress, 70, (4), 45-54.*
Lees, F. P. (1996), Loss prevention in the Process Industries—Hazard identification, assessmeand control, Vols. 1–3. London: *Butterworth Heinemann.*
Lees' F.P., (2005), *Loss Prevention in the Process Industries,* (Partially updated by S. Mannan), vols. 1–3, Elsevier/Butterworth-Heinemann, Oxford,
Lees, F. P., and Ang, M. L., (1984), Safety cases with in control of industrial major accident hazards, (CIMAH - regulation 1984), *Butterworths*, London.
Lees, F.P. (1994), Major accident inquiries - lawyers or engineers? *Process Safety Environ.,* 72B, 10.
Lees, F.P., (1980), Accident fatality number - a supplementary risk criterion, in *Loss prevention and Safety promotion 3,* 2, 426.
Lees, F.P., (1996), Loss Prevention in chemical process industries, Butterworts, London.
Lewis, D.J., (1984), Case histories of past accidents, the causes and consequences, *in Major hazard installation: Planning and assessment* (ed:Petts, J.I.), Loughborough University of Technology, UK.
Lewis, D.J., (1993), Case histories: accidents, Dept of Chem. Engng, *Loughborough University of Technol,* UK,
Lewis, D. J. (1979), *The MO/1(1 Fire, Explosion and Toxicity Index-a Development of the Dow Index.* New York: AIChE on Loss Prevention.
Loss Prevention Bulletin, List of accidents, 1981, 1982, 1983, 1984, 1986. *The Institution of Chemical Engineer*, Rugby.
Mahoney, D.G., (1990), *Large property damage losses in the hydrocarbon-chemical industries: a thirty-year review,* 13th ed., M&M Protection Consultants, New York.
Marshal, V.C., (1977), What are major hazards? *Safety Surveyor,* 5(3), 5.
Marshall, V.C., (1987), *Major Chemical Hazards,* Ellis Horwood, Chichester.
McCoy, S.A., Crawley, F., Chung, P.W.H., (2005), An expert user's trial of the Hazid system for computer-aided Hazop emulation, *Institution of Chemical Engineers Symposium Series*, pp.
McKelvey, C T., (1988), *IEEE Transactions on Reliability, 37, (2),* 167-170.
Montague, D. F., (1990), *Reliabilitv Engineering and System Safetv 29,* (I), 27-33.
Mulvihill, R. J., (1988), *IEEE Transactions on Reliability, 37,(2),* 149-153.
National Fire Protection Association. (1991), *Hazardous Chemical Data, NFPA Code-325M.* NJ: NFPA.
Nash, J.R., (1976), Darkest hours, *Nelson Hall,* Chicago.
Norstrom, G.P., (1982), Fire/explosion losses in the CPI, *Chem. Engng. Prog.,* 78(8), 80.
Ozog, H., & Stickles, R. P., (1991), *Oil and Gas Journal. January,* 81-90.
Oycleye, O. 0., & Kramer, M. A. (1988), *MChE Journal,* 34(9), 1441.
Pasman, H. J., Duxbury, H. A., & Bjordal, E. M. (1992), *Journal of Hazardous Materials, 30,* I-J8.
Parmer, J. C, & Lees, F P., (1987), *Reliability Engg.,* 7,277-303.
Pasquill, F, & Smith, F B. (1983), *Atmospheric Diffusion* (3rd cdn.), New York: John Wiley.
Palmer, K.N., (1983), Explosions in flammable smokes from smouldering fires. *Loss prevention and Safety Promotion 4,* 3, D11.

Post, R.L., Hendershot, D.C., Kers, P., (2002), *Chemical Engineering progress*, 98, pp. 60-66.

Prugh, R.W., (1991), Quantitative evaluation of BLEVE hazards, *J. of Fire Protection Engineering*, 3 (1), 9-21.

Piccinini, M., & Levy, G., (1984), *Canadaian Journal of Chemical Engineering,* 62, 547.

Popazoglou, LA" Nivoliantiou, A 0" & Christon, M, (1992), *Journal of Loss Prevention in the Process Industries,* 5(3), 181-191,

Pully, A, S, (1993), *Process Safety Progress,* 12, (2), 106-10.

Rabuse, S., (2008), Rock-solid safety, *Hydrocarbon Processing* 87, pp. 131-160.

Raghavan, K.V., and Swaminathan, G., (1996), Hazard assessment and disaster mitigation in petrochemical industries, *Oxford Publishing Company*, Chennai.

Rauzy, A (1993), *Reliability Engg, and System Safety,* 40, 203-21 I,

Rose, J, C, Wells, G, L, & Yeats, B, H, (1978), A Guide to Project Procedure*, Institution of Chemical Engineers*, London.

Roy,P.K., Bhatt, A., Rajagopal, C., (2003), Quantitative risk assessment for accidental release of titanium tetrachloride in a titanium sponge production plant, Journal of Hazardous Materials 102, (2-3), pp. 167-186.

Schwab, F., Urban, I., (2003), More effective use of FMEA [Effektiverer einsatz der FMEA], *Kunststoffe Plast Europe* 93, pp. 25-26 | 63.

Sivapirakasam, S.P., Surianarayanan, M., Swaminathan, G., (2008), Hazard assessment for the safe storage, manufacturing and handling of flash compositions, *Journal of Loss Prevention in the Process Industries.*

Sweeny, J, C., (1993), *Process Safety Progress,* 12,(2), 83-90,

Shafagi, A, & Cook, B, F., (1988), *IEEE Transactions on Reliability,* 37,(2), 161-165,

Singh, & Munday, G, (1979), *IFAL: a model for the evaluation of chemical process losses, Design* 79, London: Instn Chern, Engrs.

Scheffler, N, E., (1994), *Process Safety Progress,* 13,(4), 214-218,

Shastri, S., Lam, C.P., Werner, B., (2004), A machine learning approach to generate rules for process fault diagnosis, *Journal of Chemical Engineering of Japan* 37, pp. 691-697.

Suchman, E.A., (1961), A conceptual analysis of the accident problem, *Social Problems*, 8(3), 241-246.

Suokas, J, (1988), *Accident Analysis and Prevention,* 20(1), 67-85,

SYHI (1993), *Environment Protection Act CR-RI6735,* Washington, DC

Tanaka, H" Fan, L T" Lai, F, S" & Toguchi, K (1983), *IEEE Transactions on Reliability R-32,* 51, 453--456,

The Hindu (1997), Major fire in Vizag refinery, *The Hindu publication*, September 15, 1.

The Hindu (1997a), Blast in a factory, *The Hindu publication*, November 25, 11.

The Hindu (1997b), Blast in sterlite claims two lives, *The Hindu publication*, August 31, 15.

Tyler, B, J, (1982), *Plant/Operation Progress,* 4, 172,

Taylor, J. R., (1993), Use of simulation for emergency planning and training, *Health, Safety and Loss Prevention,* 387.

Tyler, B, J., Thomas, A. R., Doran, P., & Grieg, T, R., (1994), A toxicity hazard index. *Hazards,* XIII, 351.

Thomas, I.F., (1995), The politics of safety. *Chem Engng of Australia*, 20(1), 14.

TPL (1992), Risk analysis of storage unit of chemical process industry, *Tamilnadu Petrochemical Ltd*., Manali, Madras, India.

Ulerich, N., (1988), *IEEE Transaction on Reliability,* 37, (2), 171.

Van Sciver, G. R, (1990), *Reliability Engg and System Safety, 29,* 55-68.

Van Ulden, A. P., & Hostlag, A. A (1985), *Journal of Climate and Applied Meteorology,* 24, 1196.

Van Ulden, A. P. (1988), *The Spreading and Mixing ofa Dense Cloud in Still Air* 1, S. Puttock, Ed. Oxford, Clarendon Press.

Venkatasubramanian, V., & Vaidyanathan, R. (1994), *AIChE Journal, 40,(3), 496-505.*

Ven der Voort, M.M., Klein, A.J.J., de Maaijer, M, van den Berg, A.C., van Deursen, J.R., Versloot, N.H.A., (2007), A quantitative risk assessment tool for the external safety of industrial plants with adust explosion hazard, Journal of Loss Prevention in the Process Industries 20 (4-6), pp. 375-386.

Vera, A., Stevens, F., (2008), Is HAZOP always the method of choice for identification of major process plant hazards?, *Institution of Chemical Engineers Symposium Series* (154), pp. 603-610.

Wang, H., Jiang, N., (2005), Fault is olation and identification approach in principal component subspace, *Huagong Xuebao/Journal of Chemical Industry and Engineering (China)* 56, pp. 659-663.

Wells, G.L., Phang,C., Wardman.M., and Whetton,C., (1992), Incident scenarios: their identification and evaluation. *Process Safety Environ,* 70B**,** 179

Whitehouse, H. B. (1985), *IFAL-a New Risk Analysis Tool. The Assessment and Control of Major Hazards.* Rugby: Instn Chern. Engrs. 309.

WHO. (1984), *Major Hazard Control: A Practical Manual.* Geneva: International Labor Office.

Yang, Y., Chemg, G., Zhi, H., Yang, B., Zhao, Y., (2004), Fault tree models of flue gas energy recovery system and the fuzzy evaluation on unstable operation states, *Petroleum Refinery Engineering* 34, 00. 21-27.

Yllera, J. (1988), *Engineering Risk and Hazard Assessment* (Kandel and Avni, Eds.), Florida: CRC Press Inc.

Zhao, X., Yan, W., Shao, H., (2006), Monitoring and Fault Diagnosis for Batch Process Based on Feature Extract in Fisher Subspace 11 Supported by the National Natural Science Foundation of China (No. 60504033), *Chinese Journal of Chemical Engineering* 14, pp. 759-764.

Table 14.1. Work – related fatalities (USBL, 1999)

Cause	Number of fatalities	Percent of total
Transportation accident	2630	44
Homicide/suicide	960	16
Contact with objects and equipment	941	16
Falls	702	12
Exposure to harmful substances and environments	572	9
Fire and/or Explosion	205	3
Total fatalities	6010	100

Table 14.2. Statewise distribution of highly polluting industries in India

S. No	State	Number of units
1	Andhra Pradesh	367
2	Assam	18
3	Bihar	48
4	Chhattisgarh	25
5	Goa	11
6	Gujarat	345
7	Haryana	127
8	Himachal Pradesh	19
9	Jammu & Kashmir	11
10	Jharkhand	21
11	Karnataka	143
12	Kerala	49
13	Madhya Pradesh	79
14	Maharashtra	565
15	Meghalaya	8
16	Orissa	52
17	Punjab	101
18	Rajasthan	111
19	Tamil Nadu	220
20	Tripura	10
21	Uttaranchal	36
22	Uttar Pradesh	274
23	West Bengal	87
24	Chandigarh	1
25	Daman	3
26	Delhi	5
27	Pondicherry	8
	Total	**2744**

Table 14.3. Hazardous waste generating units in India, as per Hazardous Waste (Management & Handling) Rules, 200

Sl.No	State/ UT	Number of Units	Quantity of hazardous waste generated, TPA (tones per annum)
1.	Andhra Pradesh	1,583	4,95,985
2.	Assam	53	16,038
3.	Bihar	34	3,967
4.	Chhattisgarh	152	73,001
5.	Delhi	2,411	18,600
6.	Goa	204	10,274
7.	Gujarat	5,739	12,76,027
8.	Haryana	1,458	19,240
9.	Himachal Pradesh	880	33,517†
10.	Jammu& Kashmir	147	41,353
11.	Jharkhand	352	1,81,227
12.	Karnataka	1,621	71,625
13.	Kerala	530	88,794

Sl.No	State/ UT	Number of Units	Quantity of hazardous waste generated, TPA (tones per annum)
14.	Madhya Pradesh	1085	1,21,537
15.	Maharashtra	4,571	14,07,480
16.	Meghalaya	39	29,493
17.	Nagaland	03	11
18.	Orissa	257	83,915
19.	Punjab	2,189	1,13,248
20.	Rajasthan	588	35,57,323
21.	Tamilnadu	2,422	1,90,924
22.	Tripura*	135	264
23.	Uttaranchal	51	7,759
24.	Uttar Pradesh	1,777	1,17,572
25.	West Bengal	640	2,36,000
26.	Chandigarh	111	836
27.	Daman, Diu,& NH	598	30,862
28.	Pondicherry	86	34,768
Total		29,716	82,61,630

Table 14.4. Main causes of large fires in the chemical process industries (Khan and Abbasi 1998)

Causes	Proportion (%)
Flammable liquid or gas (release, overflow)	17.8
Overheating, hot surfaces, etc.	15.6
Pipe of fitting failure	11.1
Electrical breakdown	11.1
Cutting and welding	11.1
Arson	4.4
Others	28.7

Table 14.5. Information related to explosions in the chemical and allied industries (Less 1996, 2005)

Aspect	Proportion (%)
Main cause	
Uncontrolled chemical reaction	35
Combustion explosion in equipment	13.3
Vapour cloud explosion	10.0
Overpressure	8.3
Decomposition	5.0
Combustion sparks	5.0
Pressure vessel failure	3.3
Improper operation	3.3
Others	16.8
Frequent location of occurrence	
Enclosed process or manufacturing building	46.7
Outdoor structure	31.7

Table 14.5. (Continued)

Aspect	Proportion (%)
Yard	6.7
Tank farm	3.3
Boiler house	3.3
Others	8.3
Various contributing factors	
Rupture of equipment	26.7
Human element	18.3
Improper procedures	18.3
Faulty design	11.7
Vapour-laden superheat	11.7
Congestion	11.7
Flammable liquid	8.3
Long replacement time	6.7
Inadequate combustion control	5.0
Inadequate explosion relief	5.0

Table 14.6. Characteristics of accidental release from pipe work and in-line equipment (Bellamy et al., 1989; Less, 1996, 2005)

Aspect	Number of incidents
Location	
Chemical plant	278
Refinery	96
Factory	187
Storage depot	47
Tank yard	28
Fuel station	15
Other	38
Unknown	232
Total	921
Operation	
Normal operation	343
Storage	103
Loading/unloading	33
Maintenance	146
Modification	8
Contract work	18
Testing	5
Unknown	128
Other	40
Start-up	42
Shut-down	18
Total	884
Materials released	
Ammonia	54
Hydrocarbons (unspecified)	54

Aspect	Number of incidents
Chlorine	50
Hydrogen	37
Benzene	33
Crude oil	28
Steam	25
Natural gas	24
Propane	20
Butane	18
Fuel oil	18
Hydrochloric acid	16
Sulphuric acid	16
Ethylene	16
Hydrogen sulphide	14
Water	13
Nitrogen	13
Oxygen	13
Vinyl chloride	12
LEG	12
Styrene	11
Naphtha petroleum	10
Total	507
Material phase	
Liquid	393
Gas	260
Vapour	13
Solid	9
Liquid + gas/vapour	120
Solid + gas/vapour	3
Total	798
Unignited material dispersion	
Flammable	127
Toxic	123
Flammable/toxic	47
Corrosive	97
Irritant	1
Unignited gas	96
Vapour cloud	180
Liquid	212
Spill	186
Jet/spurt	8
Spray	10
Total	1087
Fire or explosion event	
Fire	145
Flash fire	11
Pool fire	4
Jet fire	1
Fireball	7
BLEVE[a]	4
Explosion	63

Table 14.6. (Continued)

Aspect	Number of incidents
Explosion followed by fire	77
Explosion followed by flash fire	2
Total	314

Table 14.7. Some of the most gruesome transportation accidents dealing with hazardous chemicals

Year	Location	Chemical	Incident	Fatality/injury
Pipeline transport				
1981	S. Raface, Venezuela	LPG	Explosion	18/35
1984	Cubato, Brazil	Gasoline	Fire and explosion	508/31
1984	Ghari Dhoda, Pakistan	LNG	Explosion	60/11
1988	Mexico City, Mexico	Crude oil	Fire and explosion	12/80
1989	Nizhnevartovsk, ussia	LPG	Explosion and fire	462/290
2004	Skikda, Algeria	LNG	Explosion and fire	13/79
Road transport				
1975	Texas, USA	LPG	Explosion	16/35
1978	Los Afaques, Spain	Propylene	Explosion and fire	216/400
Year	Location	Chemical	Incident	Fatality/injury
1978	Xilotopee, Mexico	Butane	Explosion	100/200
1987	Preston, UK	Diesel oil	Fire	12/16
1988	Karo, Nigeria	Petrol	Explosion and fire	15/35
1995	Madras, India	Benzene	Explosion and fire	115/10
1996	Palermo, Italy	Propane	Explosion and fire	5/25
1999	Athens, Greece	LPG	Explosion and fire	4/13
2001	Kanpur, India	LPG	Explosion and fire	12/6
2004	Balatimore, USA	Propane	Explosion and fire	10/7
Rail transport				
1974	Decatur, USA	Isobutane	Explosion	71/52
1978	Tennessee, USA	Propane	Explosion	25/50
1981	Potosi, Mexico	Chlorine	Toxic release	20/1000
1983	Pojuca, Brazil	Gasoline	Fire	10/40
1983	Dhurabai, India	Kerosine	Explosion	47/15
1988	Arzanas, Russia	Explosive	Explosion	73/230
2002	Cario, Egypt	Butane	Explosion and fire	337/7500
2004	Neyshabur, Iran	Miscellaneous	Explosion and fire	328/460

Table 14.8. Major factors leading to accidents in the chemical industries

Factor	Proportion (%)
Equipment failure	29
Operational failure	21
Inadequate material evaluation	16
Chemical process problems	11
Material movement problems	9
Ineffective loss prevention program	6
Plant site problems	4

Factor	Proportion (%)
Inadequate plant layout	2
Structures not in conformity with the user	
Requirement	2

Table 14.9. List of major accidents in chemical process industries, 1926-1969

Year	Location	Chemical	Event	Deaths/injured
1926	St. Auban, France	Chlorine	Toxic release	19/105
1928	Homburg, Germany	Phosgene	Toxic release	10/50
1929	Syracause, New York	Chlorine	Toxic release	1/100
1939	Zarnesti, Romania	Chlorine	Toxic release	60/?
1940	Mjodelana, Norway	Chlorine	Toxic release	3/34
1942	Tessenderloo, Belgium	Ammonium nitrate	Explosion	>/100
1943	Ludigshafen. Germany	Butadiene	Explosion	57/37
1943	Los Angeles, CA	Butane	Fire	5/>25
1944	Cleveland, OH	LNG	Fire and explosion	128/300
1944	Denison, TX	Butane	Fire	10/45
1947	Brest France	Ammonium nitrate	Explosion	21/?
1947	Rauma, Finland	Chlorine	Toxic release	19/200
1947	Texas City, TX	Ammonium nitrate	Explosion	552/3000
1948	Ludigshafen, Germany	Dimethyl ether	Explosion	245/2500
1949	Perth, NJ	Hydrocarbons	Fire	4/26
1950	Poza Rica, Mexico	Hydrogen sulphide	Toxic release	22/320
1952	Walsum, Germany	Chlorine	Toxic release	7/56
1954	Bitburg, Germany	Kerosine	Fire	32/16
1955	Whiting, IN	Naptha	Explosion	2/30
1958	Niagara Falls, NY	Nitromethane	Explosion	?/>200i
1958	Signal Hills, CA	Oil forth	Fire	2/34
1959	Meldrin, GA	LPG	Explosion	23/78
1959	Kansas City, MO	Gasoline	Fire	5/?
1959	Phillipsburg. NJ	Seal oil	Explosion	6/6
1959	Roseberg, OR	Ammonium nitrate	Explosion	13/74
1959	Ube, Japan	Ammonia plant	Explosion	11/40
1960	Forepart, TX	Ally l chloride	Explosion	6/14
1960	Kingsport, TN	Aniline plant	Explosion	15/55
1961	La Barre, LA	Chlorine	Toxic release	1/114
1962	Doe Run, Key	Ethylene oxide	Explosion	2/19
1962	New Belin, NY	LPG	Explosion	10/75
1962	Ras Taruna, Saudi Arabia	Propane	Fire	1/111
1962	Toledo	Acrylic polymide	Explosion	10/46
1964	Mebronville. MA	PVC	Explosion	7/27
1964	Texas, USA	Ethylene	Explosion	2/34
1965	Louisville, KY	Mono acetylene	Explosion	12/60
1965	Natchitoches, LA	Natural gas	Explosion	17/56
1966	Freyzin, France	Propane	Fire and explosion	18/83
1966	Larsoe, LA	NGL	Fire	7/20
1966	LaSallie, Quebee	Styrene	Explosion	11/10

Table 14.9. (Continued)

Year	Location	Chemical	Event	Deaths/injured
1966	West Germany	Methane	Explosion	3/83
1967	Antwerp, Belgium	VCM	Explosion	4/33
1967	Hawthorn, NJ	?	Explosion	2/16
1967	Lake Charles, LA	Isobutane	Explosion	7/14
1968	Fast Germany	VCM	Explosion	24
1968	Hull, UK	Acetic acid	Explosion	2/13
1968	Lievin, France	Ammonia	Toxic release	5/35
1968	Pernis, Netherlands	Oil (compr.)	Explosion	2/85
1969	Basel, Switzerland	Nitro liquid	Explosion	3/28
1969	Repcelak, Hungary	Carbon dioxide	Explosion	9/23
1969	Crete, NB	Ammonia	Toxic release	8/20
1969	Escombreaes	Petroleum	Explosion	4/3
1969	Laurel, MS	LPG	Explosion	2/976
1969	Puerto la Cruz	Light hydro.	Explosion	5/23
1969	Teeside, UK	Cyclohexane	Fire	2/23

Tabel 14.10. List of major accidents in chemical process industries, 1970-1989

Year	Location	Chemical	Event	Deaths/injured
1970	Philadelphia, Panama	Catl. cracker	Explosion	7/42
1971	Emmerich, Germany	Ammonia	Toxic release	4/53
1971	Houston, TX	VCM	Explosion	1/50
1971	Longview, TX	Ethylene	Explosion	4/60
1971	Netherlands	Butadiene	Explosion	8/21
1972	Rio de Janerio, Brazil	Butane	Explosion	37/53
1972	Lynchburg, VA	Propane	Fire	2/3
1972	Netherlands	Hydrogen	Explosion	4/40
1972	Weirton, WV	Coke plant	Explosion	10/10
1972	West Virginia, USA	Gas	Explosion	21/20
1973	Kingman, AZ	Propane	Fire	13/89
1973	Austine, TX	NGL	Fire	6/21
1973	Japan	VCM	Explosion	1/16
1973	Potchefstroom	Ammonia	Toxic release	18/34
1973	St. Amand L'Eaux, France	Propane	Explosion	5/45
1973	Sheffield, UK	Gas works	Explosion	4/24
1973	Staten Island, NY	LNG	Fire	40
1974	Beaumont, TX	Isoprene	Explosion	2/10
1974	Czechoslovakia	Ethylene	Explosion	14/79
1974	Decatur. IL	Propane	Explosion	7/152
1974	Flixborough, UK	Cyclohexane	Explosion	28/76
1974	Houston, TX	Butadiene	Explosion	1/235
1974	Madras, India	Potassium sol.	Hot release	9/15
1974	Wenatchee, WA	MENiterate	Explosion	2/66
1975	Antwerp, Belgium	Ethylene	Explosion	6
1975	Beek, Netherlands	Propylene	Explosion	14/108
1975	Eagle Pass, TX	Propane	Fire	16/7

Year	Location	Chemical	Event	Deaths/injured
1975	Philadelphia. Panama	Oil vapours	Explosion	8/20
1975	Scunthorpe, UK	Water-methyl	Explosion	11/15
1975	South Africa	Methane	Explosion	7/7
1976	Chalmette, LA	Ethyl benzene	Explosion	13/?
1976	Houston, TX	Ammonia	Toxic release	6/200
1976	Los Angles. CA	Gasoline	Fire	6/35
1976	Gadsden. AL	Gasoline	Fire	3/24
1976	Sandefijord, Norway	Flamm. Liquid	Explosion	6/?
1976	Serveso, Italy	TCDD	Toxic release	?/300
1977	Colombia, USA	Ammonia	Toxic release	30/22
1977	Gela, Italy	Ethylene oxide	Explosion	1/25
1977	Gujarat, India	Hydrogen	Explosion	5/35
1977	Mexico	Ammonia	Toxic release	2/102
1977	Umm Said, Qatar	LPG	Fire	7/87
1977	Westwego,LA	Explosive dust	Explosion	35/5
1978	Chicago, IL	Hydrogen	Toxic release	8/29
1978	Santa Cruz, Mexico	Sulphate	Fire	52/88
1978	St. Marys, wv	Propylene	?	51/26
1978	San Carlos, Spain	Cooling water	Explosion	211
1978	Texas City, TX	Propylene	Fire	7/11
1978	Waverly, TN	Butane	Explosion	12/21
1978	Youngestown, FL	Propane	Toxic release	8/50
1979	Banter Bay, Eire,	Chlorine	Explosion	50
1981	Montanas, Mexico	Oil	Toxic release	29/50
1982	Spencer, OK	Heated water	Burning	7/12
1983	Reserve, LA	Chlorobutadine	Fire and toxic release	3/12
1983	Houston, TX	Methyl bromide	Toxic release	2/11
1984	Brazil	Gasoline	Fire and toxic release	508/221
1984	Roeoville, IL	Propane	Explosion	15/76
1984	Mexico City, Mexico	LPG	Fire and explosion	550/23
1985	Clinton, USA	Ammonia	Toxic release	5/8
1985	Breed Ford. UK	Ammonia	Toxic release	2/13
1985	Brazil	Ammonia	Toxic release	>5000educated
1985	Illinois, USA	Naptha	Explosion	7/12
1985	Priola, Italy	Ethylene	Explosion	23/11
1985	Algerais, Spain	Naptha	Explosion and fire	18/56
1985	Mont Belyieu, TX	Propane	Fire	4/13
1986	Basel, Switzerland	Fungicide	Toxic release	Toxification
1986	Ohio, USA	HCL	Toxic release	3/26
1986	Kennedy Space Center, FL	Hydrogen	Explosion	7/119
1986	Pascagoula, MS	Aniline	Fire	3/76
1987	Grangemouth, UK	Hydrocarbon	Explosion and fire	67/21
1987	Piper Alpha	Hydrogen	Explosion	167/55
1987	Antwerp, Belgium	Ethylene oxide	Explosion	5/20
1987	Pampa, TX	Acetic acid	Explosion	3/43

Tabel 14.10. (Continued)

Year	Location	Chemical	Event	Deaths/injured
1987	Louisiana, TX	Hydrocarbon	Fire and explosion	15/21
1988	Maharashtra, India	Naptha	Fire	25/23
1988	Rafnes, Norway	Vinyl cloride	Explosion	7/13
1988	Narco, LA	Propane	Explosion	7/48
1989	Antwerp, Belgium	Aldehyde	Explosion	32/11
1989	USSR	Ammonia	Explosion and toxic release	7/57
1989	Baker, Gulf of Mexico	Natural gas	Explosion	2/24
1989	Worms, Germany	Carbon dioxide	Explosion	3/25
1989	Pasadena, TX	Ethylene	Explosion	23/314
1989	Boston Rouge, LA	Ethane	Explosion	4/12
1989	Phillips, USA	Ethylene	Explosion	23/130

Table 14.11. List of major accidents in chemical process industries (1990-2008)

Year	Location	Chemical	Event	Deaths/injured
1990	Channeiview, TX	Waste oil	Fire	5/13
1990	Rio de Janerio, Brazil	Hydrocarbon	Fire	3/?
1990	Czechoslovakia	Hydrogen	Explosion	15/26
1990	Fagaras, Romania	Explosives	Explosion	21/34
1990	Thane, India	Hydrocarbon	Fire and explosion	35/10
1990	Porto de Leixoes, Portugal	Propane	Fire and explosion	14/76
1992	Sodegraura, Japan	Hydrogen	Explosion	10/7
1993	Panipat, India	Ammonia	Explosion and toxic release	3/25
1994	Dornka, Egypt	Fuel	Fire	410/?
1995	Gujrat, India	Natural gas	Fire	?/?
1995	Ukhta, Russia	Gas	Fire	12/?
1996	Bombay, India	Hydrocarbon	Fire	2/45
1996	Weyauwega, WI, USA	Propane, LPG	a: Information unavailable	
1996	Palermo, Italy	Propane	a: Information unavailable	5/25
1997	Chennai, India	LPG	Fire	3/4
1997	Chennai, India	Molten metal	Explosion	2/5
1997	Gujarat, India	Hydrocarbon	Explosion	3/11
1997	Visag India	LPG	Fire and explosion	60/30
1997	Burnside, IL, USA	LPG	Explosion	2/ 2
1997	Japan	a	Dust explosion	1/1
1997	Blaye, France	Storage	Dust explosion	11/1
1998	Alberta City, IA, USA	Propane	Explosion	2/7
1998	Khanapara, India	LPG	Explosion	IO/12
1998	Patna, India	Gasoline	Fire	50/71
1998	Chembur, India	Oil	Fire	I/3

Year	Location	Chemical	Event	Deaths/injured
1999	Between Athens and Lamia, Greece	LPG	a: Information unavailable	4/ 13
1999	Michigan	Coal dust (cause for secondary explosion)	Powerhouse	6/14
1999	Massachusetts	Oven	Dust explosion	3/12
1999	Toronto, Canada	LPG		
1999	Quebec, Canada	Hydrocarbons	Explosion	2i, 350 evac
2000	Ohio, USA	Propane	Explosion	1/3
2000	Japan		Dust explosion	1/1
2000	Modesto California	Aluminum dust	Dust explosion	A
2000	Downey, CA, USA	Propane	Explosion	2/1
2000	Texas, USA	Propane	Explosion	2/1
2000	Palgar, India	FIRE, Explosion	Pharma Products	5/1
2001	Kanpur, India	LPG	a: Information unavailable	12/ 6
2001	Aonla, India	FIRE	Diesel	3/1
2001	Sanaswadi, India	Explosion	Fire work chemicals	II/ 10
2001	Jamnagar, India	LPG	a: Information unavailable	12/1
2001	Katpadi, India	Explosion	Fire work chemicals	25/1
2002	Mississippi	Rubber	Dust explosion	5/1
2002	Cairo, Egypt	Butane	a: Information unavailable	37/7500
2002	Ghaziabad, India	Toxic gas	Gaseous leak	4/1
2002	Agra, India		Fire	30 /1
2002	Tivissa, Spain	LNG	Explosion	½
2002	Pimpri, India	Gaseous leak	Toxic gas	I/10
2002	Gronton, CT, USA	Boranetetrahydrofur an	Explosion	2/1
2002	Vadodara, India	Chlorine	Explosion	2/18
2002	Ankleshwar, India	Ammonia gas	a: Information unavailable	2/1
2002	Koulpatti, India		Fire	6/ 50
2002	Ponda, India	Steel Plant	Explosion	5/ 6
2002	Sadasivpet, India	Omofozole	Explosion	5/ 4
2003	Kentucky, India	Resin	Dust explosion	7/1
2003	Kinston, NC	Polyethylene	Dust explosion	6/38
2003	Athanur,	Denatured alcohol	Fire, Explosion	2/3
2003	Vakaner, India	Oil	Explosion	4/15
2003	Louisville, USA	Maltodextrin and other chemicals	Dust explosion	1/1
2003	Agra, India	Chemicals	Fire	2 / 7
2003	Ludhiana, India		Explosion	9 /90
2003	Rourkela, India	Hot metal slag	Splaash	1/ 2
2003	Dighikalas, India	HCl vapours	Leak	1/ 7
2003	Punjab, India		Explosion and Fire	2/18
2003	Surat, India	Oil tanker	Explosion	1/9
2003	Attibele, India	Pharma products	Explosion	1/4
2003	Bhadravathi, India		Explosion	7/9

Table 14.11. (Continued)

Year	Location	Chemical	Event	Deaths/injured
2004	Avon, Ohio		Dust explosion	
2004	Baltimore-Washington Highway, USA	Propane	Explosion and Fire	10/1
2004	Skikida, Algeria	LNG	Explosion and Fire	13/74
2004	Mihama, Japan	Steam	Leak of super heated steam	4/7

Table 14.12. Fatal accidents in manufacturing industry in different countries

Country	Fatality rate	
	Deaths per 1000 man-years	Deaths per 100 000 workers per year
Argentina	0.020	Not available (NA)
Austria	0.142	NA
Belgium	0.140	NA
Canada	0.080	14
Czechoslovakia	0.061	NA
France	0.068	11
Germany (FRG)	0.120	17
Germany(GDR)	0.030	NA
Italy	0.080	NA
Japan	0.010	5
Netherlands	0.009	4
Norway	0.050	NA
Poland	0.066	NA
Spain	0.109	NA
Switzerland	0.080	NA
UK	0.020	4
USA	0.022	7

Table 14.13. Guide words and their physical significance

Guide word	Meaning	Parameter	Deviation
None	Negation intention	Flow	No flow
		Level	Zero level
Less	Quantitative decrease	Flow	Low flow rate
		Level	Low level
		Temperature	Low temperature
		Pressure	Low pressure
		Concentration	Low concentration
More	Quantitative increase	Flow	High flow rate
		Level	High level
		Temperature	High temperature
		Pressure	High pressure
		Concentration	High concentration
Reverse	Logical opposite	Flow	Reverse flow rate

Guide word	Meaning	Parameter	Deviation
		Pressure	Reverse pressure
Part of	Qualitative decrease	Concentration	Concentration decrease
		Flow	Flow decrease
		Level	Level decrease
As-Well-As	Qualitative increase	Concentration of impurity	Concentration increase
		Temperature of substance	Temperature increase
		Level of impurity	Level increase
		Pressure of substance	Pressure increase
		Flow of impurity	Flow increases
Other Than	Complete substitution	Concentration of desired substance	Concentration zero
		Level of desired substance	Level zero
		Flow of desired substance	Flow rate zero

Table 14.14. A typical HAZoP table for possible deviations in operating pressure in esterifier unit Poly: Poly ethylene terephthalate, Process unit: Esterifiers, Process Continuous PTA based

Deviation	Causes	Consequences	Remark
High pressure	Specific knowledge Vacuum pump not working properly; high pressure outlet from melter; out flow of methanol is low; outlet valve is choked; low temperature in esterifier; leak of vacuum from esterifier	Low product yield; poor quality of product; low methanol collection; adverse impact to poly condensation unit	Moderately hazardous
High pressure	General knowledge Leak in input line; leak in the reactor; outlet valve is not open; high inflow of reactant	Poor quality of product; high flow rate to next unit; chances of leak from the unit	Moderately hazardous
Low pressure	Specific knowledge Excess working of vacuum pump; high temperature thus fast reaction causing further vacuum; presence of impurity; low pressure input	High reaction rate causing further low vacuum; high outlet flow rate; chances of side reaction; hot spot formation; chances of esterifier damage due to excess vacuum	Hazardous
Low pressure	General knowledge Low feed flow rate; highout let flow rate; unwanted side reaction; vacuum indicator faulty	Product quality change; high outlet flow rate causing excess load to the next unit; chances of leak due over vacuum	Hazardous

Table 14.15. A typical HAZOP table for possible deviations in level of the chemical in melter unit Plant: Poly ethylene terephthalate, Process unit: Melter and mixer, Process; Continuos PTA based

Deviation (guide word)	Causes	Consequences	Remark
Low level	Specific knowledge Low flow rate of MEG; low flow rate of PTA; no raw material available at storage; high temperature in melter	Probability of running dry; overheating leading to failure; high temperature at outlet; chances of leak and subsequent fire	Hazardous

Table 14.15. (Continued)

Deviation (guide word)	Causes	Consequences	Remark
Low level	General knowledge Leak in feed line; leak in melter; inlet valve choked; non-availability of raw materials	Low outlet flow; excess heating; chances of leak	Hazardous
High level	Specific knowledge High flow rate of MEG; high flow rate of PTA; pump running over-speed; outlet valve not open or choked; low temperature in melter	Over-pressurisation of vessel that may cause leak; not proper mixing; high outlet flow rate; adverse effect to esterification unit	Moderately hazardous
High level	General knowledge Outlet valve choked; level indicator faulty; presence of foreign material	Accumulation of material; over-pressurisation which may cause adverse affect to the unit operation; poor quality at outlet;	Moderately hazardous

Table 14.16. Parameters used in the effectiveness study of various risk assessment schemes

Parameters	Detail description
A	Quantitative results
B	Inexpensive to execute (in terms of expert
	time/computational time/data requirement)
C	Sequence of steps optimal
D	In some steps numerous techniques have been clustered
	without giving criteria of which to choose in which
	situation, this may lead persons not very well-versed to
	either waste time or bypass some crucial aspects
E	Precision
F	Applicability at various stages of the project
G	Covers most of the aspects of risk study
H	Cumulative performance index

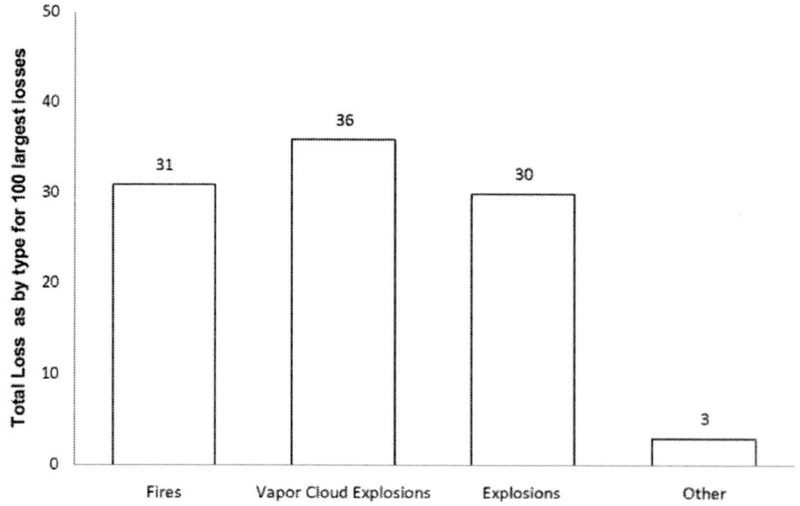

Figure 14.1. (Continued on next page).

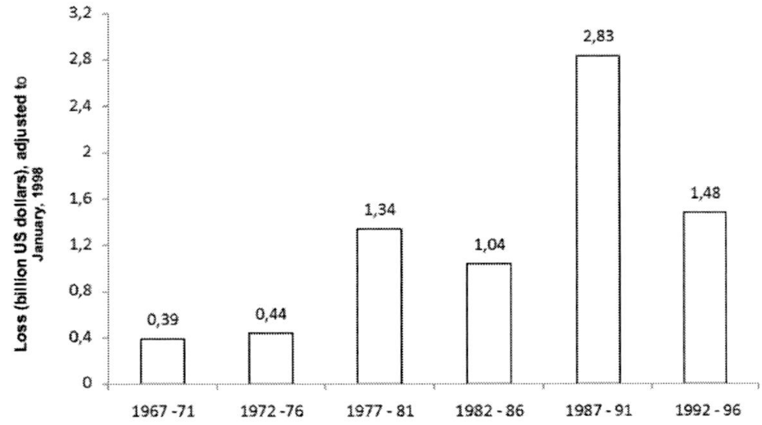

Figure 14.1. Losses incurred due to accidents in the hydrocarbon-chemical plants in the USA (Crowl 2003).

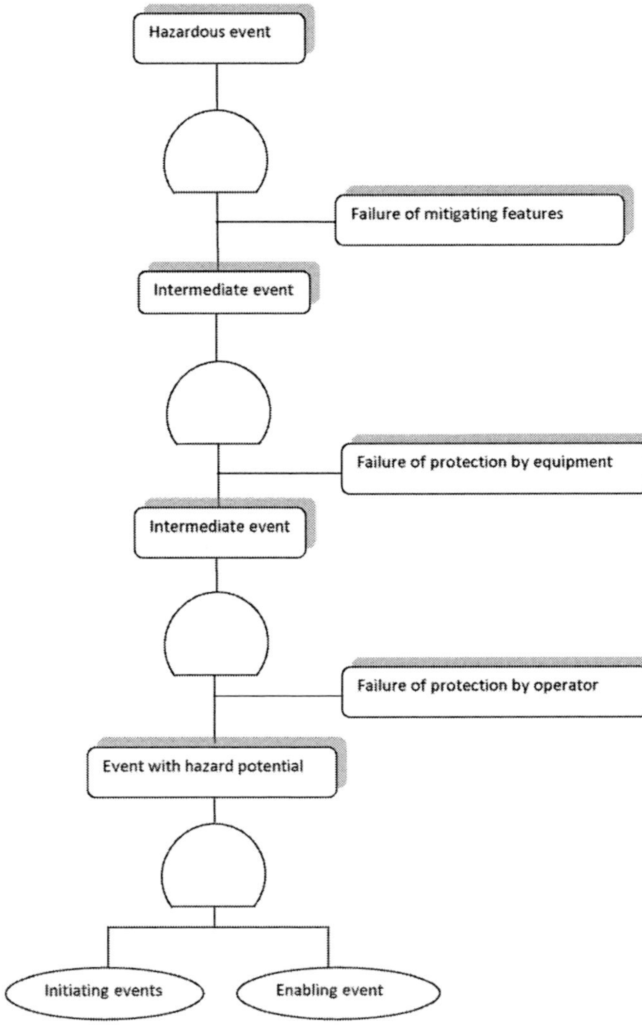

Figure 14.2. Accident fault tree model.

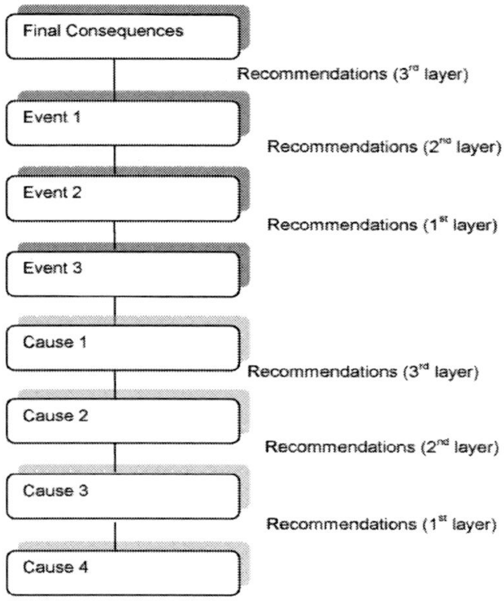

Figure 14.3. Accident model of Kletz.

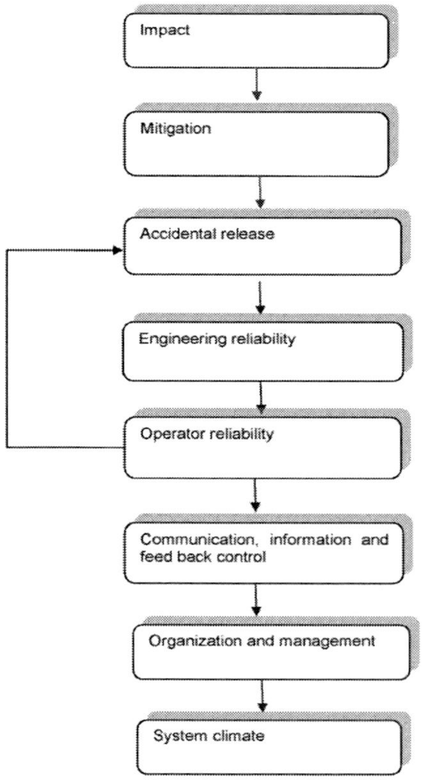

Figure 14.4. The model of Gayer and Bellamy for the accident process.

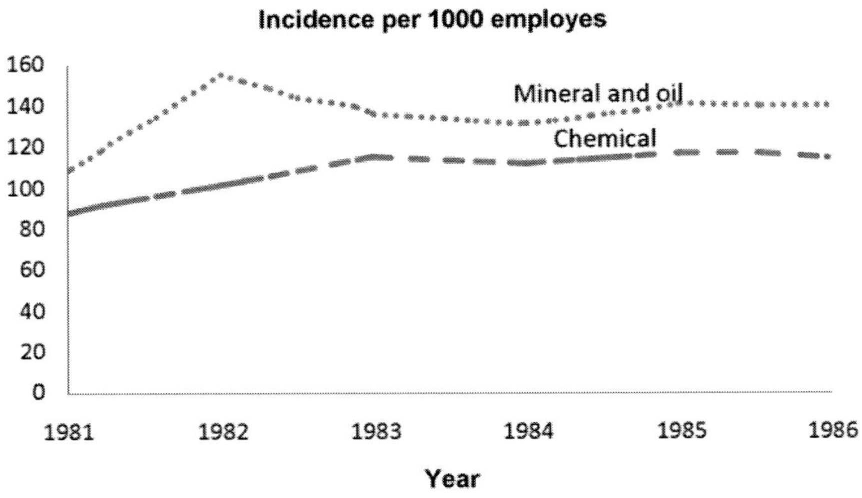

Figure 14.5. Incidence rates of fatal and major injuries in chemical and mineral oil processing industries (1981-1990).

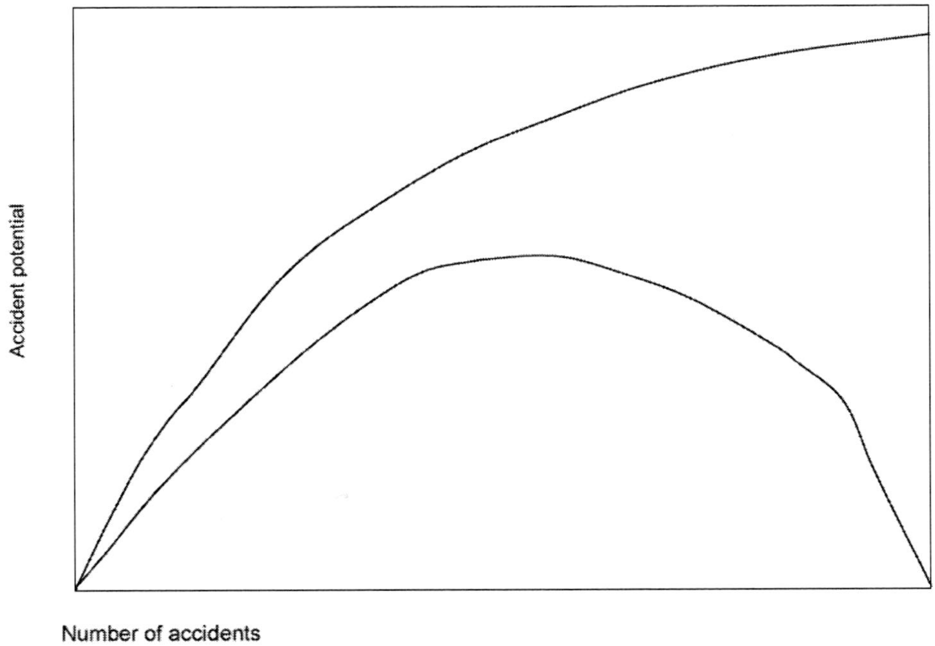

A: Actual reporting situation.
B: Ideal reporting situation.

Figure 14.6. Under-reporting of accident (Lees, 1996).

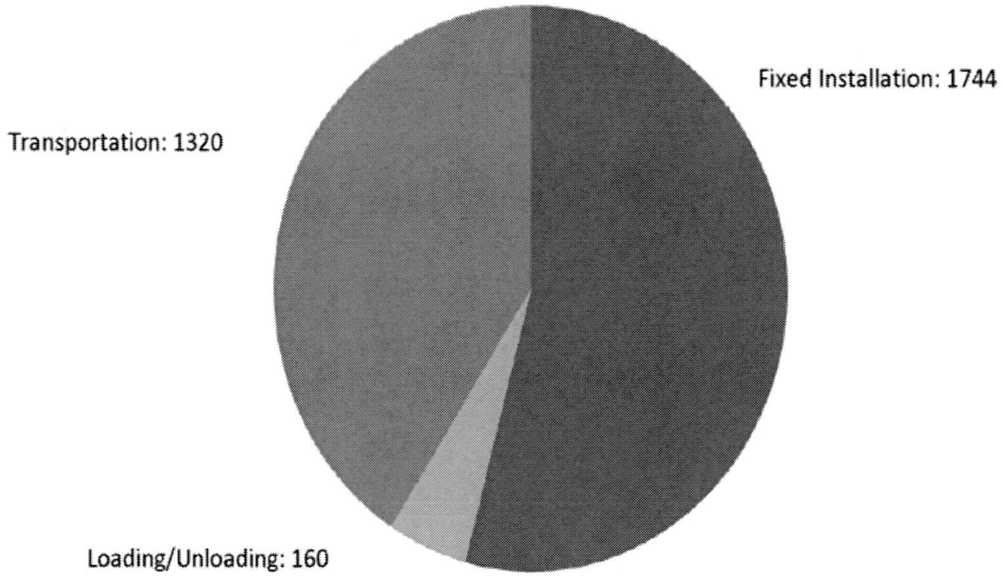

Figure 14.7. Broad classification of accident types.

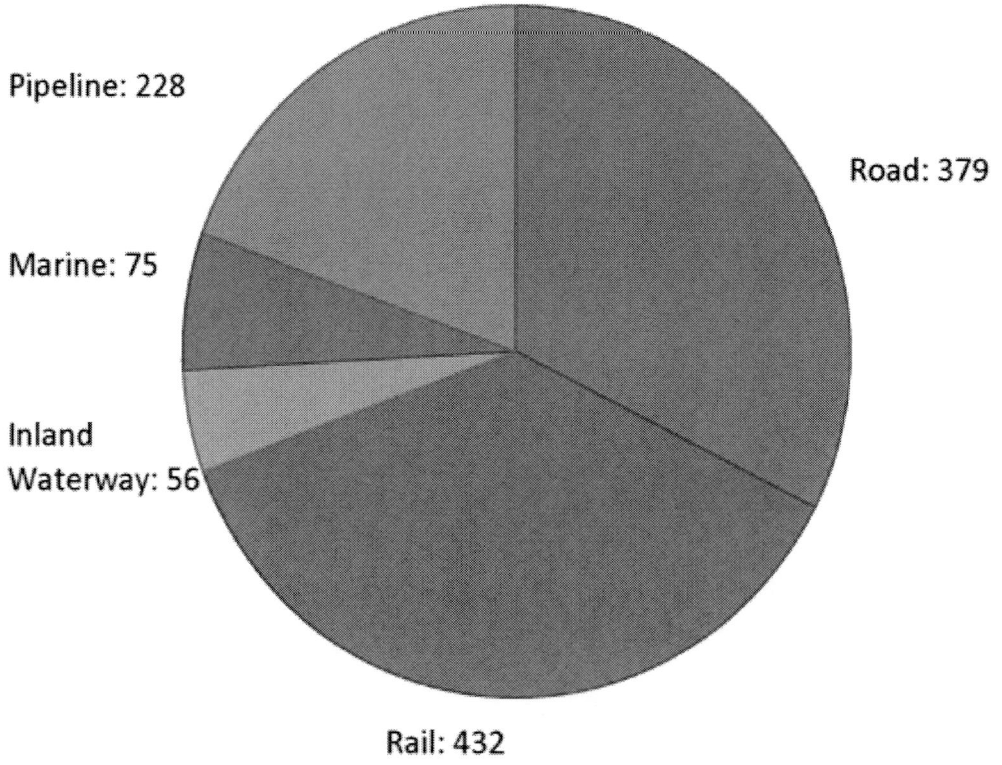

Figure 14.8. Distribution of different types of accidents.

Figure 14.9a. Arial view of the extensive damage caused in the Nypro plant and surrounding plants due the VCE that occurred at Flixborough (1974).

Figure 14.9b. Scenes of destruction at the Nypro plant from the Flixborough disaster (1974).

Figure 14.10a. Dense smoke billowing out of the PEPCON factory premises (Henderson, 1988).

Figure 14.10b. The second and largest explosion at the PEPCON plant.

Figure 14.11a. The Trans-Siberian railway accident (Siberia, 1989). The bulged-out shape of the coaches indicate that explosions happened inside as well.

Figure 14.11b. The explosion flattened large tracts of forest. In this picture, the demolished part of the forest can be seen in the foreground and the undisturbed part of the forest is visible in the distance (Siberia 1989).

Figure 14.12. Sao Paulo, Brazil (1984).

Figure 14.13a. Pictures from the pesticide and fungicide warehouse fire at Basel (1986).

Figure 14.13b. The toxic products from the warehouse fire drained into the Rhine, destroying aquatic life for miles. In this picture, fish dead on account of exposure to the toxic chemicals has washed ashore in great numbers (Basel 1986).

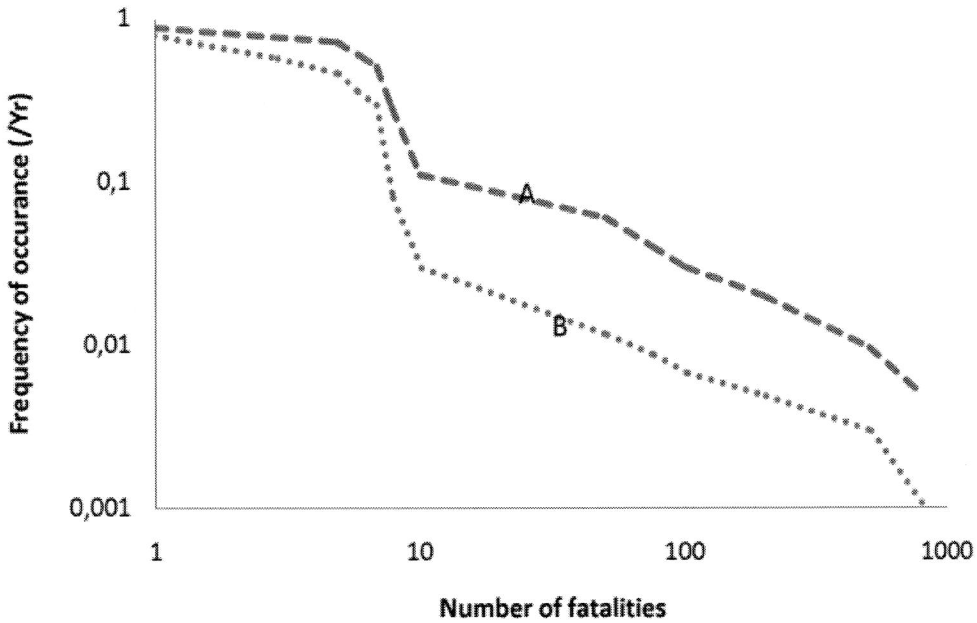

Figure 14.14. FN Curves for transportation and fixed installation accidents – A: Transportation, B: Fixed installation.

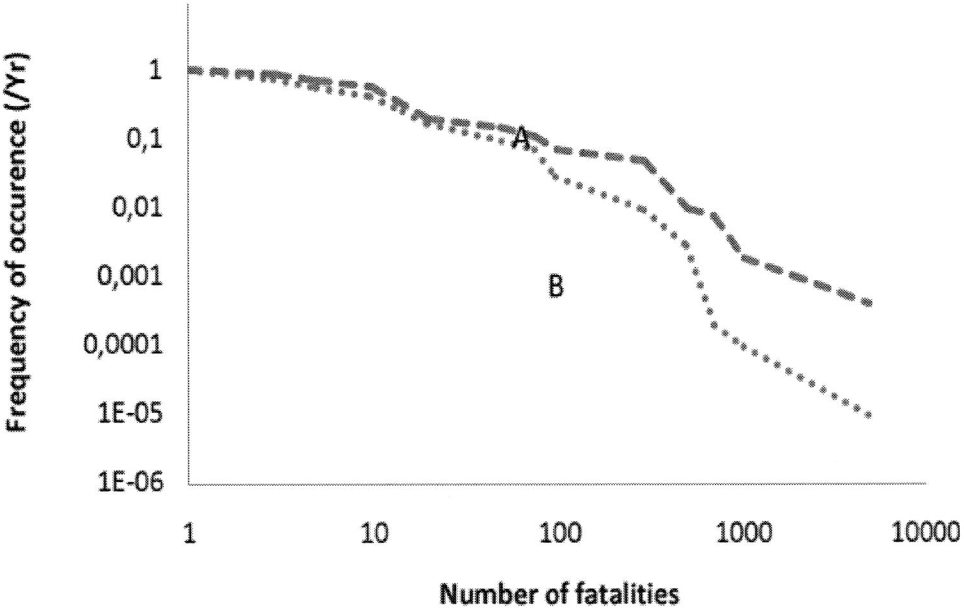

Figure 14.15. FN Curves for fire, explosion, and toxic release accidents A: Fire and explosion, B: toxic release.

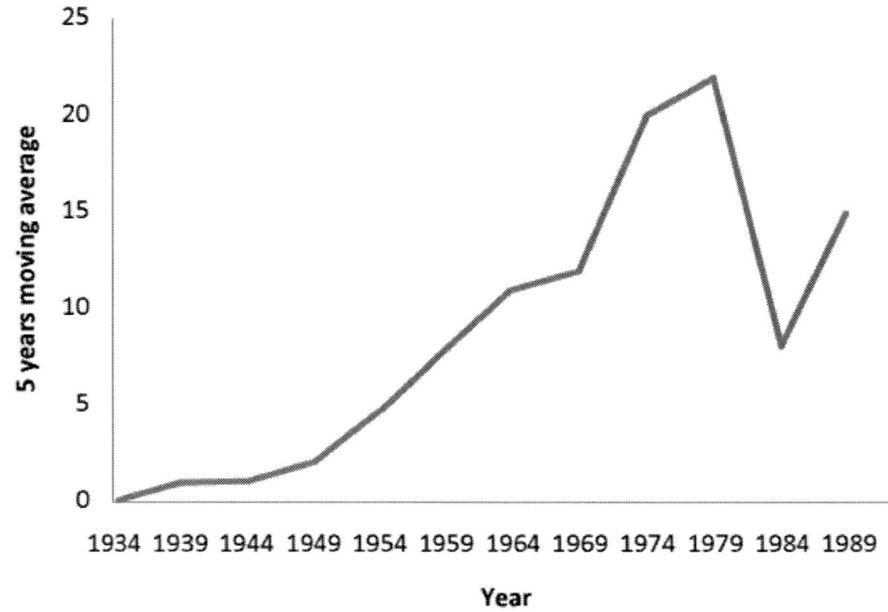

Figure 14.16. Trend of vapour cloud explosion events.

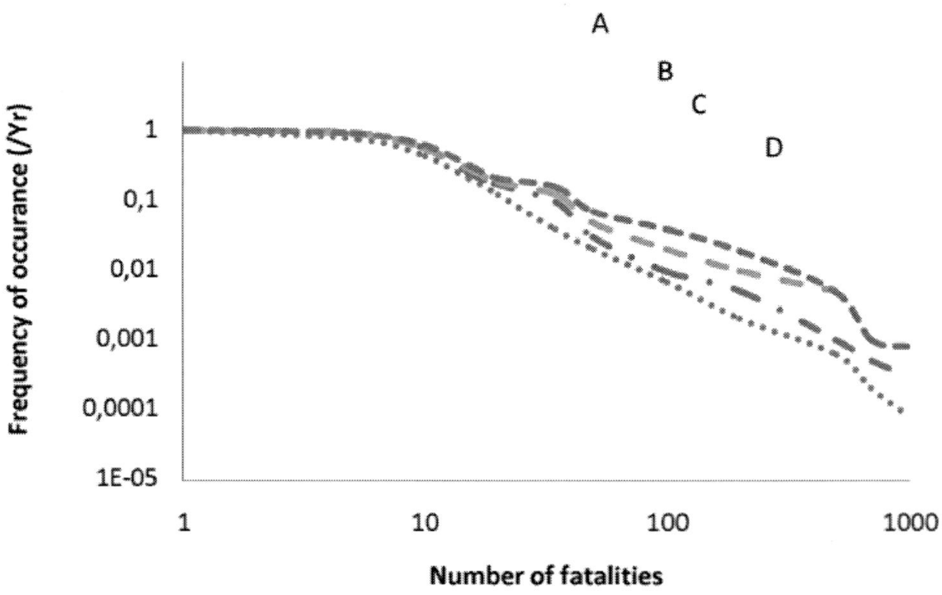

A,B,C,D,E,F and G are measures of quantifiability, cost-effectiveness, economy of steps, clarity, precision, range of applicability, and coverage respectively. H represents the cumulative index score.

Figure 14.17. FN Curves for various accidental events A: VCE, B: CVCE C: BLEVE, D: Fire.

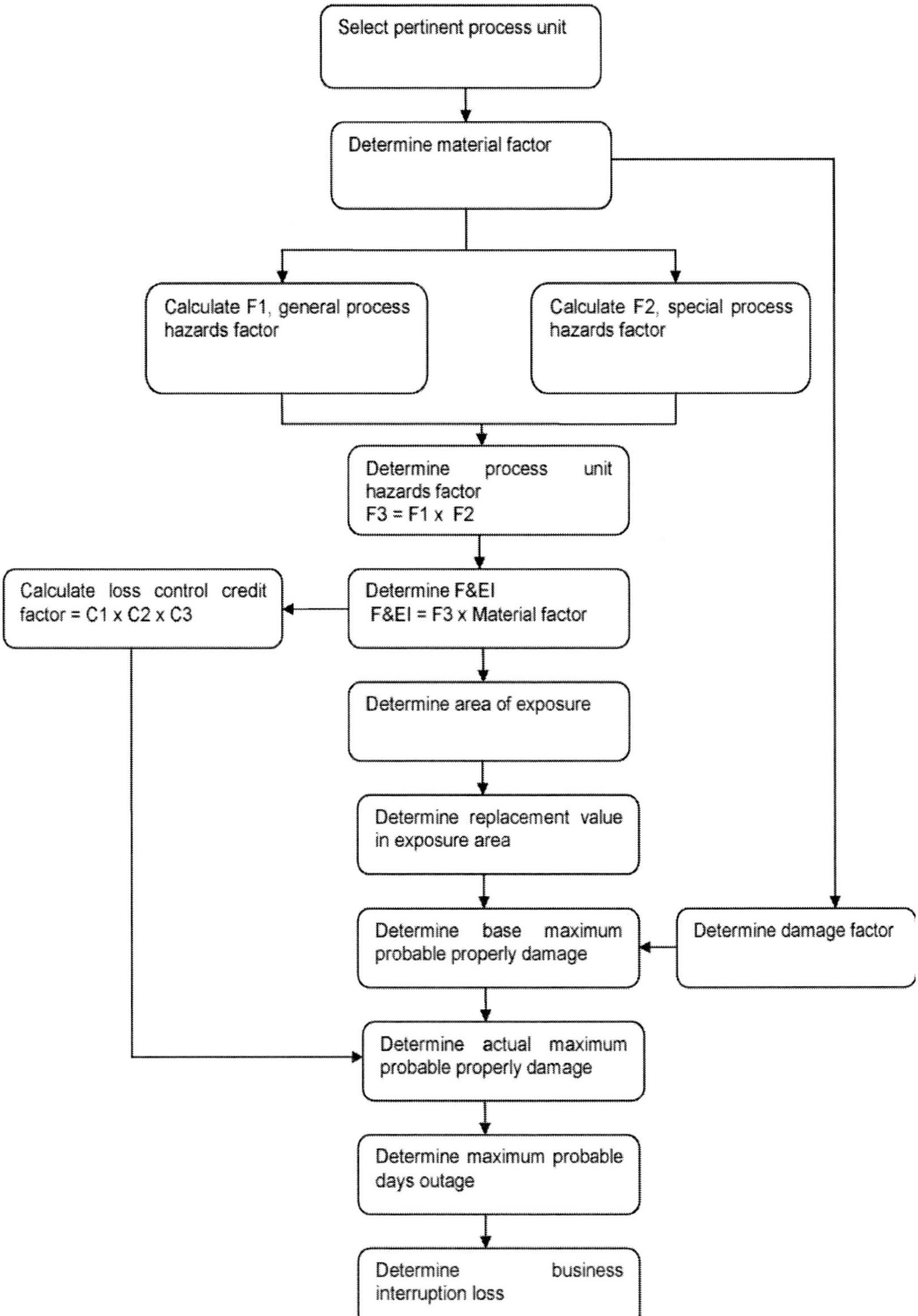

Figure 14.18. Procedure for calculating the Dow fire and explosion Index.

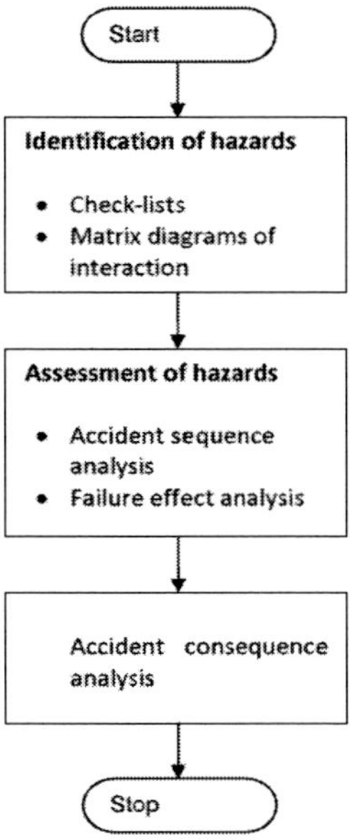

Figure 14.19. WHO's hazard assessment procedure.

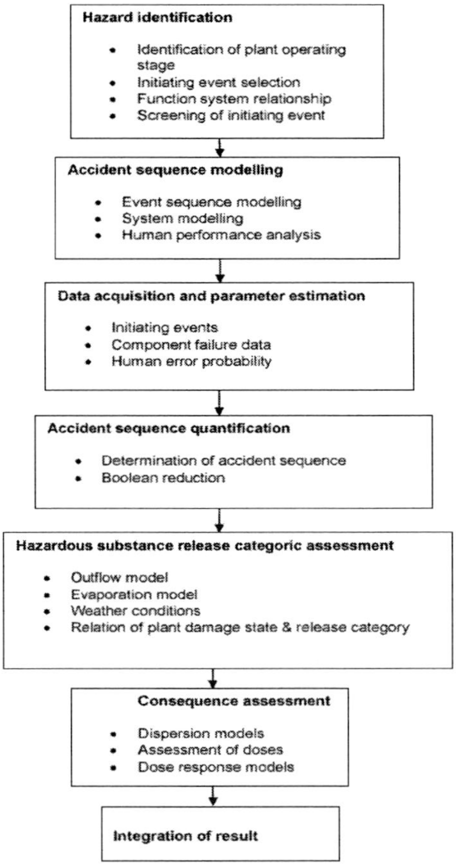

Figure 14.20. Steps to carry out probabilistic safety analysis in the chemical process industries (Papazoglou et al 1992).

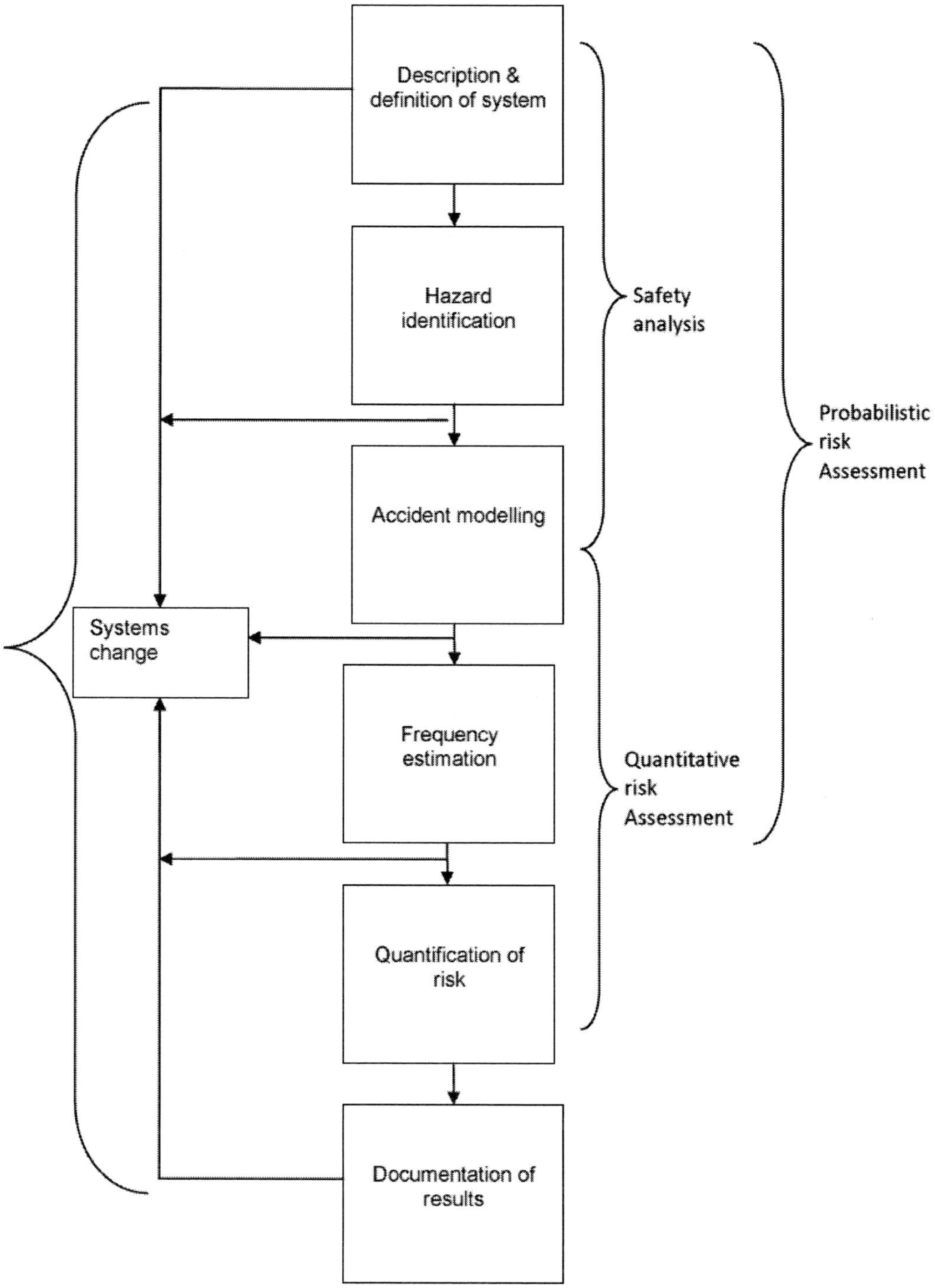

Figure 14.21. Simplified block diagram showing the main steps of different risk and safety procedures.

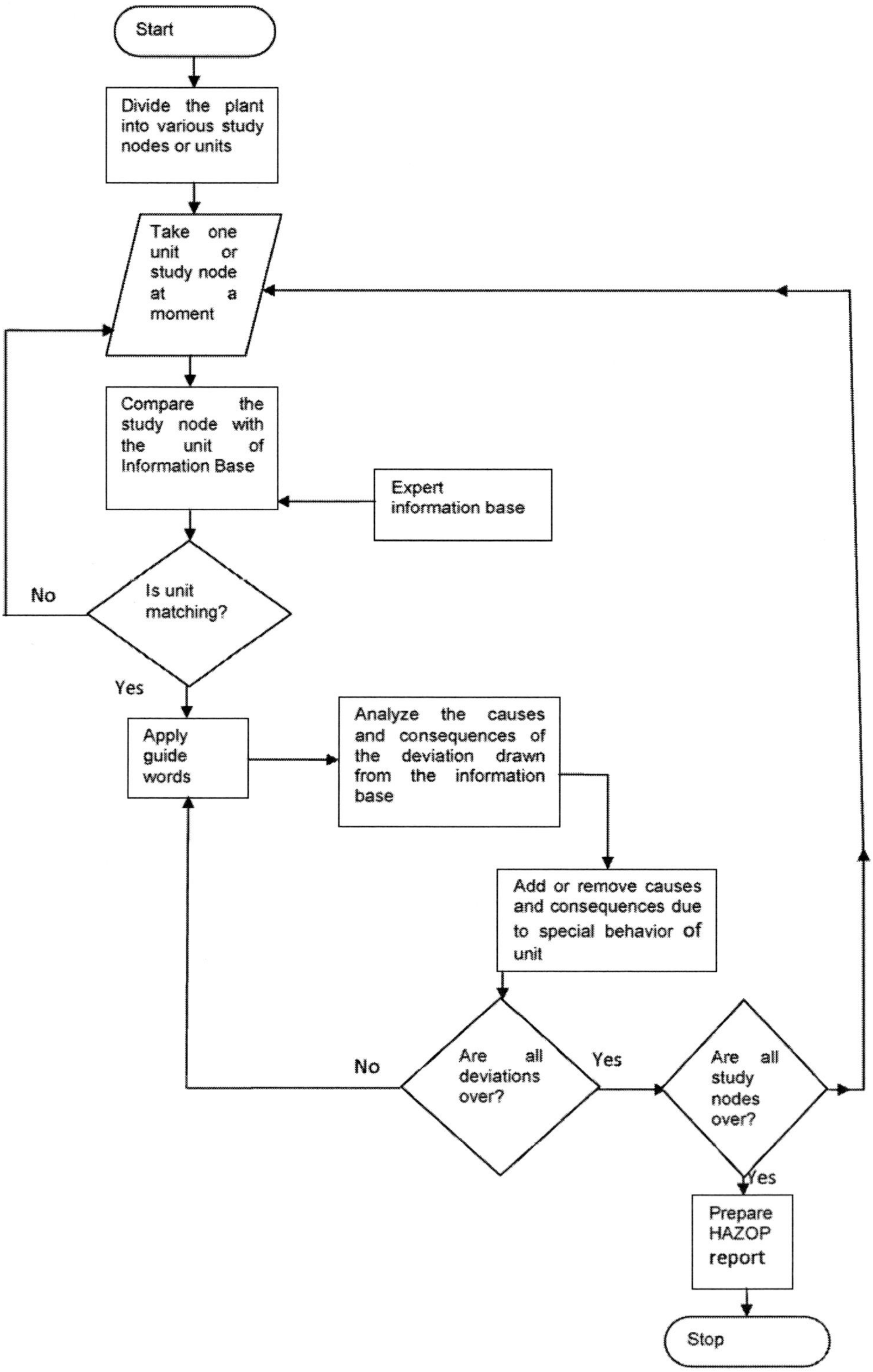

Figure 14.22. optHAZOP study procedure.

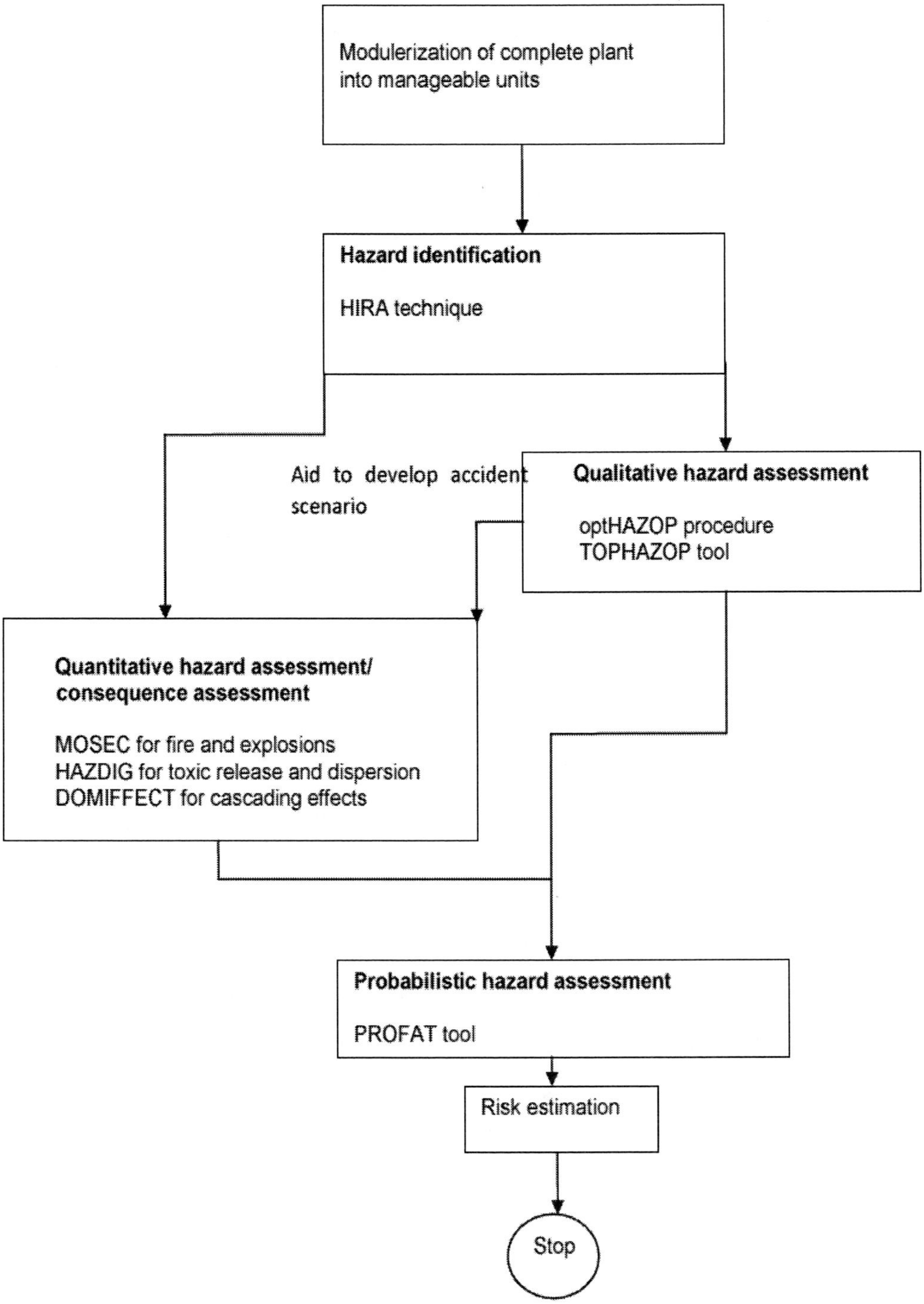

Figure 14.23. Simplified block diagram showing various steps with techniques and/or tools for conducting optimal risk analysis.

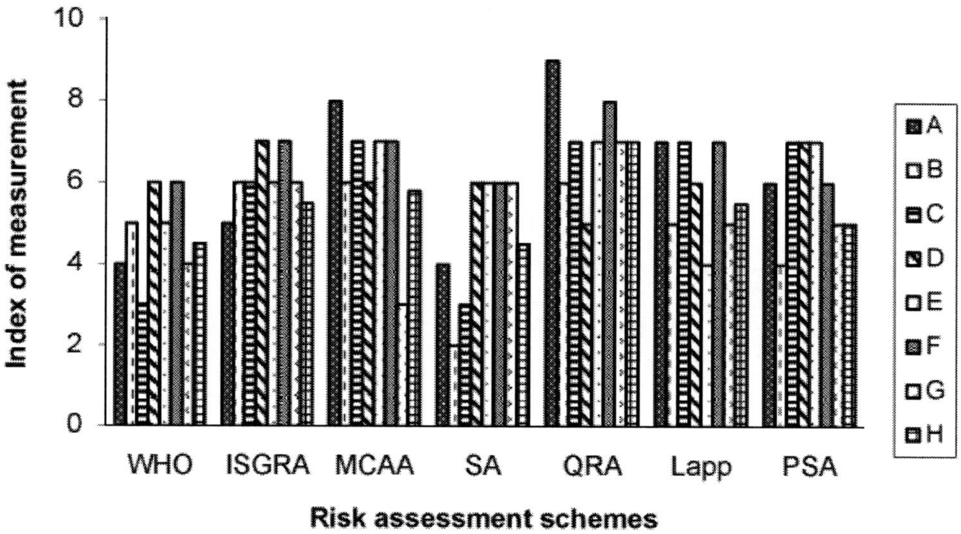

A,B,C,D,E,F and G are measures of quantifiability, cost-effectiveness, economy of steps, clarity, precision, range of applicability, and coverage respectively. H represents the cumulative index score.

Figure 14.24. Comparison of parameters for various schemes of risk assessment.

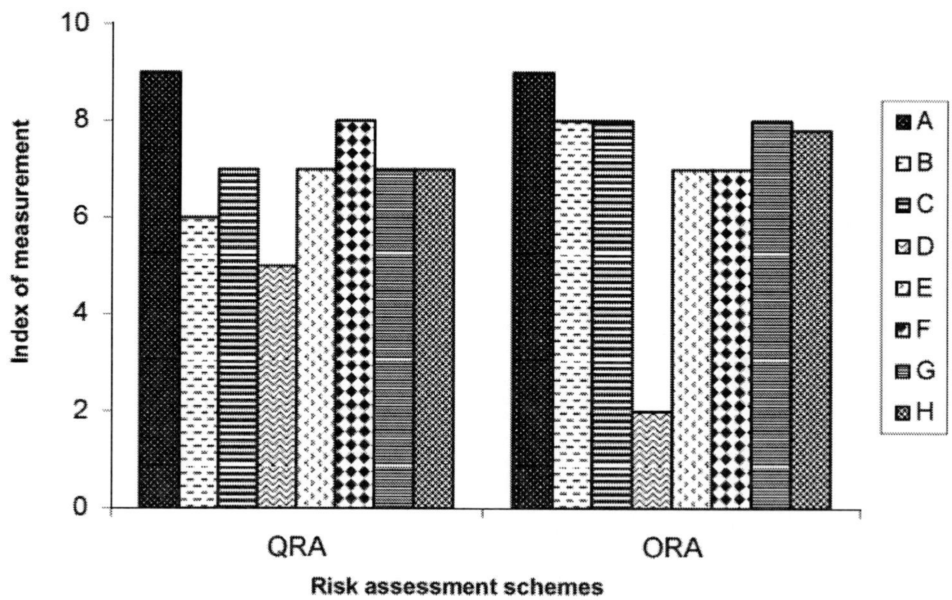

A,B,C,D,E,F and G are measures of quantifiability, cost-effectiveness, economy of steps, clarity, precision, range of applicability, and coverage respectively. H represents the cumulative index score.

Figure 14.25. Comparison of effectiveness of ORA over QRA.

Chapter 15

Illustrative Risk Assessment of MFL

15.1. Introduction

Madras Fertilizers Limited (MFL) is one of the leading NPK fertilizer producers in India with a high level of technical capability and market acceptability, especially in the southern states.

The main units of MFL are,

I. Ammonia plant
II. NPK plant
III. Urea plant

The process block diagrams are presented in Figures 15.1 to 15.3.

15.2. Hazard Identification and Ranking

The results of hazard ranking for different units of ammonia plant are presented in Table 15.1. It is seen that the components of the ammonia synthesis unit are associated with the highest values of FEDI (fire and explosion damage index) as well as TDI (toxic damage index). The next most hazardous unit is the reformer unit, closely followed by catacarb process unit.

In a similar fashion, hazard identification and ranking was conducted for the other two plants. The results are summarized in Tables 15.2 and 15.3.

15.3. Hazop Study

Ammonia and Urea Plant

A detailed and systematic HAZOP study was performed for the urea plant. Tables 15.4 to 15.8 illustrate the manner of HAZOP study with reference to one of the units (of stream

10). In the like manner HAZOP studies on all other units was conducted. The most hazardous section of the urea plant was revealed to be the synthesis and decomposition section (Section No. 14-200).

Stream<10> is the main feed stream (ammonia) to urea reactor. Any deviation in the operating parameter of this stream will strongly affect the urea reactor and also the ammonia storage tank (G-203) which may lead to harmful consequences. The most likely deviations of stream<10> are high pressure, high temperature, and low temperature. These deviations may result in explosion and toxic release. High pressure in stream<10> may induce high pressure in ammonia storage tank, which may lead to luxury release of Ammonia. The hazard potential of this pipeline is very high and hence deviations should be rigorously monitored and controlled.

Stream<9> is used for recycling unconverted ammonium carbonate. The HAZOP study suggests that the unit is most vulnerable to high temperature deviation, which may raise the temperature of the urea reactor suddenly and lead to explosions and toxic release. High temperature may also cause failure of carbonate recycle pump, J-202 A,B which can further lead to uncontrolled pressure release in the urea reactor, making the reactor vulnerable to failure. Other deviations such as high flow rate and high pressure are less hazardous. All in all a high temperature alarm system is recommended for stream<9> to mitigate the risk associated with deviations in this stream. Other deviations such as high flow rate and high pressure should also be controlled to keep up the reactor performance in safe bounds.

Compressed carbon dioxide is fed into urea reactor by stream<3>, which is operating under high pressure. The most hazardous deviation in stream<3> is high temperature, which may cause hot spot formation and damage to urea reactor. Thus explosion as well as toxic release are possible warranting, proper forewarning and control measures. High temperature alarm system may be installed and cooling water jacket may be provided for stream<3>. Other deviations such as high pressure and high flow rate are less hazardous. Yet serious enough to affect the performance and life of the reactor. So, these deviations should also be taken care of. Stream<11> is the main outlet of urea reactor. It connects both urea reactor and first flash separator, (Equipment No.G-202). Any disturbance in operating parameters will affect both the equipment. The stream is most vulnerable to deviations in pressure and flow rate. Higher than design pressure in stream <11> may exert backpressure to urea reactor and may affect the latter's performance. Serious consequences due to explosion and toxic release of ammonia may occur. Similar effect is likely in the first flash separator, as the separation process is highly exothermic. A high- pressure alarm system is recommended for stream<11> to avoid risks due to backpressures. The study also suggests that proper control measures supported by regular inspection of pressure and flow rate is warranted for stream <11>.

Ammonia and carbon dioxide supplied by stream<10> and stream<3> are made to react to form ammonium carbonate which inturn gives urea in the urea reactor. The first reaction is exothermic and the second one is endothermic. Thus, any deviation in the operating parameters shall seriously affect the reactor performance. It is also possible that beyond a point the deviations may lead to accidents. The most influential parameters in this respect are temperature and pressure. If these go beyond operational limits, explosion and toxic release may result. The HAZOP study suggests precise monitoring and control of pressure and temperature of the reactor as well as the feeder streams. The adverse impact of an accident due to deviation in any one of the operating parameters of this unit is likely to be severe enough to cause damage even beyond the boundary of the industry. Installation of high temperature and pressure alarm systems for the urea reactor as well as for the feeder streams connected to reactor is warranted. Installation of pressure relief valves and proper maintenance of cooling water and emergency water system are also recommended. Stream<13> is feed to the first decomposer (Equipment No: E-205). It connects both first flash separator (Equipment No. G-202) and first decomposer. Any significant deviation in the operability will lead to serious accidents in both the equipment. The most hazardous disturbance is high temperature. It would create high pressure and temperature in the primary decomposer. The HAZOP study suggests that high temperature may lead to toxic release in primary decomposer or in the pipeline, possibly with explosion. The study recommends installation of a high temperature switch to the pipeline. The pipeline may also be jacketed by a cooling water system. The other moderately hazardous disturbances are high pressure and high flow rate. These deviations affect the working efficiency of the decomposer. The study recommends setting up of regular monitoring and precise controls to foreclose the problems associated with these deviations.

The first decomposer (Equipment No. E-205) is useful in upgrading the urea solution by decomposing ammonium carbonate into ammonia and carbon dioxide. The most hazardous disturbances likely in this unit are high pressure and high temperature but their damage potential is lower than that of the units described above. Nevertheless similar safety precautions are recommended.

Stream<28> feeds ammonia to the first absorber. The only hazardous deviation of this unit is high pressure. This deviation may cause toxic release of ammonia. The HAZOP study suggests installation of a leak detecting system, besides a high temperature forewarning system.

Stream<31> supplies flashed and decomposed gases to first absorber. The most hazardous deviations are high pressure and high flow rate. These disturbances may cause rupture in the absorber leading to toxic release of ammonia. The HAZOP study suggests provision of appropriate alarm system in this sub-unit. The possible rupture of stream<31> can be preempted by using high pressure relief valve.

Stream<27> is the outlet stream of the first absorber. This adds ammonia to the main ammonia recycle stream<17>. The HAZOP study suggests that any disturbance may cause leak and failure of stream<17>. High pressure is the most hazardous deviation. HAZOP suggests that this deviation may cause build-up of excessively high pressure in the main pipeline and also in the ammonia storage tank, with subsequent explosive release of ammonia. The study recommends provisions of high pressure switch and alarm system to stream<27>, stream<17> and ammonia storage tank.

The first absorber is one of the most vulnerable equipment as it deals with high capacity of ammonia at extreme conditions. The most hazardous deviation is elevation in pressure of feed rate. The study suggests high pressure and high capacity level alarm system for the absorber. In addition a proper crisis management programme should be put in the place. Regular inspections and proper maintenance of the controller and other equipment can further reduce the probability of accidents. In summary, the most hazardous units are urea reactor and streams<31>, and <17>. A moderately hazardous unit is the absorber. Appropriate control, forewarning and emergency preparedness packages have been suggested.

NPK plant

The main equipment studied in the NPK plant are preneutralizer (Equipment No:05-1R-101A) and the streams <31>, <32> and <33> A&B.

Stream<31> is the feed line, which supplies Ammonia to preneutralizer through sub cooler (Equipment No: 05-1x-101A). The HAZOP study suggests that the equipment is most vulnerable to rises in flow rate, pressure and temperature. Any of these deviations may lead to failure of preneutralizer, leakage of Ammonia and rupture of the pipeline. The probability of explosion is not high but the rupture may lead to toxic release and dispersion of Ammonia in significant quantities. The study recommends installation of precise control switches of high quality and regular monitoring and maintenance of pipes and valves. To control the temperature of the stream, regular maintenance and checking of preneutralizer sub cooler is also recommended.

Stream No<32> feeds phosphoric acid to preneutralizer. Any deviation in the operating parameters would perturb the preneutralizer and the outlet stream<33> A&B. HAZOP assesses that the equipment is vulnerable to deviations in terms of rise in flow rate, rise in pressure and fall in temperature. These deviations may cause instantaneous reactions with slugs of heat release in preneutralizer leading to hot spot formation (uneven heat distribution). The failure of the unit would occur releasing highly corrosive acid. Accordingly, flow, pressure and temperature control systems supported by regular monitoring is advised. Further, high pressure and low temperature alarm systems are recommended for the stream <32>. Other deviations such as low flow rate, low pressure and high temperature are moderately hazardous in terms of low product quality and these deviations should also be avoided.

There are two outlet streams for fertilizer slurry: <33> A&B. The HAZOP study of these streams indicates that the only hazardous deviation is low temperature of the stream <33> A&B, which may lead to solidification of the slurry and the resultant choking. This would, in turn, cause high pressure development in the pre-neutralizer leading to failure of the unit. On the other hand increasing in temperature of the stream beyond the design level may lead to heating and failure of the pump. The net impact of it would be same as that of low temperature as flow in the pipeline would reduce or cease altogether regular monitoring and control of temperature is, therefore, necessary. The temperature and flow rate of cooling fluid in the cooling jacket should be kept under control. The pre-neutralizer is a reaction unit where the phosphoric acid and ammonia are made to react to form fertilizer slurry. The HAZOP study finds this unit to be highly hazardous. Any operational deviations may lead to major consequences like rupture of pre-neutralizer and toxic release of ammonia and other chemicals. The unit is the most vulnerable to these deviations: high level in pre-neutralizer, high temperature and high pressure. These parameters may lead to instantaneous reaction followed by uneven heat distribution (or hot spot formation), triggering to explosion and/or toxic release. The impact of such accidents may harm other units of the industry as well. Thus, rigorous control forewarning and crisis management procedures should be implemented to attenuate the risk associated with this unit.

15.4. Development of Accident Scenarios

Ammonia Plant

Primary Hydro-Desulfuriser: Scenario 1

The primary hydro desulfurizer unit (equipment No.H-151) deals with hydrogen and naptha. The accident scenario for this unit has been generated as BLEVE followed by fireball.

Primary Reformer: Scenario 2

The accident scenario for the primary reformer unit (equipment No.D-101) is generated as BLEVE followed by fireball.

Secondary Reformer: Scenario 3

The accident scenario for secondary reformer (Equipment No. H-101) is generated as CVCE followed by pool fire.

Ammonia Reactor: Scenario 4

The converter/reactor unit (Equipment No.H-701) transforms synthesis gas into ammonia. This is operated under high temperature and pressure, and there is probability of side reactions occurring with release of heat. The accident scenario for this unit has been developed as CVCE followed by dispersion.

CO_2 Absorber: Scenario 5

The absorber unit is a mass transfer unit dealing with a mixture of gases. The purpose of this unit is to recover CO_2 from the mixture of gases. The unit operates at normal temperature and pressure. There is less likeliness of the explosive release of the chemicals however, as it deals with hydrogen, highly flammable chemical there is high probability of flash fire. The accident scenario for this unit is envisaged as, flash fire followed by release and dispersion of gases.

Urea Plant

Urea Reactor: Scenario 6

Ammonia and carbon dioxide from ammonia plant is processed in urea reactor (Equipment No.H-201) to form ammonium carbonate and urea. The accident scenario for urea reactor is envisaged as CVCE followed by release and dispersion of toxic gases.

Primary Absorber: Scenario 7

Flashed and decomposed gases (Ammonia and carbon dioxide) are mixed with sufficient amount of Ammonia to produce weak Ammonium carbonate solution in the first absorber (Equipment No: F-201). The accident scenario for this equipment is envisaged as BLEVE with released gases.

Secondary Absorber: Scenario 8

The probability of explosion in secondary absorber is less and hence only detailed calculation for the release and dispersion of toxic gases has been done.

NPK Plant

Pre-Neutralizer: Scenario 9

The accident scenario for this unit is envisaged as instantaneous release of CVCE followed by dispersion.

Drier: Scenario 10

The accident scenario for drier (Equipment No.05-1D-101A) is forecasted as BLEVE followed by fireball and dispersion of ammonia.

15.5. Impact Forecasting: Consequence Assessment

The consequences of the scenarios described under section 4.0, are discussed below.

Ammonia Plant

The primary hydro desulfurizer unit (equipment No.H-151) deals with hydrogen and naptha. The accident scenario for this unit *scenario 1* has been generated as BLEVE followed by fire ball. Due to high temperature and pressure, there is high probability that hydrogen may escape as BLEVE, while naptha at elevated temperature and pressure on release and ignition may lead to a fire ball. The forecasts for this accident scenario are presented in Table 15.9. As the quantity of hydrogen is not very great, only a low intensity BLEVE may occur yielding an over-pressure of 2.8 KPa and a shock wave velocity of 333 m/s. These are not catastrophic. The fire ball is also not likely to cause significant damage at the point of observation (~500 meter away from the site). Table 15.10 summarizes dispersion of hydrogen over the study area, which too, isn't worrisome.

The accident scenario for the primary reformer unit (equipment No.D-101) is generated as BLEVE followed by fireball (Table 15.11). The intensity of explosion in this case is observed to be slightly higher compared to the previous case. Yet, the impacts remain within moderately hazardous levels. The dimensions of the fire ball shall be similar to the one likely in the HDS unit. The dispersion of released gases was forecast with HAZDIG. It revealed the probability of very high concentration build-up (100mg/ m^3) of the reformed gases at the point of observation (Table 15.12). The concentration exceeds the TLV (Threshold Limit Value) and may cause severe damage in terms of fatalities. In other words, the reformer unit poses moderately significant hazard due to fire and explosion but lethal toxic load over an area of 500-meter radius surrounding the reformer unit.

The accident scenario for secondary reformer (Equipment No. H-101) is generated as a CVCE followed by pool fire. As the unit deals with high capacity of reformed gases (H_2 and light hydrocarbons) under high pressure, the probability of CVCE is high. Further the released hydrocarbons may burn as a pool of fire. The forecasts for this scenario are presented in Table 15.13. An intense over-pressure load (2376.97 KPa) along with a high-velocity shock wave (1513.8 m/s) is likely around the site of accident. There shall also be lethal heat load. The pool fire would not be as damaging as the explosion. Toxic load due to burned and unburned reformer gases may occur over the study area (Table 15.14). The damaging impact of severe heat load, peak over pressure, and missiles would reach a distance upto 500 meters away from the epicenter of the explosion.

The converter/reactor unit (Equipment No.H-701) transforms synthesis gas into ammonia. This is operated under high temperature and pressure, and there is probability of side reactions occurring with release of heat. The accident scenario for this unit has been developed as CVCE followed by dispersion. The forecasts based on detailed calculations for this scenario are presented in Table 15.15. It is evident, that the intensity of explosion is very high as it would develop a peak overpressure of 20455 KPa which is far above the threshold limit (30Kpa). Simultaneously shock wave of very high velocity (3664 m/s) would be observed over an area of ~200 m radius. The damage potential of these events (shock waves and overpressure) would be severe enough to cause secondary accidents in the units placed within 500 m in all directions. The result for toxic dispersion is presented in Table 15.16. It indicates that the lethal toxic load would persist over a small area ~100m radius. It is important to note that the damage potential of toxic release and dispersion is comparatively low compared to the damage potential of explosion.

The absorber unit is a mass transfer unit dealing with a mixture of gases. The purpose of this unit is to recover CO_2 from the mixture of gases. The unit operates at normal temperature and pressure. There is less likeliness of the explosive release of the chemicals; however, as the unit involves hydrogen, which is highly flammable, there is a high probability of flash fire. The accident scenario for this unit is envisaged as, flash fire followed by release and dispersion of gases. The forecasts of flash fire are illustrated in Table 15.17. Upto a radius of 200 meters, intense heat load (sufficient to cause lethality at % probability) would persist. However, beyond 350 m radius the impact of this load would not be significant. The result for the toxic release of gases has been tabulated in Table 15.18. It is clear from the table that upto a distance of 500-m high concentration of these gases would be observed.

Urea Plant

The units that are identified hazardous in previous step by HIRA technique are studied here in detail. The forecasts based on detailed calculations for these scenarios are presented in Tables 15.19 to 15.23.

Ammonia and Carbon dioxide from ammonia plant are processed in the urea reactor (Equipment No.H-201) to form ammonium carbonate and urea. The accident scenario for urea reactor is envisaged as CVCE followed by release and dispersion of toxic gases. The forecasts for this scenario are presented in Table 15.19. The peak over pressure generated by CVCE is more than 500 KPa, which is extremely high. An area of more than 750m radius may be damaged due to this accident. Such impacts would extend beyond the industry boundary. The forecasts for release and dispersion for toxic gas reveals that an area of ~500 m radius would be under influence of lethal toxic load (Table 15.20).

Flashed and decomposed gases (ammonia and carbon dioxide) are mixed with sufficient amount of ammonia to produce weak ammonium carbonate solution in the first absorber (Equipment No: F-201). The accident scenario for this unit is envisaged as BLEVE with released gases. The results of detailed calculation for this scenario (Table 15.21) reveal that a lethal over pressure (greater than 0.3atm) would persist over an area of 50m radius. The shock wave of velocity sufficient to cause damage (>370 m/s) would be operative in the same region. The forecasts for dispersion of toxic gas (ammonia) are detailed in Table 15.22. Lethal concentrations of ammonia would encompass an area of more than 200m radius.

The probability of explosion in the secondary observer is less and hence only detailed calculation for the release and dispersion of toxic gases has been done. The forecasts for these scenarios are presented in Table 15.23. It is evident from the tables that high concentration of ammonia (4.55×10^{-07} Kg/m^3) would persist over an area of 500m. This concentration may not cause lethality but may cause non-recoverable ill effects to human health. Moreover, uptil 200m radius the concentration of the gas may be lethal.

NPK Plant

The units that pose significant hazards in the NPK plant are preneutralizer and drier. They have been analysed for detailed consequences and the results are illustrated in Tables 15.24 to 15.26.

The accident scenario for the preneutralizer unit is envisaged as instantaneous release of CVCE followed by dispersion. The forecasts of the accident are presented in Table 15.24. CVCE would generate an over pressure of ~884 KPa, that is several time higher than the lethal limit. Simultaneous shock wave of velocity 958.7 m/s would also be developed. The lethal damage potential of these effects (shock wave and over pressure) would be operative over an area of more than ~400 m. The results of dispersion of the released gas indicate comparatively low concentration of ammonia over the study zone of 500m radius. It is because; the high turbulence in atmosphere (due to over pressure and shock wave) would favour rapid dilution of the gas.

The accident scenario for drier (Equipment No.05-1D-101A) has been forecast as BLEVE followed by fire ball and dispersion of ammonia. The hazardous chemicals processed in this unit are ammonia, fertilizer slurry, naptha and LPG. The last two named chemicals are used as fuel for drying. The forecasts based on detailed calculations are presented in Table 15.26. The peak over pressure due to BLEVE would be 842 KPa, which is higher than the acceptable threshold by several orders of magnitude. Simultaneous shock wave of velocity 938.25 m/s would also be generated. The lethal impact of these events would reach an area of ~400 m radius. Heat load due to fire load would also be lethal over an area of more than ~500 m radius.

15.6. RISK ESTIMATION

Ammonia Plant

On the basis of the results presented above, and probability of occurrence of the various accident scenarios, risk factors have been computed. In order to enable these have been visualization of the accidents risk contours have been drawn over the industry layout Figures 15.4 to 15.8.

Urea Plant

On the basis of the results presented above, and probability of occurrence of the various accident scenarios, risk factors have been computed. The risk contours for urea plant have been plotted in Figures 15.9 to 15.11.

NPK Plant

On the basis of the results presented above, and probability of occurrence of the various accident scenarios, risk factors have been computed. The risk contours for the pre-neutralization unit of NPK plant are plotted in Figure 15.12.

Table 15.1. HIRA results of ammonia plant

SL. NO.	SUB UNIT	HAZADOUS CHEMICALS	FIRE AND EXPLOSION DAMAGE INDEX	TOXIC DAMAGE INDEX	REMARK
1.	**HDS unit** HDS-1	Naptha	226.31	12.41	Moderately hazardous.
		Hydrogen	34.41	25.43	
		Hydrogen sulphide	1.27	37.43	
	HDS-11	HDS-Naptha	11.14	3.75	
		Naptha	190.69	1.37	Hazardous
		Hydrogen	20.17	28.45	
	HDS stripper	Naptha	112.23	25.37	Hazardous
2.	**Reformer unit**				
	Primary reformer	Naptha	226.45	13.43	Moderately hazardous
		Hydrogen	144.76	59.7.	Moderately hazardous
		Reformer gases	294.52	78.67	
	Secondary reformer	Reformer gases (influent)	241.53	85.46	Highly hazardous
		Reformer gases (effluent)	323.86	100	
3.	**Shift converter unit**				
	Primary co converter	Reformer gases (influent)	100.32	11.27	hazardous
		Reformer gases (effluent)	85.42	10.37	
	Secondary co converter	Reformer gases (influent)	45.34	50.13	hazardous
		Reformer gases (effluent)	150.56	20.32	
4.	**Methanator**	Reformer gases (influent)	25.02	63.00	Less hazardous
		Reformer gases (effluent)	75.71	43.33	

Table 15.1. (Continued)

SL. NO.	SUB UNIT	HAZADOUS CHEMICALS	FIRE AND EXPLOSION DAMAGE INDEX	TOXIC DAMAGE INDEX	REMARK
5.	**Catacarb process**				
	Catacarb reboiler	Reformer gases (influent)	334.38	47.63	Moderately hazardous
		Reformer gases (effluent)	2.51	4.76	
	CO_2 absorber	Gas-I (influent)	334.564	100.00	Highly hazardous
		Gas-II (effluent)	84.25	78.43	
6.	**Ammonia synthesis unit**				
	Ammonia converter/reactor	Gas-I (influent)	2911.23	100.00	Extremely hazardous
		Gas-II (effluent)	3084.71	100.00	
	Ammonia absorber	Absorber gases	250.53	60.32	Moderately hazardous

Table 15.2. HIRA results of urea plant

SL. NO	SUB UNIT	HAZADOUS CHEMICALS	FIRE AND EXPLOSION DAMAGE INDEX	TOXIC DAMAGE INDEX	REMARK
1.	**Synthesis unit**				
	Urea reactor	Ammonia	1007.39	100.00	Extremely hazardous
		Carbon di oxide	74.81	5.43	
2.	**Absorber unit**				
	First absorber	Ammonia	415.7	100	Highly Hazardous
	Second absorber	Ammonia	343.2	100	Highly Hazardous
3.	**Flash separator**				
	First flash separator	Ammonia	125.43	38.76	Hazardous
	Second flash separator	Ammonia	178.42	48.95	
4.	**Decomposer unit**				
	First decomposer	Ammonia	213.45	75.43	Moderately hazardous
	Second decomposer	Ammonia	278.34	76.78	
5.	**Prill section**				
	Evaporator	Ammonia	111.78	15.625	Low
				209	hazardous

Table 15.3. HIRA results of NPK plant

SL. NO	SUB UNIT	HAZADOUS CHEMICALS	FIRE AND EXPLOSION DAMAGE INDEX	TOXIC DAMAGE INDEX	REMARK
1.	Preneutralizer	Ammonia	345.43	20.18	Highly Hazardous
2.	Blunger	Ammonia	228.73	14.41	Moderately hazardous
3.	Drier	Naptha	54.73	23.41	Extremely hazardous
		LPG	778.72	85.43	
4.	Scrubber	Fumes of H_3PO_4 and NPK	93.43	18.42	Less Hazardous
5.	cooler	NPK	75.42	12.43	Less hazardous

Table 15.4. HAZOP study of urea plant

PLANT SECTION LINE/EQUIPMENT GUIDE WORD DEVIATION	: UREA : UREA SYNTHESIS : STREAM, 10. : MORE : HIGH FLOW RATE	PFD/PEFD NO: 0-10-14-240	
CAUSES	CONSEQUENCES	REMARK	CRITICALITY RANK
* Pump J-201, ABC running over speed.	* Exceptionally high pressure at reactor.	* Chances of reactor damage.	
* Exceptionally high pressure at stream <10>.	* High concentration of ammonia at output stream<11>	* Proper maintenance and inspection of pump is recommended	
* Phase change at stream <10>	* High temperature in the reactor	* Product quality may decrease	III
* High flow rate from storage tank, G-203 to pump, J-201, ABC	* Chances of side reaction		
	* Chances of reactor failure due to high pressure and temperature due to side reaction		
	* Chances of hot spot formation in urea reactor		

Table 15.5. HAZOP study of urea plant

PLANT SECTION LINE/EQUIPMENT GUIDE WORD DEVIATION	: UREA : UREA SYNTHESIS : STREAM <10>. : MORE : HIGH FLOW RATE	PFD/PEFD NO: 0-10-14-240	
CAUSES	CONSEQUENCES	REMARK	CRITICALITY RANK
* Leak in ammonia pre-heater, E-301.	* Low reactor efficiency.	* Proper checking and maintenance of pressure control system is recommended.	
*Pump, J-201, ABC is not working properly.	* Chances of side reaction.		
* Leak in stream <10> or pump.	* Composition of stream <11> changes.	* Low reaction yield.	III
* Decrease in the level of Ammonia in storage tank.			

Table 15.6. HAZOP study of urea plant

PLANT SECTION LINE/EQUIPMENT GUIDE WORD DEVIATION	: UREA : UREA SYNTHESIS : STREAM <10>. : MORE : HIGH FLOW RATE	PFD/PEFD NO: 0-10-14-240	
CAUSES	CONSEQUENCES	REMARK	CRITICALITY RANK
* Pump, J-201, ABC runs over speed. *Phase change at stream <10>. * Flow control and/or pressure control fails.	* High pressure in reactor. * High temperature development * Uncontrolled reaction * Chances of reactor failure. * Leak at stream <10>.	* Chances of severe damage * Pressure control system and pump speed should be maintained	II

Table 15.7. HAZOP study of urea plant

PLANT SECTION LINE/EQUIPMENT GUIDE WORD DEVIATION	: UREA : UREA SYNTHESIS : STREAM <10>. : LESS : LOW TEMPERATURE	PFD/PEFD NO: 0-10-14-240	
CAUSES	CONSEQUENCES	REMARK	CRITICALITY RANK
• High flow rate from ammonia pre-heater, E-201. • Low heat transfer from ammonia pre-heater. • Stream available to pre-heater is low. • Temperature of stored ammonia is less. • Leak at stream <10> causing evaporation of ammonia.	• Instantaneous reaction. • High carbonate formation. • High heat generation. • Hot spot formation. • High pressure development. • Chances of leak or failure of reactor due to high pressure and temperature. • Flow rate of stream <11> increased. • Composition of stream <11> changes.	• Chances of severe damage • Proper control of feed flow rate is recommended. • Proper maintenance of low temperature system is recommended.	II

Table 15.8. HAZOP study of urea plant

Plant Section Line/Equipment Guide word Deviation	: Urea : Urea synthesis : stream,10. : More : High flow Rate	PFD/PEFD No: 0-10-14-240	
Causes	Consequences	Remark	Criticality rank
• High temperature steam to ammonia pre-heater	•Low reaction low urea yield	•Chances of toxic release and dispersion	
•High temperature at Ammonia storage	•Excess CO_2 and carbonate concentration in stream <11>	Proper maintenance of •pressure control system is recommended	
•Failure of pressure control system	•High temperature in the reactor		II
•Failure of pressure control system	•Chances of ammonia leak due to high temperature		
	•High temperature in stream<11>		
	•High concentration of ammonia in stream<11>		

Table 15.9. Consequence analysis for *scenario 1* HDS-1

Parameter	Value
Explosion: BLEVE	
SHOCK WAVE CHARACTERISTICS:	
Peak over pressure (kPa)	=2.847
Duration of shock wave (sec)	=41.098
Variation of over pressure (kPa/sec)	=6.36693E-06
Shock velocity in air (m/s)	=333.93
Partile velocity behind the shock wave (m/s)	= 6.53857E+00
Dynamic pressure (Dimension less)	=2.84364E-02
Reflected pressure (kPa)	=5.76388E+00
Over pressure impulse (kPa/sec)	=8.51556E+00
MISSILE WILL NOT EXIST	
FIRE BALL: RESULT	
Radius of fire ball (m)	=66.50
Duration of fire ball (sec)	=27.174
Energy released by fire ball (KJ)	=242965120.0
Area of fire ball (m^2)	=55546.164
Head flux at distance 500.0m (kj/sq.m)	=3.71E+01

Table 15.10. Consequence analysis for *scenario 1*

Parameters	Values, units
GASEOUS RELEASE	
PUFF RADIUS	
Total mass in the vessel	2.645600E+02 Kg
Mass of vapor flashed	2.645600E+02 Kg
Radius of puff is	8.919664E+00 m
VAPOR GENERATION RATE	
PASQUILL-GIFFORD INSTANTANEOUS MODEL	
Evaporation of liquid pool for 1 hour taken as a basis ELEVATED SOURCE	
Concentration at 500.00 m is	1.488658E-09 Kg/m^3
PUFF CHARACTERISTICS	
Concentration at ground level is	7.548594E-10 Kg/m^3
Concentration at cloud axis is	3.264465E-05 Kg/m^3
Concentration at cloud edge	3.264465E-06 Kg/m^3
Cloud radius	2.285779E+02 m
Maximum distance travelled by the cloud	1.150000E+03 m
Dosage on downwind dist 500.00 is	2.540853E-03 Kg/m^3

Table 15.11. consequence analysis for *scenario 2* Primary Reformer

Parameter	Value
Explosion: BLEVE	
SHOCK WAVE CHARACTERISTICS:	
Peak over pressure (kPa)	= 16.245081
Duration of shock wave (sec)	= 41.098106
Variation of over pressure (kPa/sec)	= 3.63195E-05
Shock velocity in air (m/s)	= 351.833160
Particle velocity behind the shock wave (m/s)	= 3.53578E+01
Dynamic pressure (Dimension less)	= 9.07300E-01
Reflected pressure (kPa)	= 3.46677E+01
Over pressure impulse (kPa/sec)	= 4.85762E+01
MISSILE WILL NOT EXIST	
FIRE BALL: RESULT	
Radius of fire ball (m)	= 64.53
Duration of fire ball (sec)	= 26.371
Energy released by fire ball (kj)	= 204317808.00
Area of fire ball (m^2)	= 52310.91
Head flux at distance 500.0m (kj/sq. m.)	= 3.12333E+01

Table 15.12. Consequence analysis for *scenario 2*

Parameters	Values, unit
GASEOUS RELEASE	
PUFF RADIUS	
Total mass in the vessel	4.348300E-04 Kg
Mass of vapor flashed	4.348300E-04 Kg
Radius of puff is	2.448709E+01 m
LIQUID EMISSION RATE	
Area of puncture	6.361726E-03 m^2
Liquid emission rate through the puncture	1.959806E-01 Kg/sec
POOL AREA	
Total time of spill	2.218740E+05 se
Area of liquid pool after 60.00 is	1.268118E+05 m^2
VAPOR GENERATION RATE	
Area of liquid pool	1.268118E+05 sq.m
Evaporation due to forced circulation	9.898561E+00 Kg/sec
PASQUILL-GIFFORD CONTINOUS MODEL	
ELEVATED SOURCE	
Concentration at 500.00 m is	1.00262E-04 Kg/m^3
PLUME CHARACTERISTICS	
Ground level conc. of plume	1.012213E-04 Kg/m^3
Ground level conc. of plume of axis	1.109109E-04 Kg/m^3
Ground level conc. of plume at edge	1.109109E-05 Kg/m^3
Width of plume	2.509362E+02 m
Maximum ground level concentration	2.140633E-02 Kg/m^3
Distance at which maximum conc. occur	2.200000E+01 m

Table 15.13. Consequence analysis for *scenario 3*
Secondary reformer

Parameter	Value
CONFINED VAPOUR CLOUD EXPLOSION	
SHOCK WAVE CHARACTERISTICS:	
Peak over pressure (kPa)	= 2376.97
Duration of shock wave (sec)	= 41.098106
Variation of over pressure (kPa/sec)	= 2.37697E-03
Shock velocity in air (m/s)	= 1513.80
Particle velocity behind the shock wave (m/s)	= 1.20342E+03
Dynamic pressure (Dimension less)	= 4.57650E+03
Reflected pressure (kPa)	= 1.57375E+04
Over pressure impulse (kPa/sec)	= 7.10765E+03
EXPLOSION CHARCTERISTICS	
Total energy available (KJ)	= 1.83285E+10
Adiabatic energy lost due to expansion of gas (KJ)	= 7.39905E+07
Shock wave energy change (KJ)	= 8.30553E09+
Energy lost in fracture/plastic deformation of material (KJ)	= 2.04511E+08
Energy available for bursting	= 1.38426E+10
Energy released by CVCE explosion (KJ)	= 9744501993.65
Heat intensity (KJ/sq.m) at distance 500.0 (m)	= 4.34468E+03

Table 15.13. (Continued)

Parameter	Value
MISSELE CHARACTERISTICS	
Initial velocity of fragment (m/s)	= 1.53488E+03
Kinetic energy associated with fragments (KJ)	= 1.17793E+06
Fragment velocity (m/s) at distance 500.0 (m)	= 2.61014E+02
Penetration ability (based on empirical model):	
In concrete structure at a distance (m) 500.0 (m)	= 7.59046E-02
In brick structure at a distant (m) 500.0 (m)	= 9.69893E-02
In MS structure at a distance(m) 500.0 (m)	= 1.56608E-02
POOL FIRE	
INSTANTAEOUS	
Amount of material engage in fire (Kg)	= 75548.79
Pool fire radius (m)	= 5.000000
Area fire radius (m)	= 78.500000
Burning rate (m/s)	= 24.652662
Damage distance of 100% lethality (m)	= 665.939026
Heat intensity (kw/sq.m) at a distance 500.0m is	= 1.24375

Table 15.14. The output of TORAP for Scenario 3

Parameter	Value
Toxic release and dispersion	
Box instantaneous model: elevated source	
Concentration at distance of 200 m (kg/m^3)	: 1.95×10^{-5}
Heavy gas puff characteristics	
Ground-level concentration of puff (kg/m^3)	: 0.854×10^{-5}
Ground-level concentration on puff axis (kg/m^3)	: 0.854×10^{-4}
Cloud radius (m)	: 2.958×10^{2}
Maximum distance travelled by the cloud (m)	: 1.345×10^{2}
Maximum ground-level concentration (kg/m^3)	: 2.134×10^{-2}

Table 15.15. Consequence analysis for *scenario 4* Ammonia reactor

Parameter	Value
CONFINED VAPOUR CLOUD EXPLOSION	
SHOCK WAVE CHARACTERISTICS:	
Peak over pressure (kPa)	= 20454.8
Duration of shock wave (sec)	= 41.4
Variation of over pressure (kPa/sec)	= 10475.7
Shock velocity in air (m/s)	= 3663.8
Particle velocity behind the shock wave (m/s)	= 2.731E+03
Dynamic pressure (Dimension less)	= 2.78E+04
Reflected pressure (kPa)	= 9.13E+04
Over pressure impulse (kPa/sec)	= 3.53E+04
EXPLOSION CHARCTERISTICS	
Total energy available (KJ)	= 3.77E+11

Parameter	Value
Adiabatic energy lost due to expansion of gas (KJ)	= 2.69E+08
Shock wave energy change (KJ)	= 1.59E+11
Energy lost in fracture/plastic deformation of material (KJ)	= 2.06E+10
Energy available for bursting	= 2.66E+11
Energy released by CVCE explosion (KJ)	= 1.96E+11
Heat intensity (KJ/sq.m) at distance 500.0 (m)	= 8.74E+4
MISSELE CHARACTERISTICS	
Initial velocity of fragment (m/s)	= 7253.2
Kinetic energy associated with fragments (KJ)	= 1.31E+08
Fragment velocity (m/s) at distance 500.0 (m)	= 5.46E+02
Penetration ability (based on empirical model):	
In concrete structure at a distance (m) 500.0 (m)	= 0.50
In brick structure at a distant (m) 500.0 (m)	= 0.68
In MS structure at a distance(m) 500.0 (m)	= 0.10

Table 15.16. Consequence analysis for *scenario 4*

Parameters	Values, units
TWO PHASE RELEASE	
PUFF RADIUS	
Total mass in the vessel	1.000000E+05 Kg
Mass of vapor flashed	4.255378E+04 Kg
Radius of puff is	2.362941E+01 m
VAPOR GENERATION RATE	
Area of liquid pool	5.000E+00 sq. m.
Evaporation due to forced circulation	1.6500E-01 Kg/sec
PASQUILL-GIFFORD INSTANTANEOUS MODEL	
Evaporation of liquid pool for 1 hour taken as a basis	
ELEVATED SOURCE	
Concentration at 500.00 m is	2.442806E-06 Kg/m^3
PUFF CHARACTERISTICS	
Concentration at ground level is	1.214172E-07 Kg/m^3
Concentration at cloud axis is	5.250805E-03 Kg/m^3
Concentration at cloud edge	5.250805E-04 Kg/m^3
Cloud radius	2.285449E+02 m
Maximum distance travelled by the cloud	1.150000E+03 m
Dosage on downwind dist 500.00 is	4.086895E-01 Kg/m^3

Table 15.17. Consequence analysis for *scenario 5* CO_2 Absorber

Parameter	Value
FLASH FIRE :RESULT	
Volume of vapour cloud (m^3)	= 437277.46
Area under fire effect (m^2)	= 508282.40
Total heat radiated (kj)	= 292243.50
Half life time of fire (sec)	= 18.25
Effective time of fire (sec)	= 54.77
Heat intensity at a distance 200.0m is (KW/m^2)	= 0.81

Table 15.18. Consequence analysis for *scenario 5*

Parameters	Values, unit
GASEOUS RELEASE	
PUFF RADIUS	
Total mass in the vessel	8.872260E+04 Kg
Mass of vapor flashed	8.872260+04 Kg
Radius of puff is	3.091307E+01 m
LIQUID EMISSION RATE	
Area of puncture	9.852035E-03 m^2
Liquid emission rate through the puncture	2.923498E-01 Kg/sec
POOL AREA	
Total time of spill	3.034810E+05 sec
Area of liquid pool afer 60.00 is	1.645039E+05 m^2
VAPOR GENERATION RATE	
Area of liquid pool	1.645039E+01 sq.m
Evaporation due to forced circulation	5.153172E+00 Kh/sec
PASQUILL-GIFFORD CONTINOUS MODEL	
ELEVATED SOURCE	
Concentration at 500.00 m is	5.597928E-05 Kg/m^3
PLUME CHARACTERISTICS	
Ground level conc. of plume	5.645961E-05 Kg/m^3
Ground level conc. of plume of axis	6.186428E-05 Kg/m^3
Ground level conc. of plume at edge	6.186428E-06 Kg/m^3
Width of plume	2.509362E+02 m
Maximum ground level concentration	2.436755E-02 Kg/m^3
Distance at which maximum conc. occur	1.400000E+01 m

Table 15.19. Consequence analysis for *scenario 6* Urea reactor

Parameter	Value
CONFINED VAPOUR CLOUD EXPLOSION	
SHOCK WAVE CHARACTERISTICS:	
Peak over pressure (kPa)	= 5065.67
Duration of shock wave (sec)	= 41.09
Variation of over pressure (kPa/sec)	= 1.88E-03
Shock velocity in air (m/s)	= 2179.88
Particle velocity behind the shock wave (m/s)	= 177E+03
Dynamic pressure (Dimension less)	= 1.11E+04
Reflected pressure (kPa)	= 3.68E+04
Over pressure impulse (kPa/sec)	= 1.51E+04
EXPLOSION CHARCTERISTICS	
Total energy available (KJ)	= 6.93E+10
Adiabatic energy lost due to expansion of gas (KJ)	= 4.71E+07
Shock wave energy change (KJ)	= 3.36E+10
Energy lost in fracture/plastic deformation of material (KJ)	= 5.56E+09
Energy available for bursting	= 5.59E+10
Energy released by CVCE explosion (KJ)	= 3.01E+10
Heat intensity (KJ/sq. m) at distance 500.0 (m)	= 1.34E+02
MISSELE CHARACTERISTICS	

Parameter	Value
Initial velocity of fragment (m/s)	= 7.27E+02
Kinetic energy associated with fragments (KJ)	= 5.29E+05
Fragment velocity (m/s) at distance 500.0 (m)	= 1.77E+02
Penetration ability (based on empirical model):	
In concrete structure at a distance (m) 500.0 (m)	= 5.62E-02
In brick structure at a distant (m) 500.0 (m)	= 7.19E-02
In MS structure at a distance(m) 500.0 (m)	= 4.17E-02

Table 15.20. consequence analysis for *scenario 6*

Parameter	Values, unit
TWO PHASE RELEASE	
PUFF RADIUS	
Total mass in the vessel	6.140800E+04 Kg
Mass of vapor flashed	5.498489E+04 Kg
Radius of puff is	2.573781E+01 m
VAPOR GENERATION RATE	
Area of liquid pool	5.000000E+00 sq. m.
Evaporation due to forced circulation	8.19977E-05 Kg/sec
PASQUILL-GIFFORD INSTANTANEOUS MODEL	
Evaporation of liquid pool for 1 hour taken as a basis	
ELEVATED SOURCE	
Concentration at 500.00 m is	2.974495E-07 Kg/m^3
PUFF CHARACTERISTICS	
Concentration at ground level is	1.568864E-07 Kg/m^3
Concentration at cloud axis is	6.784708E-03 Kg/m^3
Concentration at cloud edge	6.784708E-04 Kg/m^3
Cloud radius	2.285449E+02 m
Maximum distance travelled by the cloud	1.150000E+03 m
Dosage on downwind dist 500.00 is	7.393102E-01 Kg/m^3

Table 15.21. Consequence analysis for *scenario 7*
Primary Absorber

Parameter	Value
Explosion: BLEVE	
SHOCK WAVE CHARACTERISTICS:	
Peak over pressure (kPa)	=13.19
Duration of shock wave (sec)	=41.09
Variation of over pressure (kPa/sec)	=4.13E-03
Shock velocity in air (m/s)	=347.84
Particle velocity behind the shock wave (m/s)	=2.9E+01
Dynamic pressure (Dimension less)	=6.0E+01
Reflected pressure (kPa)	=2.78E+01
Over pressure impulse (kPa/sec)	=3.94E+01

Table 15.22. Consequence analysis for *scenario 7*

Parameter	Value
GAS EMISSION RATE	
Gas release rate	2.684581E-01 Kg/s
PASQUILL-GIFFORD CONTINOUS MODEL	
ELEVATED SOURCE	
Concentration at 500.00 m is	3.245270E-06 Kg/m^3
PLUME CHARACTERISTICS	
Ground level conc. of plume	3.294259E-06 Kg/m^3
Ground level conc. of plume of axis	3.609607E-06 Kg/m^3
Ground level conc. of plume at edge	3.609607E-07 Kg/m^3
Width of plume	2.509362E+02 m
Maximum ground level concentration	1.777041E-04 Kg/m^3
Distance at which maximum conc. occur	4.700000E+01 m

Table 15.23. consequence analysis for *scenario 8*

Parameters	Values, unit
TWO PHASE RELEASE	
PUFF RADIUS	
Total mass in the vessel	6.938000E+03 Kg
Mass of vapor flashed	3.901951E+03 Kg
Radius of puff is	1.065534E+01 m
LIQUID EMISSION RATE	
Area of puncture	1.963495E-03 m^2
Liquid emission rate through the puncture	4.208295E+00 Kg/sec
POOL AREA	
Total time of spill	1.648649E+03 sec
Area of liquid pool after 60.00 is	2.231919E+03 m^2
VAPOR GENERATION RATE	
Area of liquid pool	2.231919E+03 sq. m.
Evaporation due to forced circulation	3.770305E-02 Kg/sec
PASQUILL-GIFFORD CONTINOUS MODEL	
ELEVATED SOURCE	
Concentration at 500.00 m is	4.557753E-07Kg/m^3
PLUME CHARACTERISTICS	
Ground level conc. of plume	4.626555E-07 Kg/m^3
Ground level conc. of plume of axis	5.069438E-07 Kg/m^3
Ground level conc. of plume at edge	5.069438E-08 Kg/m^3
Width of plume	2.509362E+02 m
Maximum ground level concentration	2.495729E-05 Kg/m^3
Distance at which maximum conc. occur	4.700000E+01 m

Table 15.24. Consequence analysis for *scenario 9*
Pre-neutralizer

Parameter	Value
CONFINED VAPOUR CLOUD EXPLOSION	
SHOCK WAVE CHARACTERISTICS:	
Peak over pressure (kPa)	= 884.97
Duration of shock wave (sec)	= 41.09
Variation of over pressure (kPa/sec)	= 69.51
Shock velocity in air (m/s)	= 958.72

Parameter	Value
Particle velocity behind the shock wave (m/s)	= 7.05E+02
Dynamic pressure (Dimension less)	= 1226.38
Reflected pressure (kPa)	= 4.71E+03
Over pressure impulse (kPa/sec)	= 2.91E+01
EXPLOSION CHARCTERISTICS	
Total energy available (KJ)	= 2.8E+09
Adiabatic energy lost due to expansion of gas (KJ)	= 8.71E+05
Shock wave energy change (KJ)	= 1.34E+09
Energy lost in fracture/plastic deformation of material (KJ)	= 2.71E+06
Energy available for bursting	= 2.23E+09
Energy released by CVCE explosion (KJ)	= 1.45E+09
Heat intensity (KJ/sq. m.) at distance 500.0 (m)	= 1.85E+03
MISSELE CHARACTERISTICS	
Initial velocity of fragment (m/s)	= 2.41E+02
Kinetic energy associated with fragments (KJ)	= 1.16E+05
Fragment velocity (m/s) at distance 500.0 (m)	= 0.0
Penetration ability (based on empirical model):	
In concrete structure at a distance (m) 500.0 (m)	= 0.0
In brick structure at a distant (m) 500.0 (m)	= 0.0
In MS structure at a distance(m) 500.0 (m)	= 0.0

Table 15.25. Consequence analysis for *scenario 9*

Parameter	Value
PUFF RADIUS	
Total mass in the vessel	1.400000E+03 Kg
Radius of puff is	7.571517E+00 m
PASQUILL-GIFFORD INSTANTANEOUS MODEL	
ELEVATED SOURCE	
Concentration at 500.00 m is	7.869155E-09 Kg/m^3
PUFF CHARACTERISTICS	
Concentration at ground level is	3.994569E-09 Kg/m^3
Concentration at cloud axis is	1.727491E-04 Kg/m^3
Concentration at cloud edge	1.727491E-05 Kg/m^3
Cloud radius	2.285449E+02 m
Maximum distance travelled by the cloud	1.150000E+03 m
Dosage on downwind dist 500.00 is	1.882398E-02 Kg/m^3

Table 15.26. Consequence analysis for scenario 10
Drier

Parameter	Value
Explosion: BLEVE	
SHOCK WAVE CHARACTERISTICS:	
Peak over pressure (kPa)	= 842.73
Duration of shock wave (sec)	= 41.09
Variation of over pressure (kPa/sec)	= 47.53
Shock velocity in air (m/s)	= 938.25
Particle velocity behind the shock wave (m/s)	= 6.86E+02
Dynamic pressure (Dimension less)	= 1.14E+03
Reflected pressure (kPa)	= 4.43E+03
Over pressure impulse (kPa/sec)	= 2.77E+10
FIRE BALL: RESULT	
Radius of fire ball (m)	= 137.229172

Table 15.26. (Continued)

Parameter	Value
Duration of fire ball (sec)	= 56.076008
Energy released by fire ball (kj)	= 1.6E+10
Area of fire ball (m^2)	= 3.14E+06
Head flux at distance 500.0m (kj/sq. m.)	= 2103.86

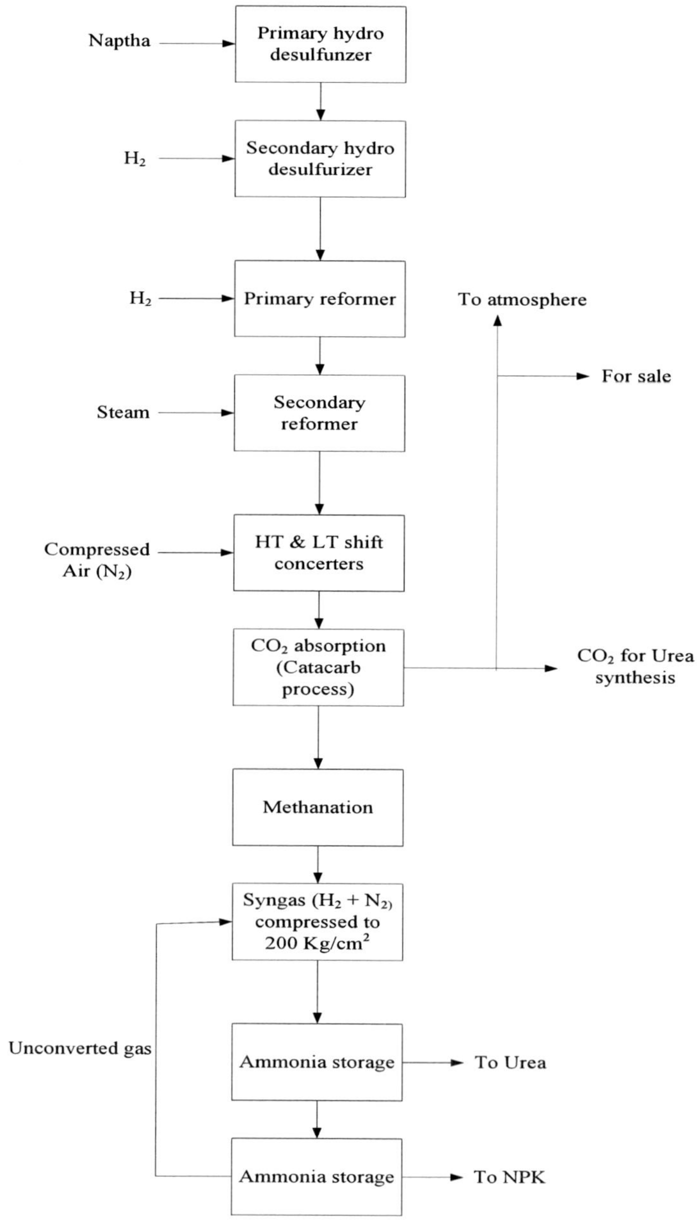

Figure 15.1. Ammonia plant.

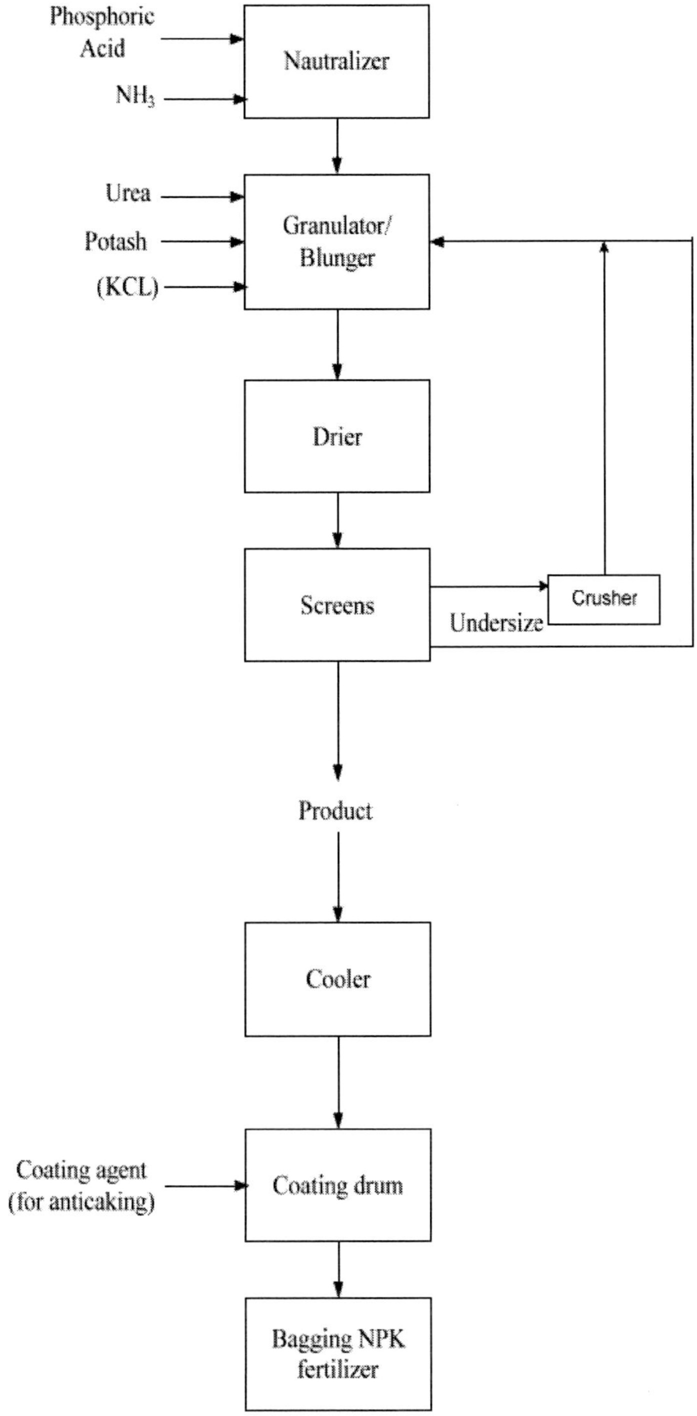

Figure 15.2. NPK plant.

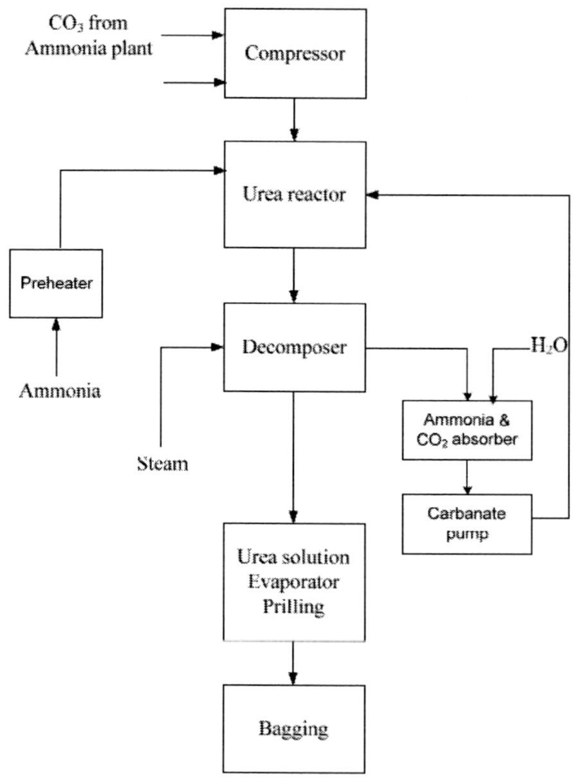

Figure 15.3. Urea plant.

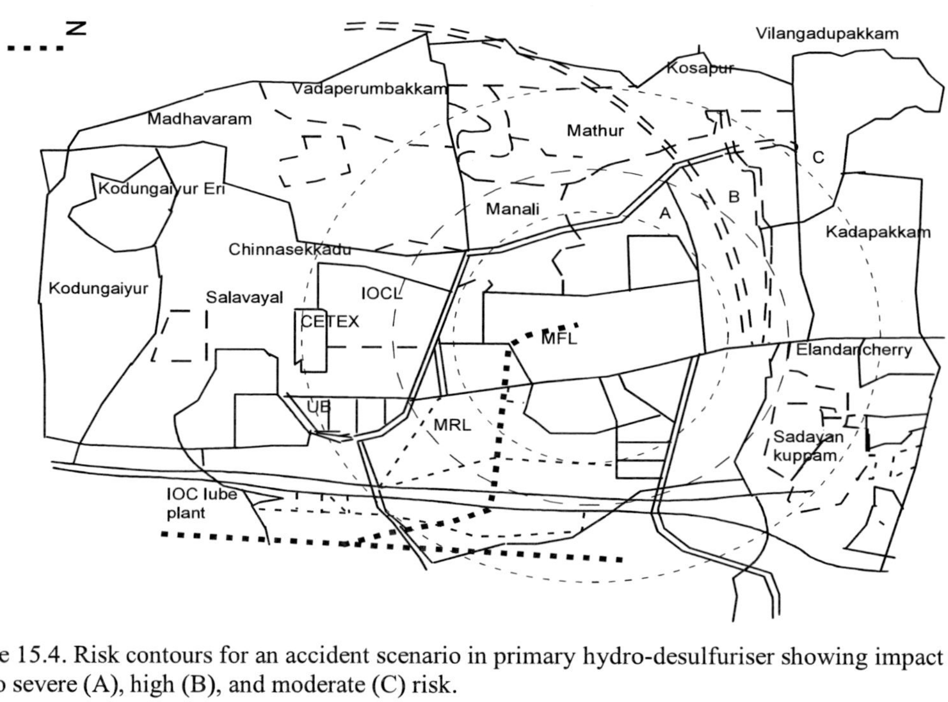

Figure 15.4. Risk contours for an accident scenario in primary hydro-desulfuriser showing impact area due to severe (A), high (B), and moderate (C) risk.

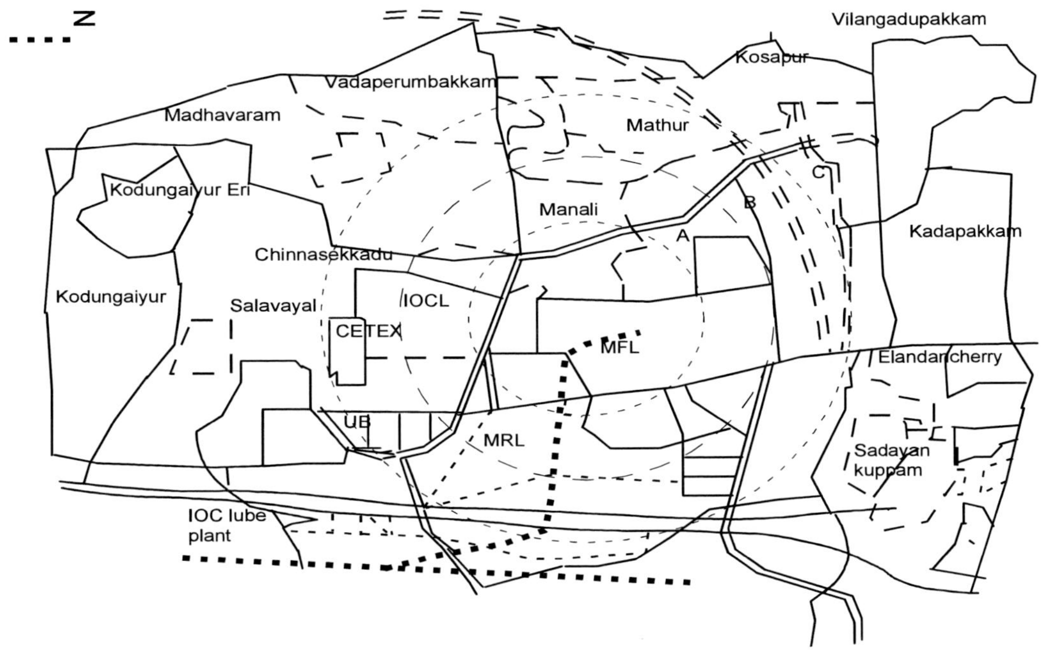

Figure 15.5. Risk contours for an accident scenario in primary reformer showing impact area due to severe (A), high (B), and moderate (C) risk.

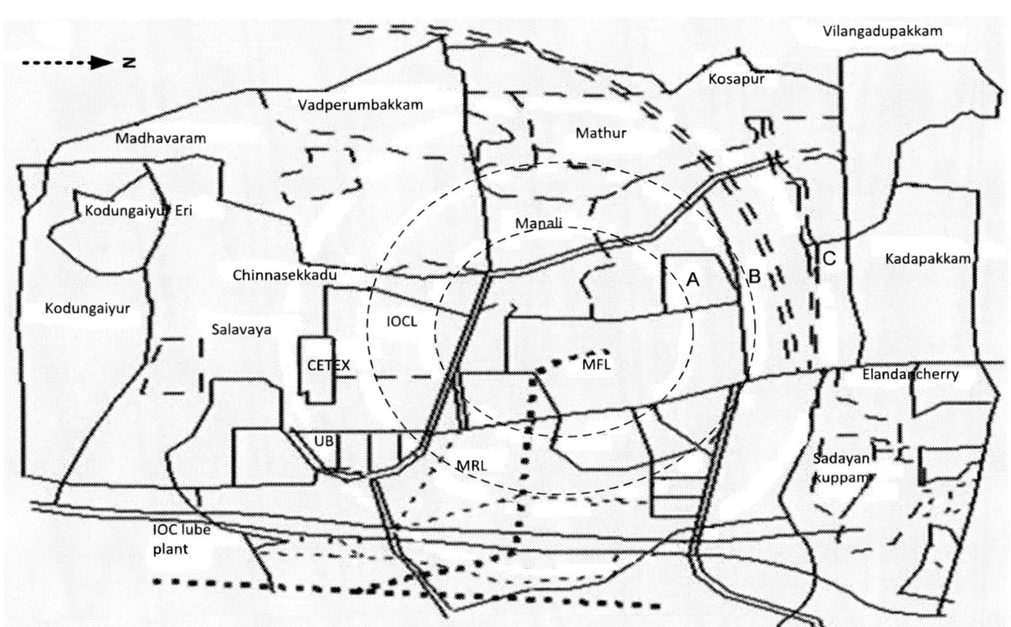

Figure 15.6. Risk contours for an accident scenario in secondary reformer showing impact area due to severe (A), high (B), and moderate (C) risk.

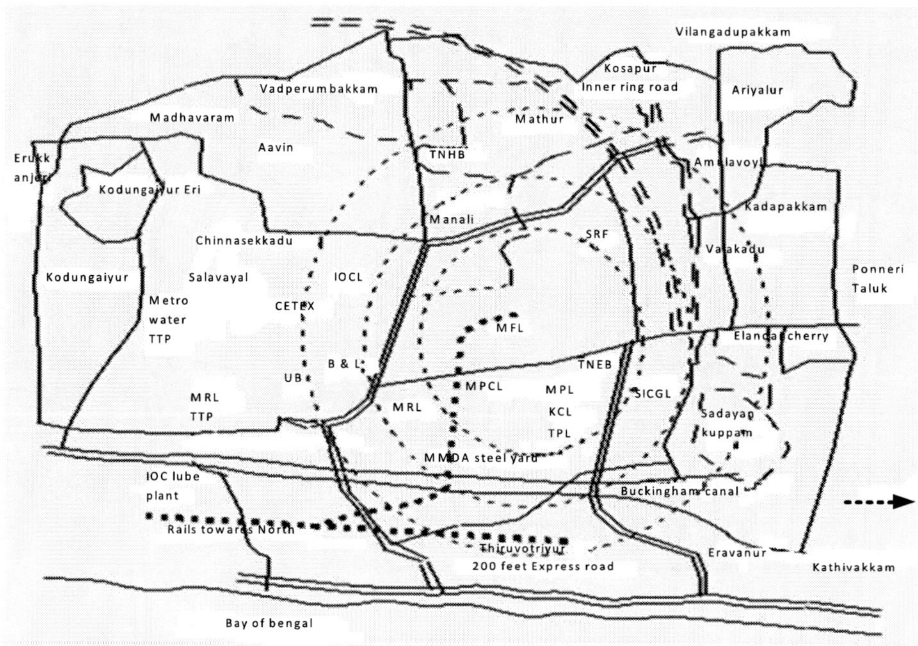

Figure 15.7. Risk contours for an accident scenario in Ammonia reactor showing impacts area due to severe (A), high (B), and moderate (C) risk.

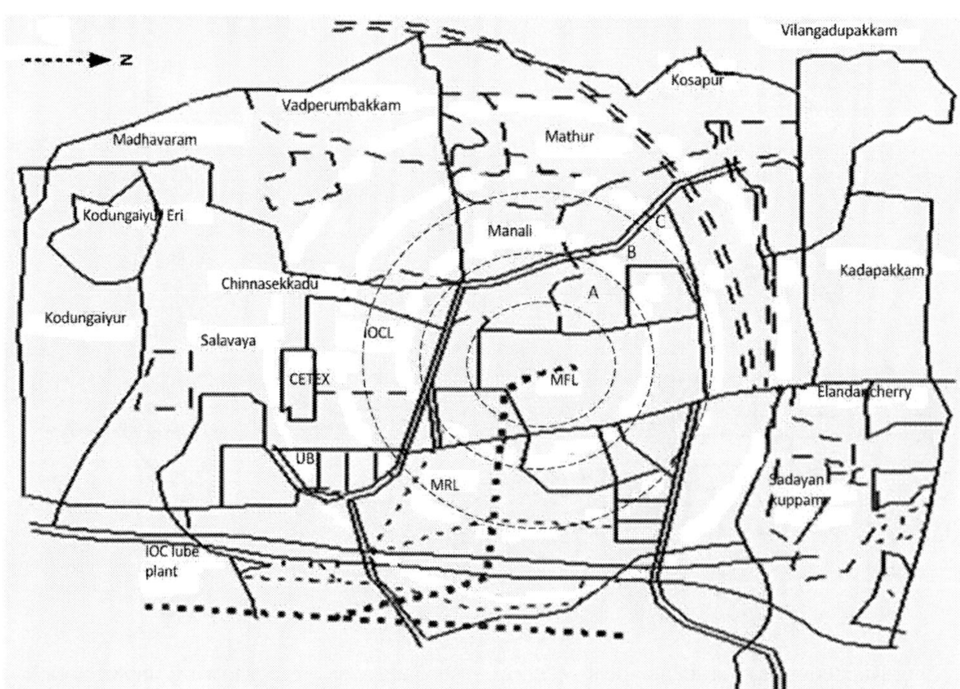

Figure 15.8. Risk contours for an accident scenario in absorber unit showing impact area due to severe (A), high (B), and moderate (C) risk.

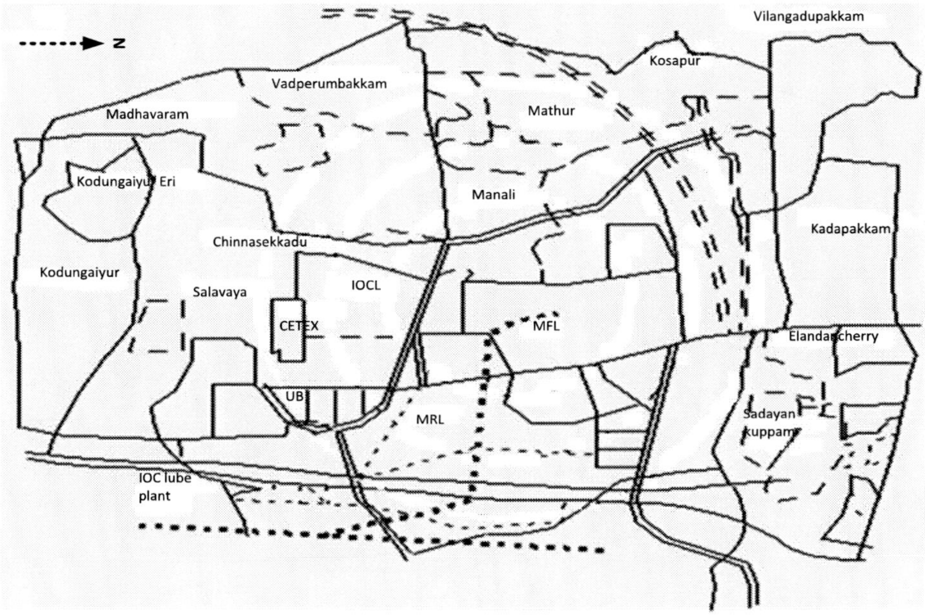

Figure 15.9. Risk contours for an accident scenario in urea reactor showing impact area due to severe (A), high (B), and moderate (C) risk.

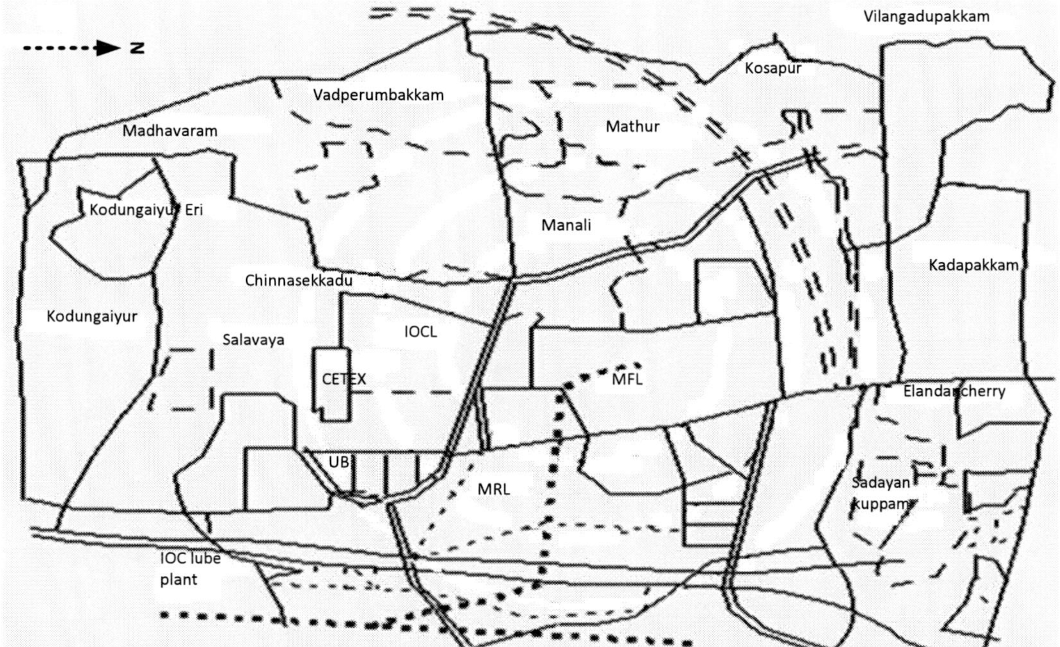

Figure 15.10. Risk contours for an accident scenario in primary absorber showing impact area due to severe (A), high (B), and moderate (C) risk.

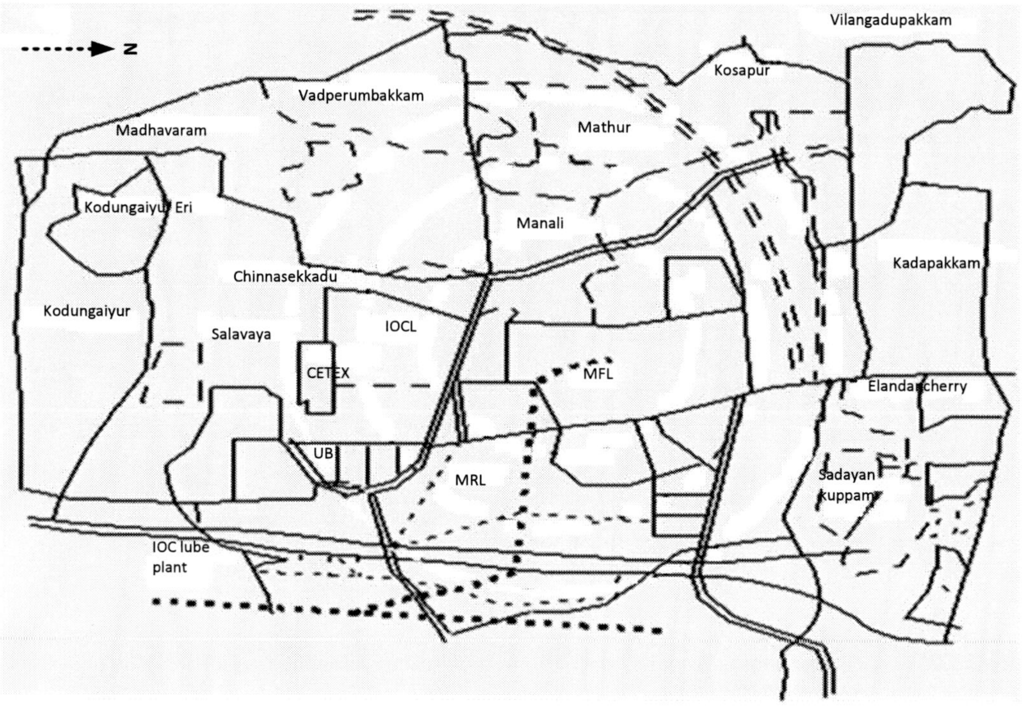

Figure 15.11. Risk contours for an accident scenario in secondary absorber showing impact area due to severe (A), high (B), and moderate (C) risk.

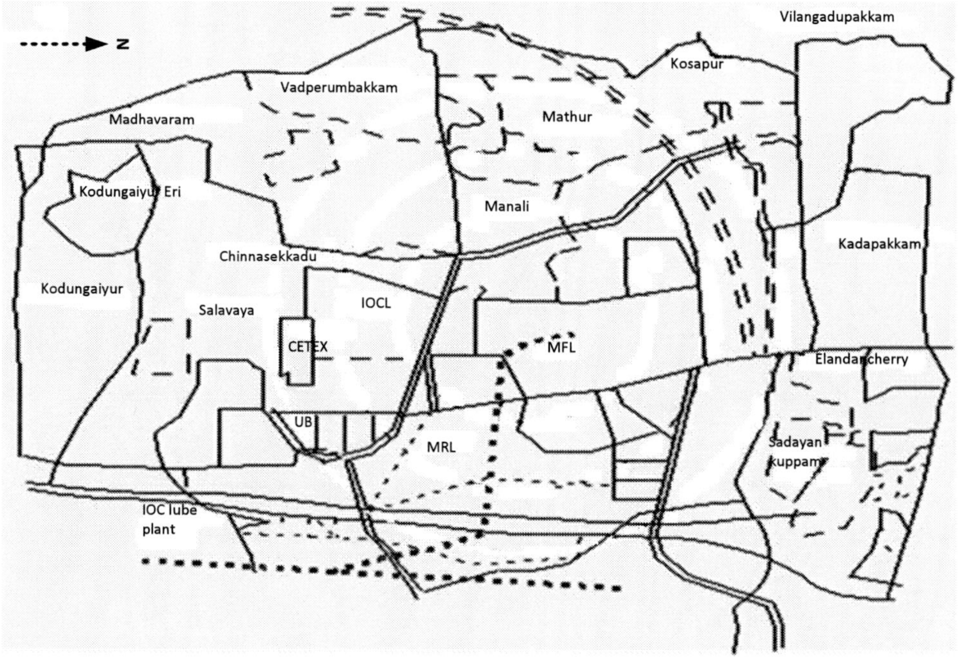

Figure 15.12. Risk contours for an accident scenario in pre-neutralizer showing impact area due to severe (A), high (B), and moderate (C) risk.

Chapter 16

Illustrative Risk Assessment of IOCL

16.1. Introduction

IOCL manufactures polyester staple fibre (2500 MT/month) and polyester yarn (1250 MT/month).

The main units of IOCL are,

I. Melter
II. Trans-esterifier
III. Polycondensation autoclave (PCAC)
IV. Drier
V. MEG (mono ethylene glycol) storage
VI. Methanol storage
VII. Ammonia storage

The simplified process block diagram of the plant is presented in Figures 16.1 and 16.2.

16.2. Hazard Identification and Ranking

As a prelude to more detailed risk assessment, all the units mentioned above were screened and ranked using fire and explosion damage index (FEDI) and toxic damage index (TDI). The results are summarized in Tables 16.1 and 16.2. Whereas the first four units do not pose much risk, the three storage units carry high risks of fire and explosion, besides toxic release.

16.3. Hazop Study

The result of an illustrative HAZOP study is presented in Tables 16.3 – 16.9. It comprises of a scrutiny of the trans-esterifier. Similar studies were conducted on all other units.

The melter is used to get a uniform mixture of dimethyl terephthalate and mono ethylene glycol. The moderately hazardous deviations are high temperature, high pressure, high and low level of reactants. The first three types of deviation may cause over pressure in the vessel, which may lead to leak and fire. A high pressure in the outlet streams may also lead to leak or bursting of the pipeline. A deviation causing low level of reactant may generate heat shock and the vessel may get overheated and fail. These forecasts suggest that appropriate alarms for high temperature and pressure may be provided and extra care may be bestowed upon the maintenance of this unit.

In the trans-esterifier unit, the reactants and products are in semisolid state and the most expected hazard is fire. The unit operates under low pressure and high temperature (140-230°C). The unit is most vulnerable to deviations in feed rate, temperature, pressure, and presence of unidentified foreign material (Tables 16.6 -16.9). Low feed rate may reduce the pressure in the reactor, making the latter crack. The resulting spillage may cause fire. Since the reaction is endothermic, operational deviating of higher-than-design temperature and low pressure may lead to violent reactions and hence lower-than-design crumpling of the reactor. Presence of foreign material may lead to unforeseeable side reaction, which may lead to either sudden rise in temperature or excess cooling of reactor (hot spot formation). The HAZOP study suggests installation and proper maintenance of feed flow rate control system, high temperature alarm, and inert gas supply system (to keep the pressure optimal) in this unit.

Polycondensation autoclave where polymerization takes place operates in vacuum (1 Torr) and at a high temperature (230°C). HAZOP study of this unit reveals the unit is most vulnerable to deviations of high temperature and low pressure. These may lead to violent reactions. The resulting hot spots may cause cracks in the reactors and lead to spillage of flammable chemicals and fire. In view of this, adequate process control systems, fire extinguishing systems and high temperature alarm systems are warranted for the safety of this unit.

HAZOP study suggests that the drier unit is most vulnerable to deviations in the form of high temperature and low pressure. Since the drying process is done in vacuum, these deviations may cause crumpling of the drier unit, leading to spillage and fire. Hence high temperature control and pressure backup systems are recommended for the drier unit. A fire in the drier may raise the pressure in the pipe line (used for drying), making it burst.

16.4. Development of Accident Scenarios

Melter, Trans Esterifier, and Esterifier Unit: Scenario 1

The melter and trans esterifier of DMT-based process and esterifier of PTA based process deal with approximately same amount of MEG at atmospheric pressure. The accident scenario for these units is developed as pool fire.

Polymerization Reactor: Scenario 2

The most credible accident scenario for this unit is generated as pool fire.

Methanol Storage: Scenario 3

The accident scenario for this unit is developed as flash fire.

Polymerization Autoclave: Scenario 4

The release of chemical, either due to crumpling of the autoclave, or a crack in the autoclave, would lead to a pool fire.

Ammonia Storage: Scenario 5

Explosion as well as toxic release threaten this unit. The most credible accident scenario envisaged in this unit is explosive release of ammonia followed by dispersion.

MEG Storage Unit : Scenario 6

A pool fire is envisaged as the most credible accident in this unit.

16.5. IMPACT FORECASTING: CONSEQUENCE ASSESSMENT

The forecasts for *scenario 1*(Pool fire) are presented in Tables 16.10 to 16.12. An area of ~100 m radius faces 100% probability of fatality due to heat load. As several other units are close to these units, secondary accidents may also occur causing further damage. However considering the distances and the types of nearby units, the probability of such domino effect is not very high.

The polymerization reactor of the PTA based process deals with the monomer, bis (hydroxy ethyl) terephthalate and the most credible accident scenario is generated as pool fire (scenario 2). The forecasts based on detailed calculations for this scenario are presented in Table 16.13. Lethal heat load (50% probability of causing fatality) would be observed over and area of 290 m radius.

Scenario 3 includes two separate methanol storage units, one of pure chemical and the other of impure one. An accident in the former to shall generate strong heat load over an area of ~500 m radius. An accident in the lather unit shall also impact the same area. Both shall have above 50%(but less than 100%) probability of causing lethality. The forecasts based on detailed calculations for this scenario are presented in Table 16.14.

The accident scenario for polymerization autoclave is envisaged as pool fire(Scenario 4). It would occur due to the release of the chemical either through crumpling of the auto clave or due to crack in the autoclave. The forecasts based on detailed calculations are presented in Table 16.15. Severe heat load (50% probability of causing fatality) would be operative over an ~90 m radius.

Ammonia storage poses both toxic and explosive release hazards (scenario 5). The forecasts based on detailed calculations for an accident scenario in this unit are enumerated in Table 16.16. Ammonia concentration well above lethal value would built – up over an area of more than ~200 m radius. The instantaneous release of ammonia as BLEVE would cause over pressure (850.68 kpa) and shock waves over an area of ~200 m radius.

The forecasts of *scenario 6* (MEG storage Table 16.17) indicate that severe heat load (50% probability of causing fatality) would occur over an area of ~50 m radius.

16.6. Risk Estimation

In order to enable swift visualization of accident scenarios, their risk contours have been drawn over the industrial layout (Figures 16.3-16.8). It may be seen that the risk contours for *scenarios 3* and *5* (Figures 16.5 and 16.7) extend beyond the boundaries of the industry and overlap the boundaries of nearby and populated areas. The risk contours for other scenarios are confined to the campus of the IOCL but the risk contour for *scenarios 6* reaches other storage vessels in the factory (Figures 16.8). This may cause secondary accidents, the impacts of which may go beyond the factory confines.

Table 16.1. Preliminary hazard assessment: hazard identification and ranking of IOCL' s process equipment

S.No	Sub unit	Hazardous chemicals	Fire and explosion damage index	Toxic damage index	Remark
1.	Melter PTA based	MEG	97	-	Less hazardous
2.	Esterification reactor PTA based	MEG	81	-	Less hazardous
3.	Polymerization reactor PTA based	Bis (hydroxyethy1) Terephthalate Monomer	91	-	Less hazardous
4.	Trans esterifier DMT based	MEG	81	-	Less hazardous
5.	Polymerization autoclave DMT based	Bis (hydroxyethy1)Terephthalate Monomer	97	-	Less hazardous

Table 16.2. Preliminary hazard assessment: hazard identification and ranking of IOCL's storage units

S.No	Sub unit	Hazardous chemicals	Fire and explosion damage index	Toxic damage index	Remark
1.	MEG storage	MEG	688	-	Extremely hazardous
2.	Methanol storage	Methanol	270	25	Moderately hazardous
3.	Ammonia storage	Ammonia	160	100	Extremely hazardous operation

Table 16.3. HAZOP study of IOCL

INDUSTRY PLANT EQUIPMENT PARAMETER	: IOCL : DMT BASED BATCH PROCESS : MELTING SYSTEM : TEMPERATURE		
Causes	*Consequences*	*Remark*	*Criticality Rank*
Guide word: Less * Low temperature feed. * Failure of heating system * Leak in the pipe line of MEG leading to evaporation and hence cooling of the stream.	* No feed to transesterifier & other units.	* No severe harm.	IV
Guide word: More * High temperature storage of raw materials.	* Crack in the melter leading to fire hazard.	* May cause severe damage	
* Overheating.	* Leak in the feed stream due to high pressure development in MEG stream which also may lead to fire hazard.	* Hazardous due to fire accidents.	III
* Irregular fuel/power supply to the heating system	* High outlet flow rate.		

Table 16.4. HAZOP study of IOCL

INDUSTRY PLANT EQUIPMENT PARAMETER	: IOCL : DMT BASED BATCH PROCESS : MELTING SYSTEM : PRESSURE		
Causes	*Consequences*	*Remark*	*Criticality Rank*
Guide word: Less * Low feed to feed. * Low heating/failure of heating system.	* No severe harm.	* No severe harm.	IV
Guide word: More * High feed rate. * High heating. * High MEG feed followed by high heating up.	* May lead to leak in MEG stream. * May cause back pressure in MEG stream. * High outlet flow rate.	* high pressure is recommended for melter.	III

Table 16.5. HAZOP study of IOCL

INDUSTRY PLANT EQUIPMENT PARAMETER	: IOCL : DMT BASED BATCH PROCESS : MELTING SYSTEM : LEVEL OF REACTANTS		
Causes	*Consequences*	*Remark*	*Criticality Rank*
Guide word: Less * Low flow rate in feed lines. * Leak in feed streams. * Leak in melter.	* Chances to run dry. * Low flow rate at outlet. * High temperature outlet.	* Reactant level indicator may be installed.	III
Guide word: More * High flow rate in feed lines. * Outlet valve closed. * Blockage in outlet pipeline.	* High heat requirement * Low quality output due to improper melting. * High pressure in MEG pipeline & outlet.	* Reactant level indicator should be installed.	III

Table 16.6. HAZOP study of ICOL

INDUSTRY PLANT EQUIPMENT PARAMETER	: IOCL : DMT BASED BATCH PROCESS : MELTING SYSTEM : LEVEL OF REACTANTS		
Causes	*Consequences*	*Remark*	*Criticality Rank*
Guide word: Less * Low quantity of raw material storage	* Low product quality	* Harmful in terms of low product quality	
* Leak in feed lines.	* Low reaction rate	* Hazardous in terms reactor failure	
* Failure of melter	* Excess high vaccum creation leading to crupling of the equipment		III
* High back pressure in feed line	* High reaction rate	* No severe hazard	
* Guide word: More High temperature/ pressure/reactant level in melter	* High methanol production	* Harmful in terms of low product quality	IV
* High working of feed pump	* High flow rate in outlet stream	* Flow control system should be installed to feed line	
* High pressure in feed line	* Low product quality		

Table 16.7. HAZOP study of ICOL

INDUSTRY PLANT EQUIPMENT PARAMETER	: IOCL : DMT BASED BATCH PROCESS : MELTING SYSTEM : LEVEL OF REACTANTS		
Causes	*Consequences*	*Remark*	*Criticality Rank*
Guide word: less * Failure of heating system	* Low product quality		
* Low temperature output from melter	* Solidification of product	* No severe harm	IV
* High reaction rate	* Low output		
* High methanol formation	* High reaction rate		
* Guide word: More High temperature outlet from melter	* High temperature and vaccum in esterifier may lead to collapse of the system	* Severe harm in terms of economic loss	III
* High working of heating system			
* Low feed level in tran-esterifier			

Table 16.8. HAZOP study of ICOL

INDUSTRY PLANT EQUIPMENT PARAMETER	: IOCL : DMT BASED BATCH PROCESS : MELTING SYSTEM : LEVEL OF REACTANTS		
Causes	*Consequences*	*Remark*	*Criticality Rank*
Guide word:More * High feed to trans-esterifies	* Low product yield	* Vacuum sealing may be provided for trans-estifier	
* Leak in the equipment and hence inflow of outer air	* Low quality of the product	* Proper monitoring and control of pressure is recommended	IV
* Low methanol outflow	* Low methanol collection		
Guide world: Less * Over working of vacuum pump	* Low feed to poly condensation auto calve	* Severe hazards may result due to failure of the unit	
	* High reaction rate		III
* High feed to tran-esterifier	* High flow rate at outlet		
	* High cooling of reactor * Hot spot formation * Crumpling of trans-esterifier	* Harmful in terms economic loss	

Table 16.9. HAZOP study of ICOL

INDUSTRY PLANT EQUIPMENT PARAMETER	: IOCL : DMT BASED BATCH PROCESS : MELTING SYSTEM : LEVEL OF REACTANTS		
Causes	*Consequences*	*Remark*	*Criticality Rank*
Guide word: Other than * Impure feed to melter	* Solidification of material in transesterifier	* No severe hazard	
* Leak melter/feed line of trans-esterifier	* Uncontrolled reaction	* Harmful in terms of high economic loss	III
* Manual mistake	* Uneven heat distribution		

Table 16.10. Consequence analysis for scenario 1 Esterifier

Parameter	Value
POOL FIRE	
CONTINOUS	
Amount of material engage in fire (kg)	=2986.11
Burning area (sq.m.)	=78.5
Mass relapse rate (kg/s)	=394.14
Evaporation rate due to conduction (kg/s)	=0.000139
Total evaporation rate (kg/s)	=0.000761
Burning rate (kg/s)	=1.21
Damage distance of 100% lethality (m)	=105
Heat intensity (KW/sq.m) at a distance 200.0m is	=0.96

Table 16.11. Consequence analysis for scenario 1 Melter

Parameter	Value
POOL FIRE	
CONTINOUS	
Amount of material engage in fire (kg)	= 2986.11
Burning area (sq.m.)	= 78.5
Mass relapse rate (kg/s)	= 394.14
Evaporation rate due to conduction (kg/s)	= 0.000139
Total evaporation rate (kg/s)	= 0.000761
Burning rate (kg/s)	= 1.21
Damage distance of 100% lethality (m)	= 105xff
Heat intensity (KW/sq.m) at a distance 200.0m is	= 0.96

Table 16.12. Consequence analysis for scenario 1 Trans-esterifier

Parameter	Value
POOL FIRE	
CONTINOUS	
Amount of material engage in fire (kg)	= 2986.11
Burning area (sq.m)	= 78.5
Mass relapse rate (kg/s)	= 394.14
Evaporation rate due to conduction (kg/s)	= 0.000139
Total evaporation rate (kg/s)	= 0.000761
Burning rate (kg/s)	= 1.21
Damage distance of 100% lethality (m)	= 105.95
Heat intensity (KW/sq.m) at a distance 200.0m is	= 0.96

Table 16.13. Consequence analysis for scenario 2 Polymerisation reactor

Parameter	Value
POOL FIRE	
CONTINOUS	
Amount of material engage in fire (kg)	= 5065.45
Burning area (sq.m)	= 176.62
Mass relapse rate (kg/s)	= 491.66
Evaporation rate due to conduction (kg/s)	= 0.000233
Total evaporation rate (kg/s)	= 0.005761
Burning rate (kg/s)	= 4.01
Damage distance of 100% lethality (m)	= 291
Heat intensity (KW/sq.m) at a distance 200.0m is	= 8.0

Table 16.14. Consequence analysis for scenario 3 Methanol storage (Pure)

Parameter	Value
FLASH FIRE: RESULT	
Volume of vapour cloud (m^3)	= 1.1E+06.5
Area under fire effect (m^2)	= 2.9E+06
Total heat radiated (kJ)	= 3.6E+07
Half life time of fire (sec)	= 2.21
Effective time of fire (sec)	= 6.62
Heat intensity at a distance 200.0m is (KW/m^2)	= 49.19
Methanol storage (impure)	
FLASH FIRE: RESULT	
Volume of vapour cloud (m^3)	= 6.2E+05
Area under fire effect (m^2)	= 9.2E+05
Total heat radiated (kJ)	= 1.2E+07
Half life time of fire (sec)	= 1.95
Effective time of fire (sec)	= 5.87
Heat intensity at a distance 200.0m is (KW/m^2)	= 15.55

Table 16.15. Consequence analysis for scenario 4 Polymerisation reactor 2 (polymerization autoclave)

Parameter	Value
POOL FIRE	
CONTINOUS	
Amount of material engage in fire (kg)	= 5065.45
Burning area (sq.m)	= 176.62
Mass relapse rate (kg/s)	= 491.66
Evaporation rate due to conduction (kg/s)	= 0.000233
Total evaporation rate (kg/s)	= 0.005761
Burning rate (kg/s)	= 4.01
Damage distance of 100% lethality (m)	= 291
Heat intensity (KW/sq.m) at a distance 200.0m is	= 8.0

Table 16.16. Consequence analysis for scenario 5

Parameter	Value
Explosion: BLEVE	
SHOCK WAVE CHARACTERISTICS:	
Peak over pressure (kPa)	= 850.68
Duration of shock wave (sec)	= 41.09
Variation of over pressure (kPa/sec)	= 4.5E-04
Shock velocity in air (m/s)	= 515.43
Particle velocity behind the shock wave (m/s)	= 2.5E+02
Dynamic pressure (Dimension less)	= 4.17E+02
Reflected pressure (kPa)	= 2.6E+03
Over pressure impulse (kPa/sec)	= 2.5E+03
Dispersion	
TWO PHASE RELEASE	
PUFF RADIUS	
Total mass in the vessel	1.000000E+05 Kg
Mass of vapor flashed	4.2553784+04 Kg
Radius of puff	2.362941E+01 m
PASQUILL-GIFFORD INSTANTANEOUS MODEL	
Evaporation of liquid pool for 1 hour taken as a basis	
ELEVATED SOURCE	
Concentration at 500.00 m	2.442826E-07 Kg/m^3
PUFF CHARACTERISTICS	
Concentration at ground level	1.214172E-07 Kg/m^3
Concentration at cloud axis	5.250805E-03 Kg/m^3
Concentration at cloud edge	5.250805E-04 Kg/m^3
Cloud radius	2.285449E+02 m
Maximum distance travelled by the cloud	1.150000E+03 m
Dosage on downwind dist 500.00	4.086895E-01 Kg/m^3

Table 16.17. Consequence analysis for scenario 6 MEG storage unit

Parameter	Value
POOL FIRE	
CONTINOUS	
Amount of material engage in fire (kg)	= 145644.0
Pool fire radius (m)	= 5.0000
Area of fire (sq.m)	= 78.5
Burning rate (kg/s)	= 2.69
Damage distance of 100% lethality (m)	= 56
Heat intensity (KW/sq.m) at a distance 200.0m is	= 0.058

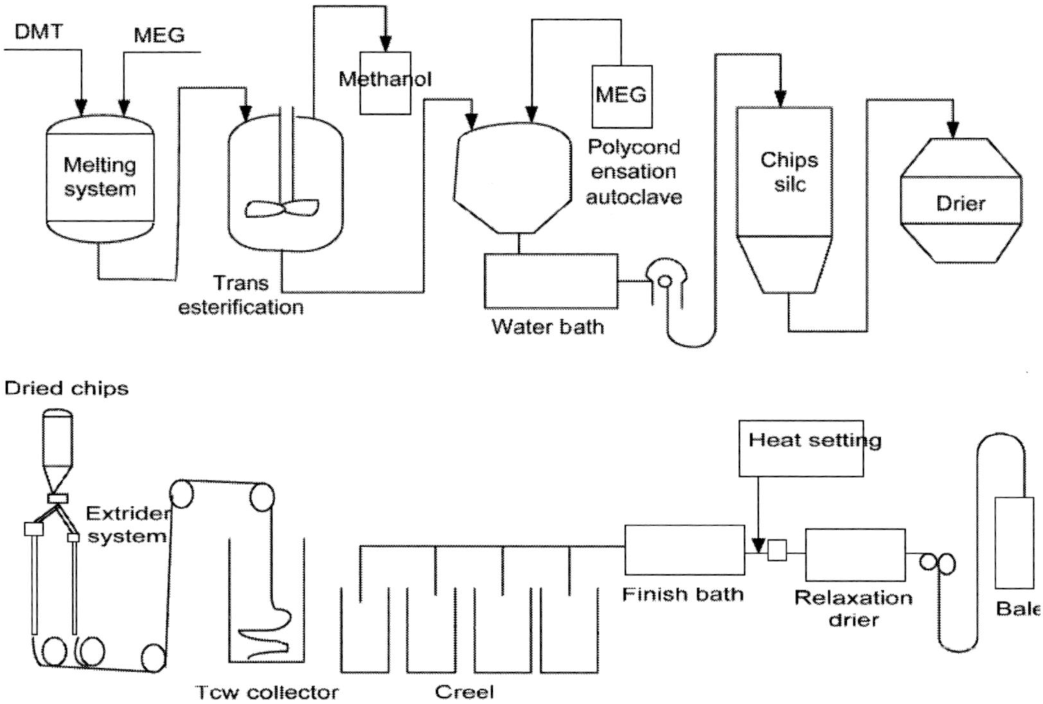

Figure 16.1. Process flow diagram for DMT process (batch process).

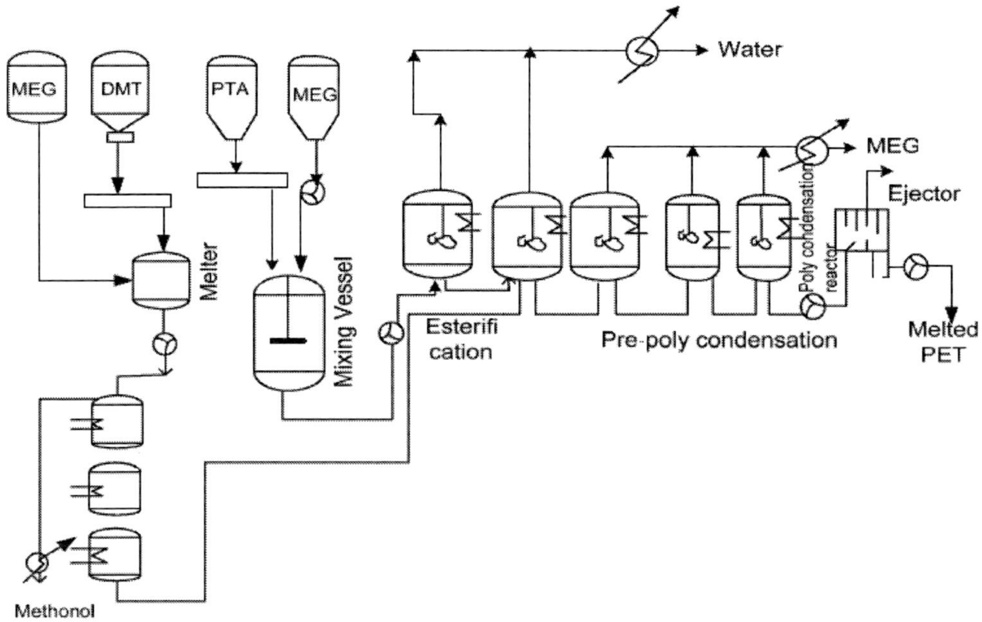

Figure 16.2. The process flow diagrams for poly ethylene terephthalate plant process (continuous process).

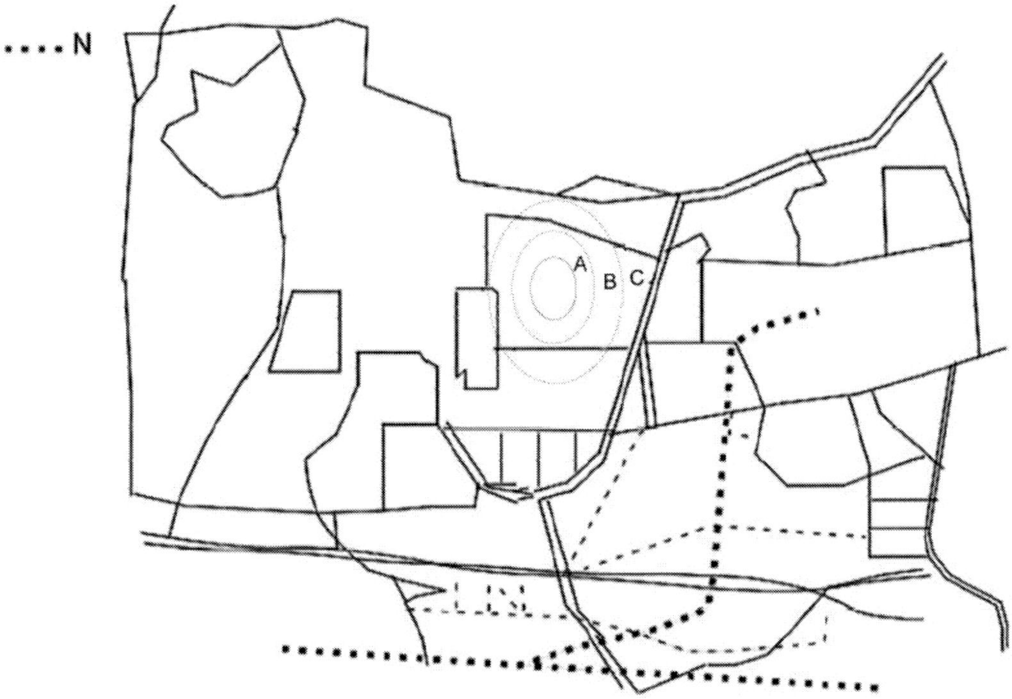

Figure 16.3. Risk contours for an accident scenario in melter, esterifies, trans-esterifies (reactor) showing impact area due to severe (A), high (B), and moderate risk (C).

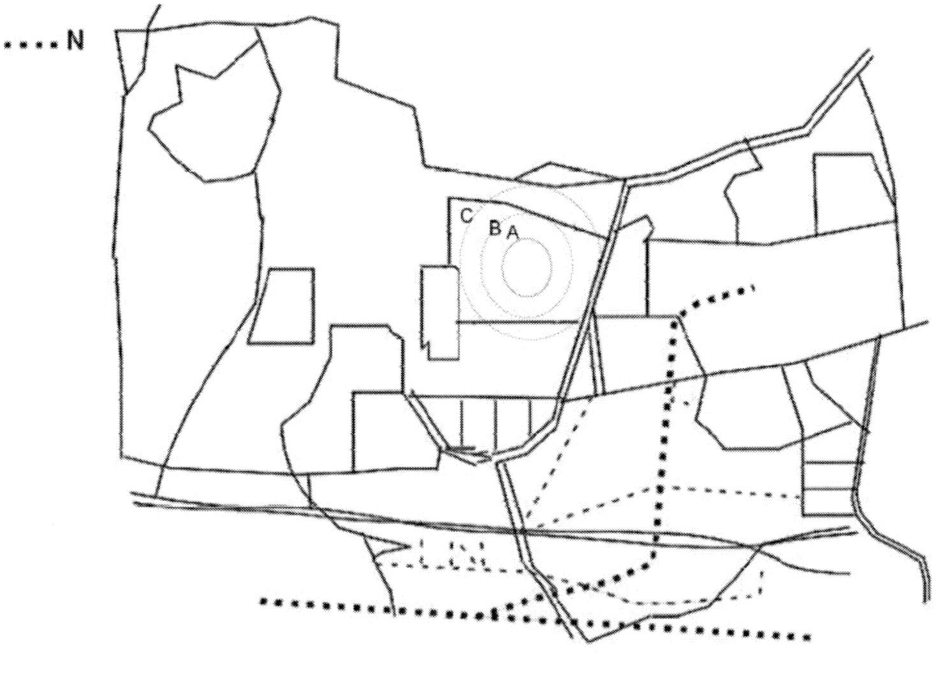

Figure 16.4. Risk contours for an accident scenario in polymerization unit showing impact area due to severe (A), high (B), and moderate risk (C).

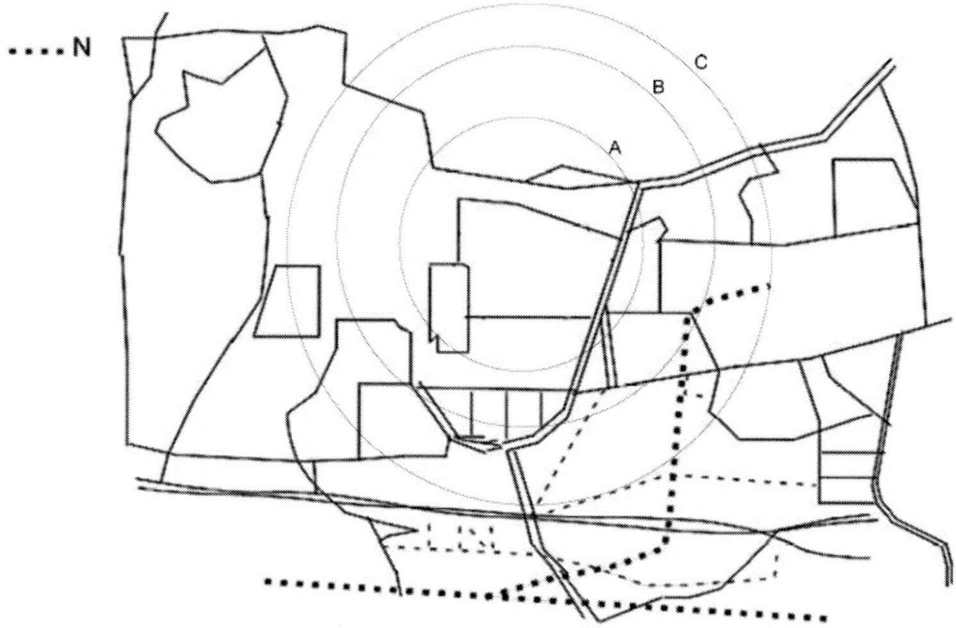

Figure 16.5. Risk contours for an accident scenario in methanol showing impact area due to severe (A), high (B), and moderate risk (C).

Figure 16.6. Risk contours for an accident scenario in Polymerisation autoclave showing impact area due to severe (A) high, (B), and moderate risk (C).

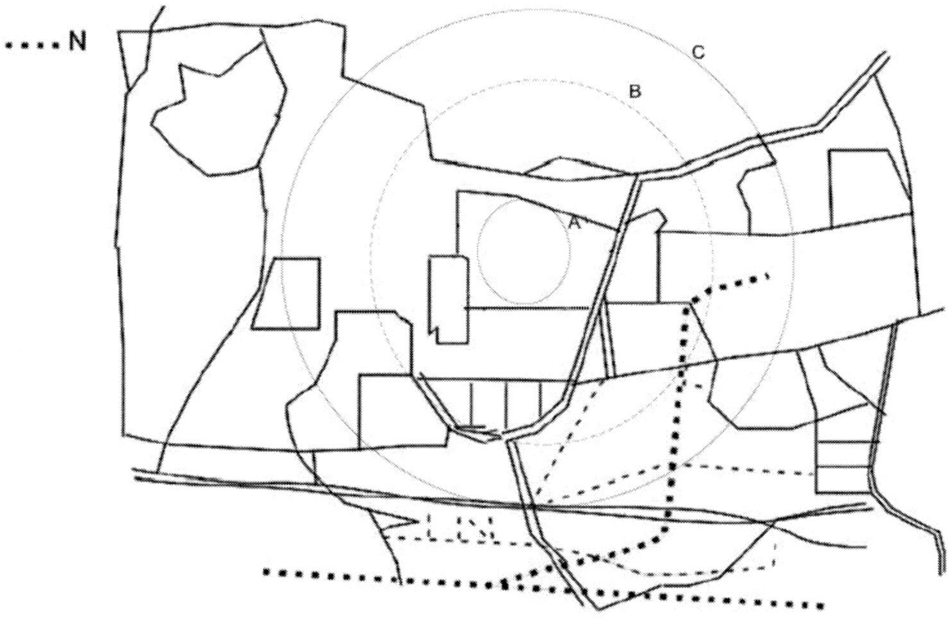

Figure 16.7. Risk contours for an accident scenario in ammonia storage showing impact area due to severe (A), high (B), and moderate risk (C).

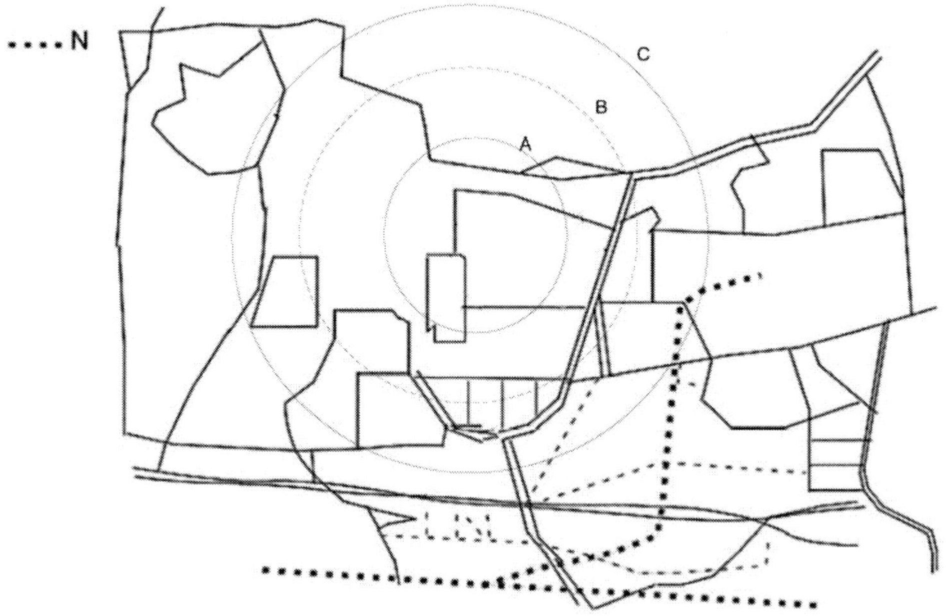

Figure 16.8. Risk contours for an accident scenario in monoethylene glycol storage showing impact area due to severe (A), high (B), and moderate risk (C).

Chapter 17

Rapid Risk Assessment of SPIC – HCD

17.1. Introduction

SPIC Heavy Chemicals Division was commissioned in 1989. The different *products manufactured by this plant are:

1) Caustic soda
2) Liquid chlorine
3) Hydrochloric acid (33% concentration)
4) Ammonium chloride.

The simplified process block diagram for various plants is presented in Figures 17.1 to 17.5.

17.2. Hazard Identification and Ranking

All the units of the SPIC's plant (Table 17.1) were screened by the Hazard Identification & Ranking Analysis (HIRA) system. The index scores in terms of toxic damage hazard and fire/explosion damage hazard are presented in Figures 17.6 and 17.7 respectively. Seven out of 17 units have TDI values above the threshold and 4 have FEDI values above the threshold. The chlorine plant's compressor, and hydrogen storage units have both indices values above the dangerous threshold.

17.3. Development of Accident Scenarios

For the units shortlisted as above, accident scenarios were developed and the impacts were forecast. Typical credible scenarios are described below.

Electrolysis Unit : Scenario 1

This unit is associated with hydrogen and chlorine gases and thus poses hazards of toxic dispersion as well as explosions/fires. One of the most credible accidents in this unit is likely to be formation of a fire ball accompanied by toxic release of chlorine.

Chlorine Preparation Unit : Scenario 2

A continuous release and dispersion of toxic gas is envisaged.

Compression Unit: Scenario 3

An instantaneous release of the compressed chlorine may occur as BLEVE (boiling liquid expanding vapour explosion). This would generate shock waves as well as toxic load.

Liquefication Unit: Scenario 4

The most probable accident in this unit is continuous release of chlorine followed by toxic dispersion.

Hydrogen Treatment Unit: Scenario 5

This unit handles hydrogen, which is highly flammable. Overpressurization or an accidental leak of hydrogen may cause BLEVE. The released chemical on meeting an ignition source would cause fireball.

Chlorine Storage Unit: Scenario 6

This unit stores chlorine in liquefied condition. Overpressurization and/or accidental leak would cause release of chlorine as BLEVE, followed by toxic dispersion.

Caustic Soda Storage Unit: Scenario 7

Release of caustic soda, which is stored in liquid state, from the vessel would cause corrosive spill and contamination of air and soil.

Hydrogen Storage Unit: Scenario 8

Hydrogen, which is stored under pressure, may leak out due to failure of the vessel or the valve. This would lead to CVCE and fireball.

Scenario 9: Accident in the Ammonia Storage Vessel

Vessel failure leading to BLEVE and toxic dispersion unit.

17.4. Impact Forecasting: Consequence Assessment

The forecast of *scenario 1* – fire ball followed by the release and dispersion of toxic chlorine – are presented in Table 17.2. The fire ball shall be formed when the vapor cloud of the released hydrogen shall meet an ignition source. The resultant radiation shall carry heat load capable of causing fatal injuries at 50 % probability over an area of 10m radius from the point of accident. This is not a very serious consequence but it shall be accompanied by toxic heat load due to the dispersion of chlorine. HAZDIG estimates that this may adversely impact an area of ~3700 m radius. Not only the premises of SPIC and nearby industries but also residents of villages are likely to come under the impact zone.

The likely impacts of *Scenario 2* – continuous release of chlorine – are enumerated in Table 17.3. Concentrations of the gas high enough to cause mortality (at 50% probability) shall envelope an area of over 5600 m radius, encompassing four populous villages (over 10,000 residents). As this unit also involves other chlorinated compounds, which may be photochemically, active, other possibilities of adding toxicity to the environment also exist.

Scenario 3, which envisages a BLEVE accompanied by dispersion of toxic chlorine, is likely to have the impacts presented in Table 17.4. The shock waves caused by the explosion may have destructive intensity (at 50% probability) upto ~37m from the point of their origin. But the impact of chlorine toxicity may reach a much larger area, upto about ~4700 m from the accident epicenter, and may cause harm to a sizable populace, including two residential colonies and two villages.

Scenario 4: The credible accident in liquefication unit has been envisaged as continuous release of chlorine followed by dispersion *(scenario 4).* The likely impact of such an accident is presented in Table 17.5. It is evident that toxic effect of chlorine would persist over an area of ~6550 m radius. As with other units involving chlorine an accident in this unit would also threaten a large area encompassing sizeable human population.

As per *scenario 5* BLEVE would generate shock waves as well as missiles. In addition, there may be secondary impacts if the released material gets ignited and forms a fire-ball thereby generating additional heat load. The forecast for this scenario is summarized in Table 17.6. The impact of the lethal shock waves may be felt in an area of ~35 m radius, whereas the impact of missile is likely to cover an area of ~1618 m radius.

In the chlorine unit chlorine is stored in large quantity under pressure. *Scenario 6* envisages an instantaneous release of chlorine as BLEVE. This would generate shock waves, missiles, and toxicity built-up (Table 17.7). The damage radii of all these possibilities are high making this scenario as the most disastrous of all.

The forecasts of *scenario 7* are presented in Table 17.8. Caustic soda is stored in liquid state; the chemical being highly corrosive, its release would cause contamination of soil and air. It would also corrode and weaken other units. Evaporation of a part this liquid would release highly corrosive vapour, influencing living and non living objects over an area of about 60 m radius.

The confined vapor cloud explosion in hydrogen tank as per *scenario 8* would generate severe shock waves and missiles effects (Table 17.9). Of all the scenarios developed for this industry, *scenario 8* has most serious portants in terms of shock wave and missile impacts. The heat load generated on ignition of the hydrogen vapour cloud may also scald a sizeable area (~100 m radius).

A BLEVE as per scenario *9* would lead to overpressure, and instantaneous toxic release. As the chemical is non-flammable no heat load would be observed but toxic dispersion would occur in (Table 17.10). The overpressure would have 50% probability of causing human death over an area of ~195 m radius, while lethal toxic load at the same probability of fatality would be observed over an area of ~3800 m radius. The hazard posed by this unit is comparatively less, because ammonia is easier to disperse and also less toxic than chlorine.

17.5. Risk Estimation

On the basis of the results presented above, and probability of occurrence of the various accident scenarios, risk factors have been computed. In order to enable these have been visualization of the accidents risk contours have been drawn over the industry layout (Figures 17.8 to 17.16).

Table 17.1. Different plants of SPIC heavy chemical industry

Plant Process units	Legends used in Figures 19.6 &19.7
Caustic soda plant	
Brine purification	A
Electrolysis B	B
Chlorine preparation	C
Chlorine Plant	
Chlorine treatment	D
Compressors	E
Liquefication unit	F
Hydrochloric Plant	
Hydrogen treatment	G
HCL synthesis	H
Absorption unit	I
Ammonium Chloride plant	
NH_4 Cl reactor	J
Centrifugation	K
Storages	
Chlorine	L
Caustic soda	M
Hydrogen	N

Plant Process units	Legends used in Figures 19.6 &19.7
Liquid Ammonia	O
NH_4Cl	P
HCl	Q

Table 17.2. Result of consequence analysis for scenario1; accident in Electrolysis unit; Fire Ball/dispersion of toxic chlorine

Electrolysis unit; Fire ball/dispersion of toxic chlorine

Parameters	Values
Fire: Fire Ball	
Radius of the fire ball	(m) :26.00
Duration of the fire ball	(s) :11.00
Energy released by fire ball	(kJ) :7.32E+05
Radiation heat flux	(kJ/sq.m) :280.00
Damage Radii (DR) due to thermal load	
DR for 100% fatality/damage	(m) :26
DR for 50% fatality/damage	(m): 37
DR for 100% third degree of burn	(m) :48
DR for 50 % third degree of burn	(m) :67
Toxic Release & Dispersion	
Heavy Gas Dispersion characteristics	
Box Instantaneous: Model	
Concentration at cloud axis	(kg/cu.m) :6.94E-03
Value of source height	(m) :7.0000
Damage radii (DR) for various degree of damage:	
DR for 100% lethality	(m) :2879
DR for 50% lethality	(m) :3700
DR for 10% lethality	(m) :6629
Puff characteristics:	
Puff concentration at centre of cloud	(kg/cu.m) :3.88E-04
Concentration at cloud edges	(kg/cu.m) :3.87E-04
Distance along downwind	(m) :200.0000
Dosage at study point	(kg/cu.m): 0.0052

Table 17.3. Result of consequence analysis for scenario 2; accident in chlorine preparation unit; continuous release of chlorine

Chlorine preparation unit
Continuous release of chlorine

Parameters	Values
Toxic Release & Dispersion	
Box continuous :Model	
Value of wind speed	(m/s) :3.5000
Ground level X concentration	(kg/cu.m) :0.0016
Damage radii (DR) for various degree of damage	
DR for 100% lethality	(m) :3947
DR for 50% lethality	(m) :5601
DR for 10% lethality	(m) :12625
Plume characteristics	
Ground level concentration of plume on axis	(kg/cu.m) :0.0032
Width of plume	(m) :112.0
Cloud concentration	(kg/cu.m) :0.0032
Maximum ground level concentration	(kg/cu.m) :0.0425

Table 17.4. Result of consequence analysis for scenario 3; accident in compressor unit; BLEVE followed by dispersion

Compressor unit
BLEVE followed by dispersion

Parameters	Values
Explosion ; BLEVE	
Total energy released	(kJ) :1.4E+07
Peak over pressure	(kPa) :3.00
Variation of over pressure in air	(kPa/s): 3.00
Shock velocity of air	(m/s) :333.00
No missile effect	
Damage radii (DR) for various degree of damage due to over pressure	
DR for 100% fatality/damage	(m) :25
DR for 100% fatality or 50% damage	(m) :37
DR for 50 % fatality or 25% damage	(m) :55
Toxic Release & Dispersion	
Heavy Gas Dispersion characteristics	
Box Instantaneous: Model	
Concentration at cloud axis	(kg/cu.m) :1.39E-02
Value of source height	(m) :4.0000
Damage radii (DR) for various degree of damage	
DR for 100% lethality	(m) :3700
DR for 50% lethality	(m) :4757
DR for 10% lethality	(m) :8522
Puff characteristics	
Puff concentration at centre of cloud	(kg/cu.m) :7.77E-04
Concentration at cloud edges	(kg/cu.m) :7.74E-04
Distance along downwind	(m) :200.0000
Dosage at study point	(kg/cu.m) :0.0104

Table 17.5. Result of consequence analysis for scenario 4; accident In Liquefication unit; continuous release of chlorine

Liquefication unit
Continuous release of chlorine

Parameters	Values
Toxic Release & Dispersion	
Box continuous :Model	
Value of wind speed	(m/s) :3.5000
Ground level X concentration	(kg/cu.m) :0.0022
Damage radii (DR) for various degree of damage	
DR for 100% lethality	(m) :4616
DR for 50% lethality	(m) :6550
DR for 10% lethality	(m) :14766
Plume characteristics	
Ground level concentration of plume on axis	(kg/cu.m) :0.0044
Width of plume	(m) :112.0
Cloud concentration	(kg/cu.m) :0.0044
Maximum ground level concentration	(kg/cu.m): 0.1776

Table 17.6. Result of consequence analysis for scenario 5; accident In Hydrogen treatment unit; BLEVE + fire ball

Hydrogen treatment unit; BLEVE + fire ball

Parameters	Values
Explosion ; BLEVE	
Total energy released	(kJ) : 1.20E+05
Peak over pressure	(kPa) : 242.00
Variation of over pressure in air	(kPa/s) : 169.00
Shock velocity of air	(m/s) : 342.00
Duration of shock wave	(m/s) : 39.0
Missile characteristics	
Initial velocity	(m/s) : 115.4
Kinetic energy of fragment	(kJ) : 2.11E+05
Fragment velocity at study point	(m/s) : 45.60
Penetration ability at study point (based on empirical models	
Concrete structure	(m) : 0.0052
Brick structure	(m) : 0.0066
Steel structure	(m) : 0.0029
Damage radii (DR) for various degree of damage due to over pressure	
DR for 100% complete damage	(m) : 22
DR for 100% fatality or 50% complete damage	(m) : 34
DR for 50% fatality or 25% complete damage	(m) : 49
Damage radii (DR) for various degree of damage due to missile	
DR for 100% damage or 100% fatality	(m) : 1502
DR for 50% damage or 100% fatality	(m) : 1618
DR for 100% fatality or 10% damage	(m) :1721
Fire: Fire Ball	
Radius of the fire ball	(m) :75.60
Duration of the fire ball	(s) :31.50
Energy released by fire ball	(kJ) :6.52E+07
Radiation heat flux	(kJ/sq.m) :5441.6
Damage Radii (DR) due to thermal load	
DR for 100% fatality/damage	(m) :81
DR for 50% fatality/damage	(m) :93
DR for 100% third degree of burn	(m) :104
DR for 50% third degree of burn:	(m) :121

Table 17.7. Result of consequence analysis for scenario 6; accident in chlorine storage vessel; BLEVE followed by toxic release and dispersion

Storage chlorine; BLEVE followed by toxic release and dispersion

Parameters	Values
Explosion : BLEVE	
Total energy released	(kJ) : 3.5E+07
Peak over pressure	(kPa) : 477.00
Variation of over pressure in air	(kPa/s) : 353.00
Shock velocity of air	(m/s) : 687.00
Duration of shock wave	(m/s) : 47.0
Missile characteristics	
Initial velocity	(m/s) :55.00
Kinetic energy of fragment	(kJ) : 7.37E+03
Fragment velocity at study point	(m/s) : 54.00
Penetration ability at study point (based on empirical models)	
Concrete structure	(m) : 0.0133

Table 17.7. (Continued)

Parameters	Values
Brick structure	(m) : 0.0170
Steel structure	(m) : 0.0054
Damage radii (DR) for various degree of damage due to over pressure	
DR for 100% complete damage	(m) : 33
DR for 100% fatality or 50% complete damage	(m) : 51
DR for 50% fatality or 25% complete damage	(m) : 75
Damage radii (DR) for various degree of damage due to missile	
DR for 100% damage or 100% fatality	(m) : 1802
DR for 50% damage or 100% fatality	(m) : 1918
DR for 100% fatality or 10% damage	(m) : 2021
Toxic Release & Dispersion	
Heavy Gas Dispersion characteristics	
Box Instantaneous: Model	
Concentration at cloud axis	(kg/cu.m) : 1.08E+03
Value of source height	(m) : 4.0000
Damage radii (DR) for various degree of damage	
DR for 100% lethality	(m) : 10954
DR for 50% lethality	(m) : 14082
DR for 10% lethality	(m) : 25229
Puff characteristics	
Puff concentration at centre of cloud	(kg/cu.m) : 2.91E+01
Concentration at cloud edges	(kg/cu.m) : 6.03E+01
Distance along downwind	(m) : 10.0000

Table 17.8. Result of consequence analysis for scenario 7; accident in caustic soda storage unit; Toxic release and dispersion

Storage caustic soda; Toxic release and dispersion

Parameters	Values
Toxic Release & Dispersion	
Box continuous: Model	
Value of wind speed	(m/s) : 3.5000
Ground level concentration	(kg/cu.m) : 0.0022
Damage radii (DR) for various degree of damage	
DR for 100% lethality	(m) : 43
DR for 50% lethality	(m) : 60
DR for 10% lethality	(m) : 136
Plume characteristics	
Ground level concentration of plume on axis	(kg/cu.m) : 0.0044
Width of plume	(m) : 8.0
Maximum ground level concentration	(kg/cu.m) : 0.0002

Table 17.9. Result of consequence analysis for scenario 8; accident in Hydrogen storage unit; CVCE + fire ball

Storage hydrogen; CVCE + fire ball

Parameters	Values
Explosion ; CVCE	
Energy released during explosion	(kJ) : 5.45E+09
Peak over pressure	(kPa) : 22550.1
Variation of over pressure in air	(kPa/s) : 15475.5

Parameters	Values
Shock velocity of air	(m/s) : 4255.8
Duration of shock wave	(m/s) : 106.0
Missile characteristics	
Initial velocity	(m/s) : 8415.3
Kinetic energy of fragment	(kJ) : 2.40E+08
Fragment velocity at study point	(m/s) : 655.00
Penetration ability at study point (based on empirical models)	
Concrete structure	(m) : 0.58
Brick structure	(m) : 0.78
Steel structure	(m) : 0.17
Damage radii (DR) for various degree of damage due to over pressure	
DR for 100% complete damage	(m) : 339
DR for 100% fatality or 50% complete damage	(m) : 518
DR for 50% fatality or 25% complete damage	(m) : 765
Damage radii (DR) for various degree of damage due to missile	
DR for 100% damage or 100% fatality	(m) : 2832
DR for 50% damage or 100% fatality	(m) : 2948
DR for 100% fatality or 10% damage	(m) : 3052
Fire: Fire Ball	
Radius of the fire ball	(m) : 215.4
Duration of the fire ball	(s) : 128.6
Energy released by fire ball	(kJ) : 9.54E+09
Radiation heat flux	(kJ/sq.m) : 755150.3
Damage Radii (DR) due to thermal load	
DR for 100% fatality/damage	(m) : 186
DR for 50% fatality/damage	(m) : 204
DR for 100% third degree of burn	(m) : 244
DR for 50 % third degree of burn	(m) : 269

Table 17.10. Result of consequence analysis for scenario 9; accident in Liquid ammonia storage unit; BLEVE toxic release and dispersion

Storage Liquid ammonia
BLEVE toxic release and dispersion

Parameters	Values
Explosion ; CVCE	
Total energy released	(kJ) : 8.33E+07
Peak over pressure	(kPa) : 8764.3
Variation of over pressure in air	(kPa/s) : 4010.5
Shock velocity of air	(m/s) : 1205.2
Duration of shock wave	(m/s) : 41.5
Missile characteristics	
Initial velocity	(m/s) : 1065.4
Kinetic energy of fragment	(kJ) : 1.91E+07
Fragment velocity at study point	(m/s) : 715.60
Penetration ability at study point (based on empirical models	
Concrete structure	(m) : 0.58
Brick structure	(m) : 0.78
Steel structure	(m) : 0.17
Damage radii (DR) for various degree of damage due to over pressure	
DR for 100% complete damage	(m) : 128
DR for 100% fatality or 50% complete damage	(m) : 195

Table 17.10. (Continued)

Parameters	Values
DR for 50% fatality or 25% complete damage	(m) : 288
Toxic Release & Dispersion	
Heavy Gas Dispersion characteristics	
Box Instantaneous: Model	
Concentration at cloud axis	(kg/cu.m) : 1.39E+01
Value of source height	(m) : 5.000
Damage radii (DR) for various degree of damage	
DR for 100% lethality	(m) : 3023
DR for 50% lethality	(m) : 3887
DR for 10% lethality	(m) : 6964
Puff characteristics	
Puff concentration at centre of cloud	(kg/cu.m) : 7.77E-01
Concentration at cloud edges	(kg/cu.m) : 7.74E-01
Distance along downwind	(m) : 200.0000
Dosage at study point	(kg/cu.m) : 10.4108

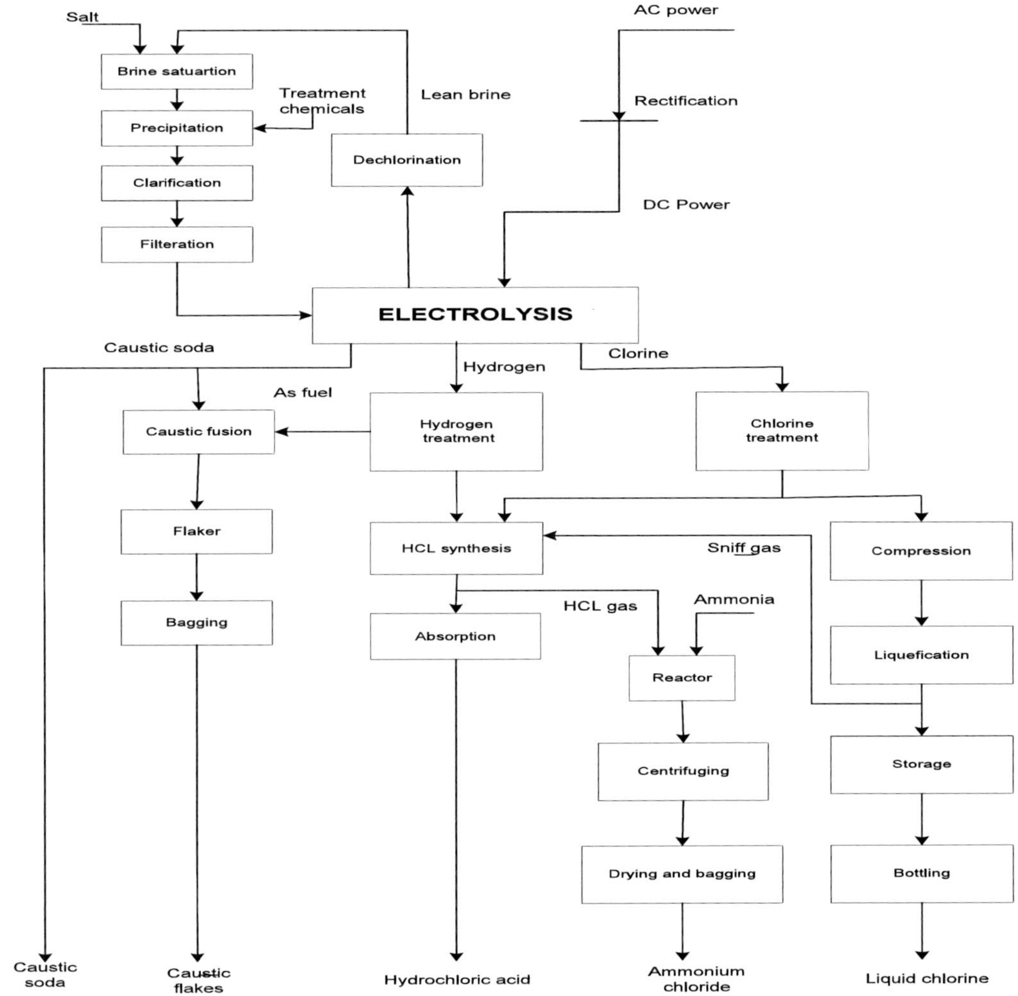

Figure 17.1. The block diagram of complete plant process for SPIC plant.

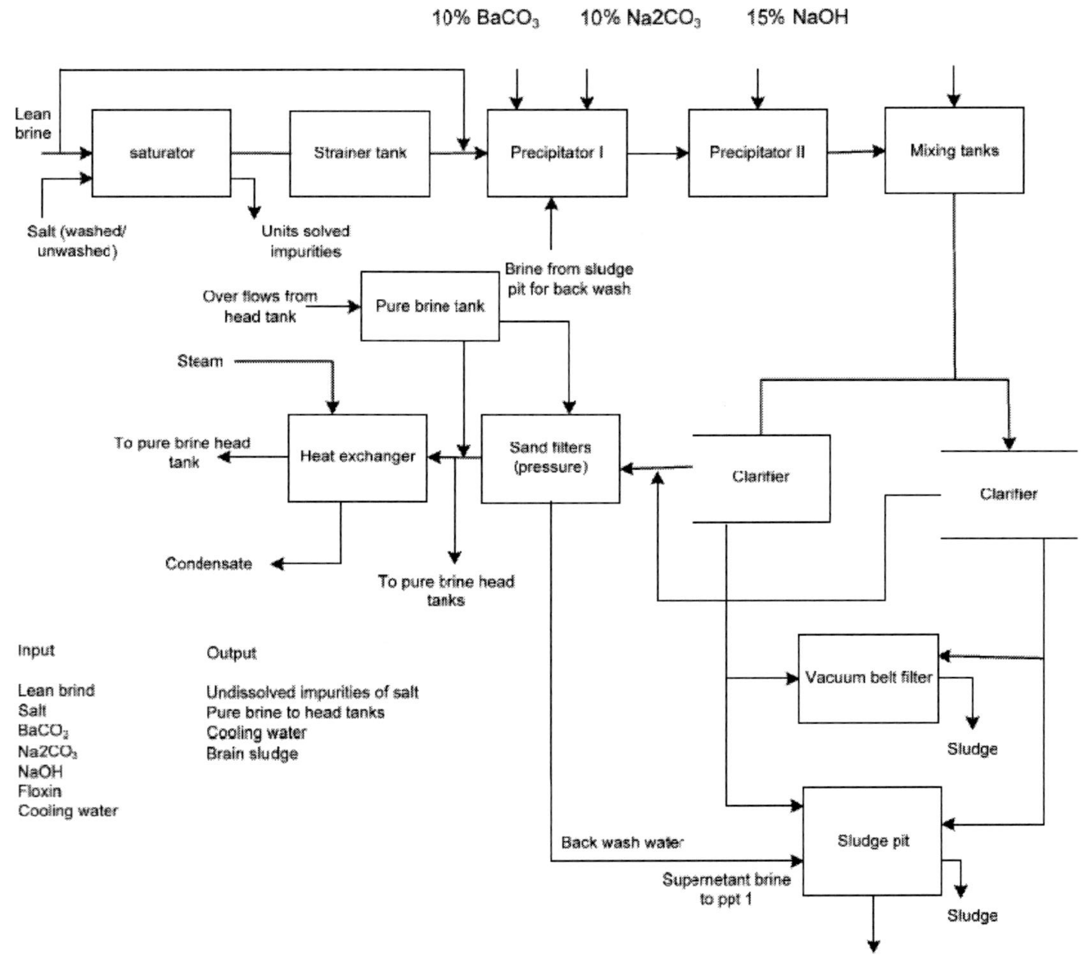

Figure 17.2. Process block diagram of brine house.

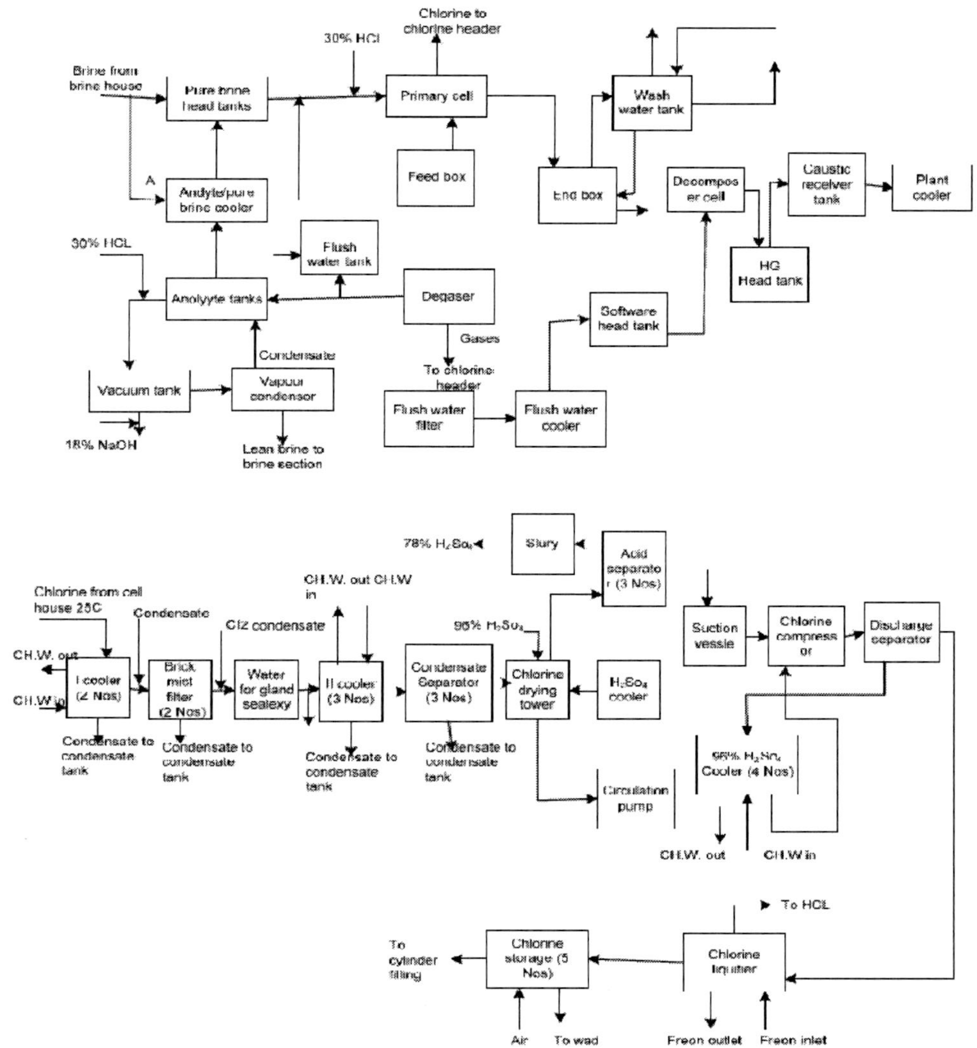

Figure 17.3. Process flow diagram of Chlorine section.

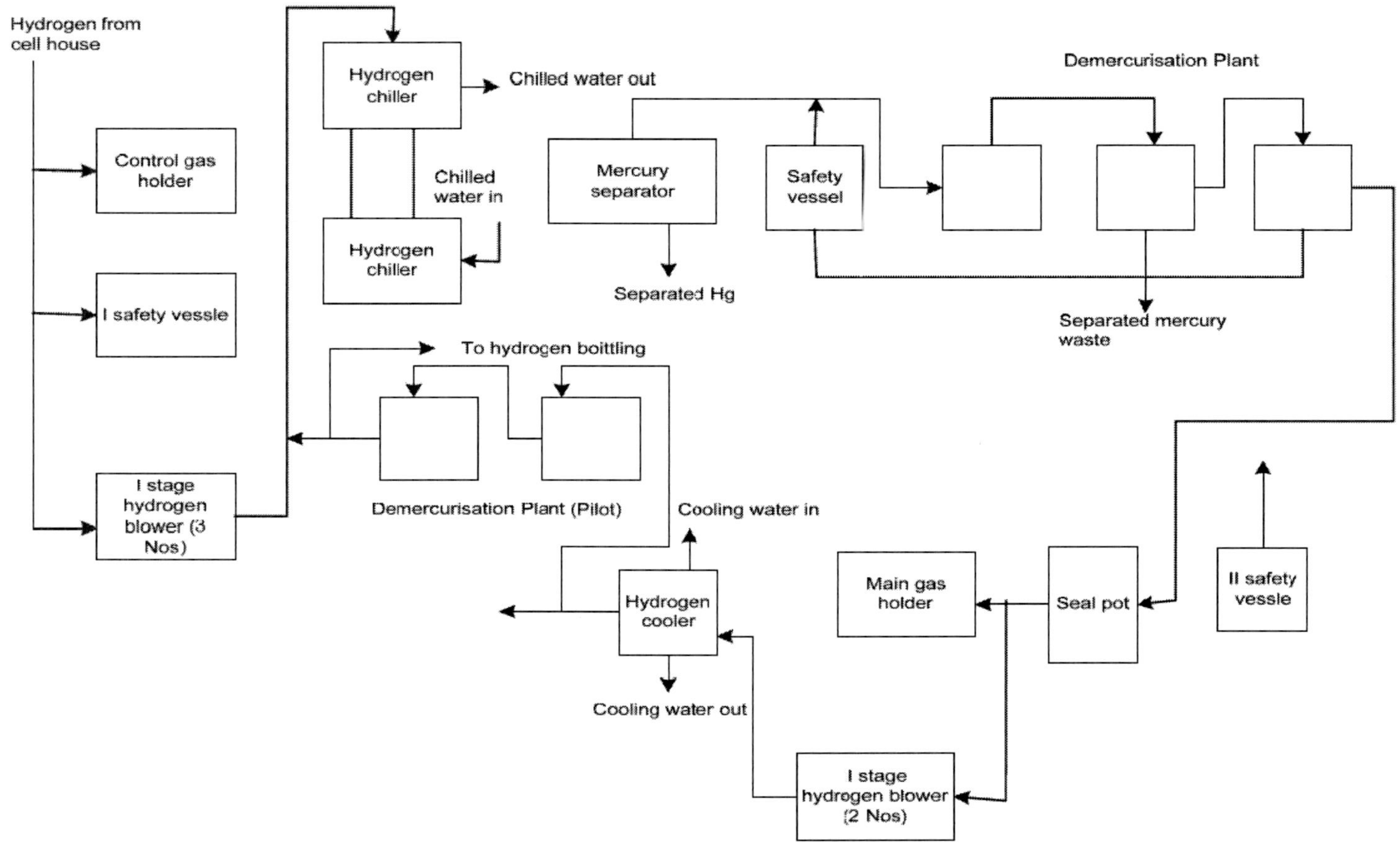

Figure 17.4. Process flow diagram of Hydrogen section.

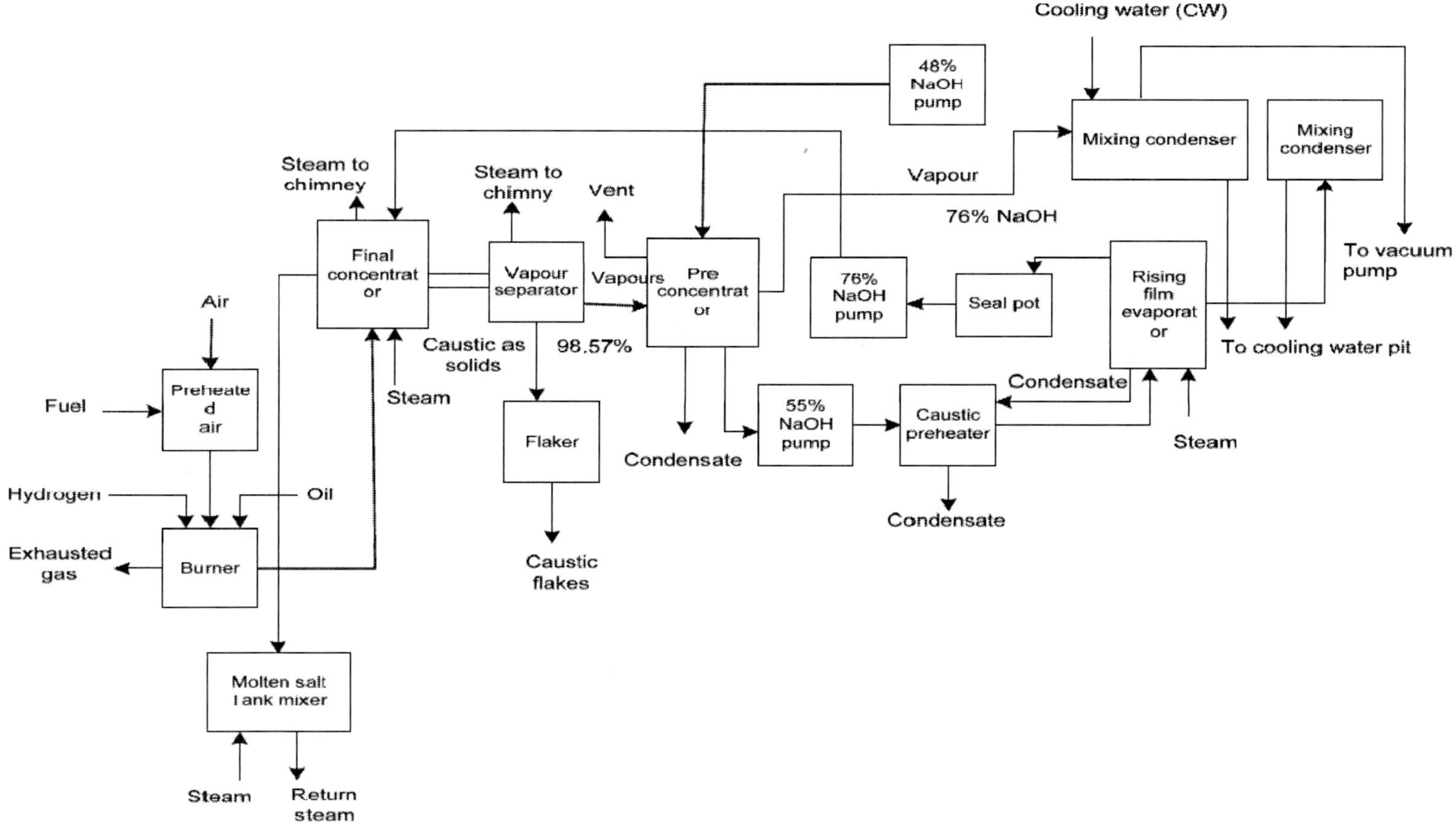

Figure 17.5. Block diagram of Caustic flake unit.

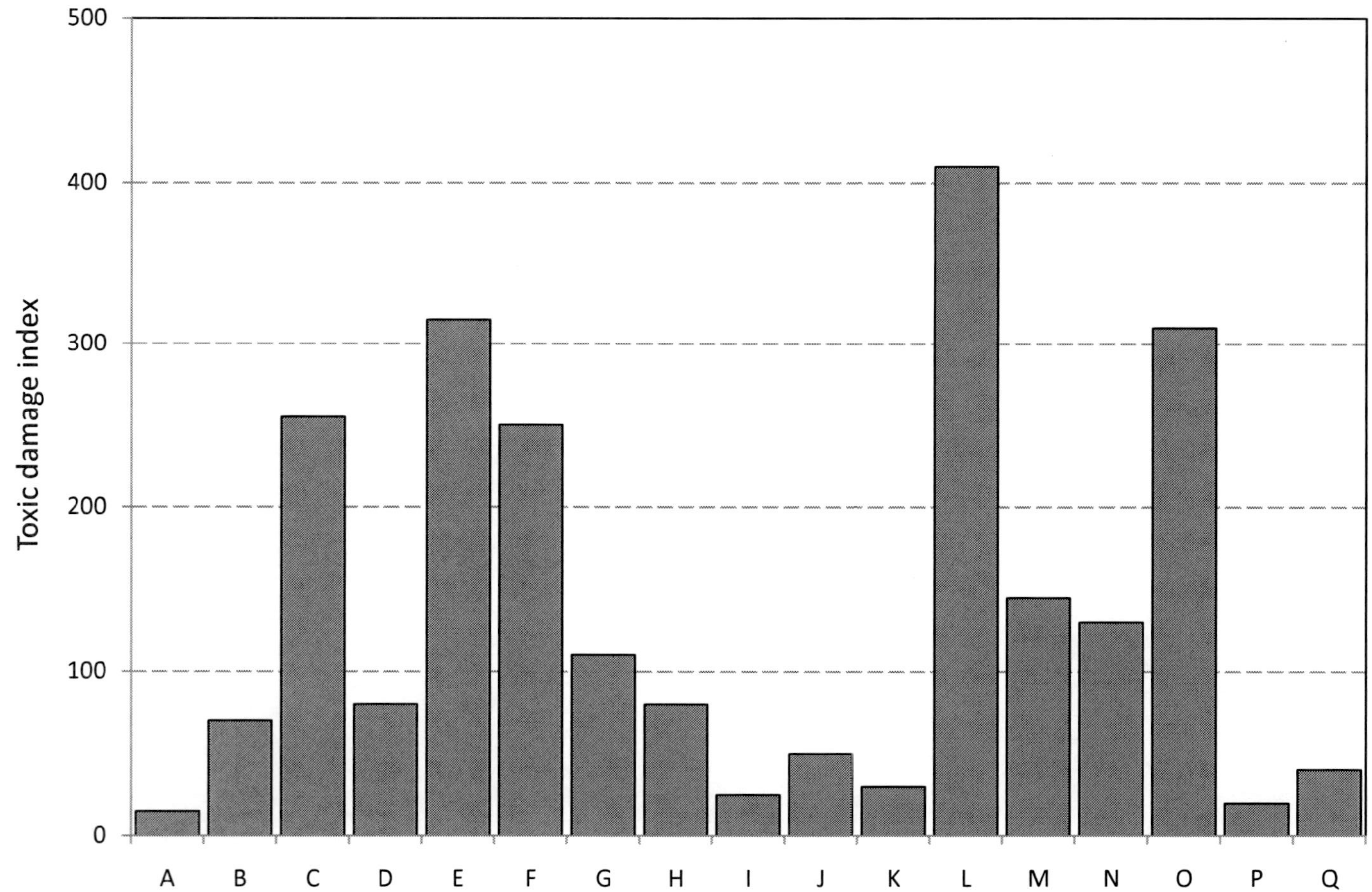

Figure 17.6. Toxic damage index for various units in SPIC.

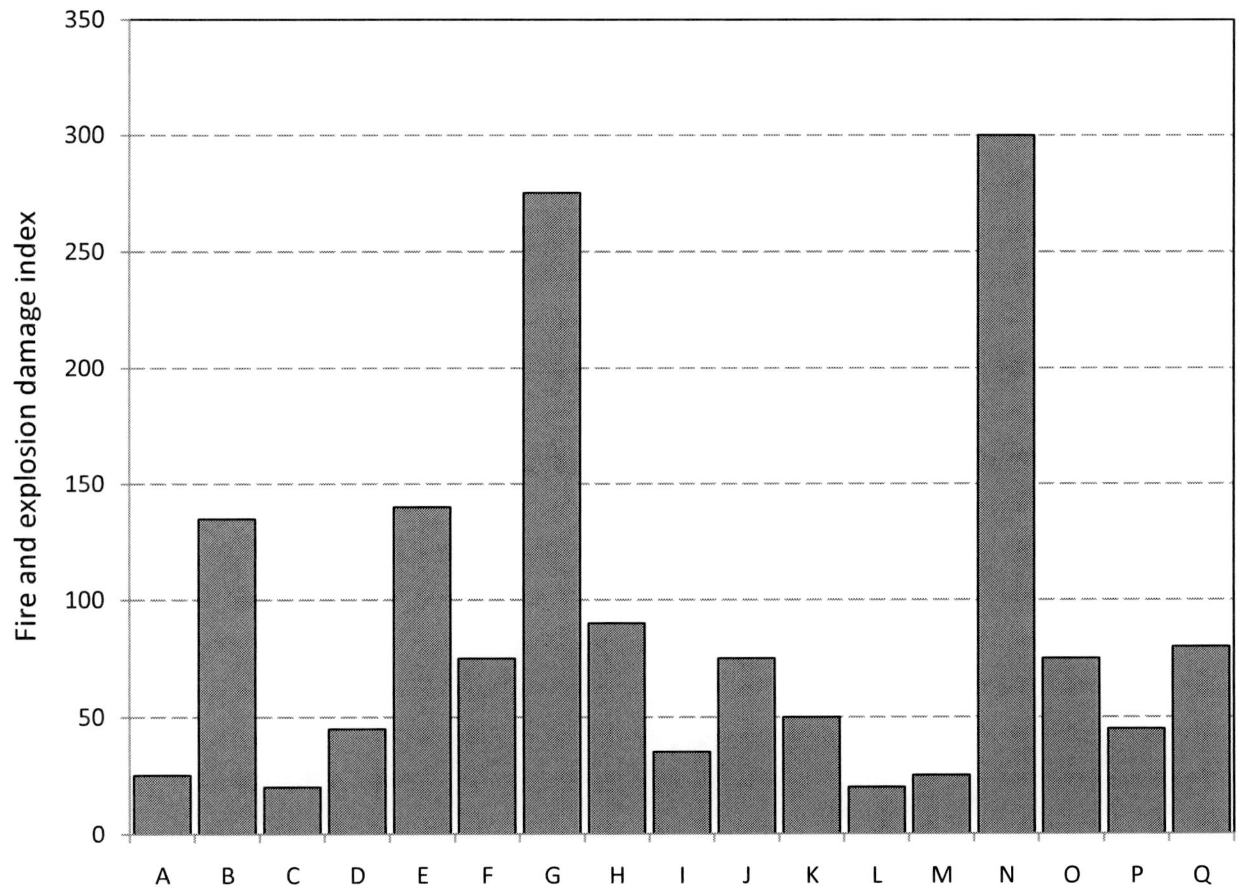

Figure 17.7. Fire and explosion damage index for various units in SPIC.

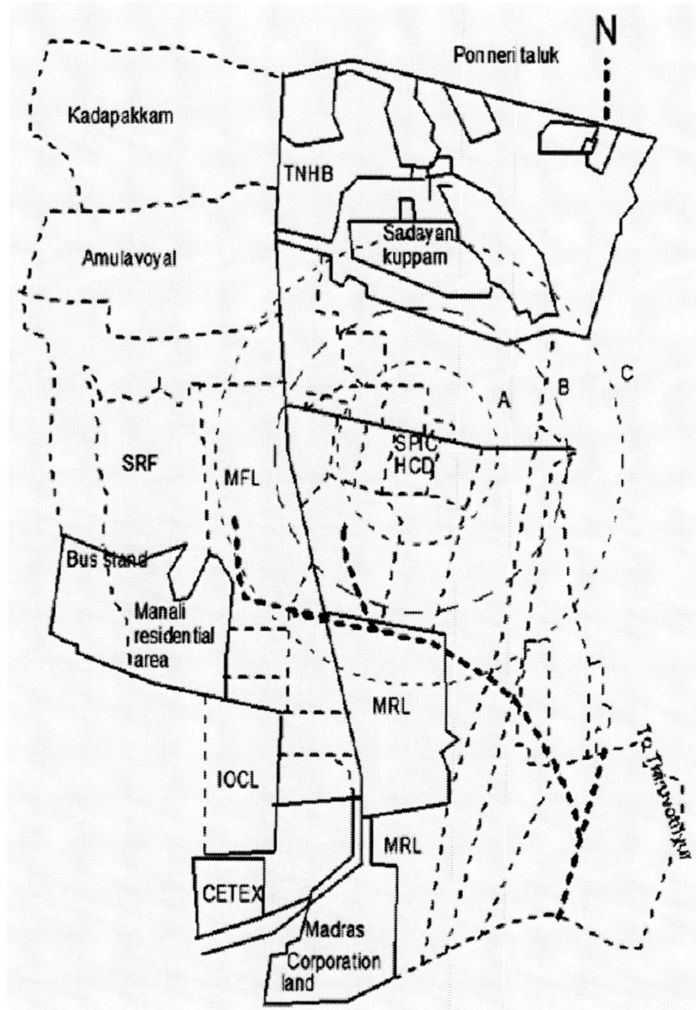

Figure 17.8. Risk contours for scenario 1 showing impact area due to severe (A), high (B) and moderate (C) risk.

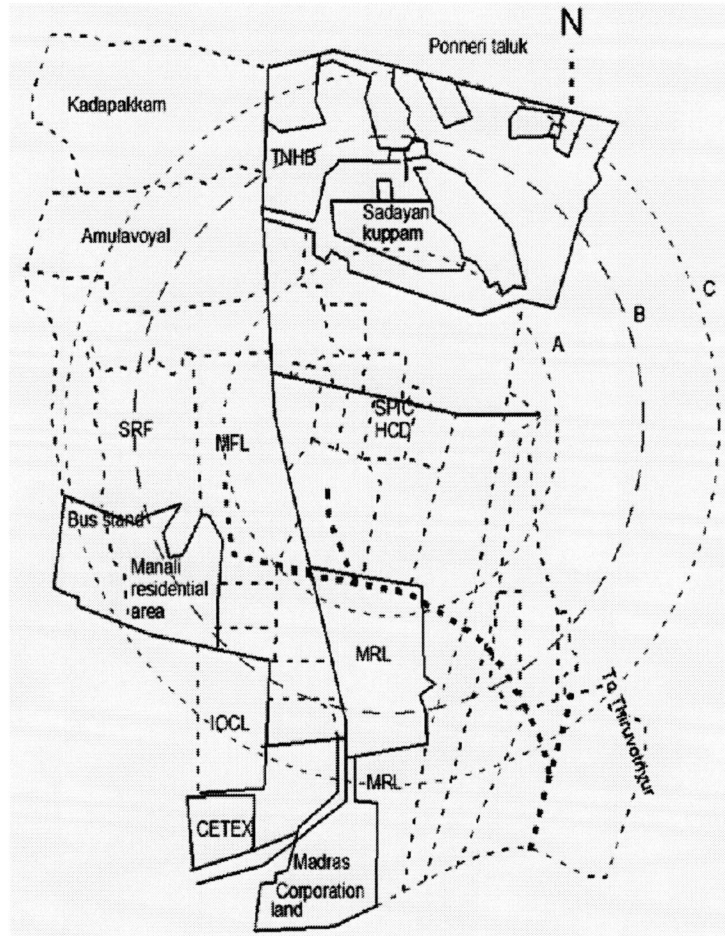

Figure 17.9. Risk contours for scenario 2 showing impact area due to severe (A), high (B) and moderate (C) risk.

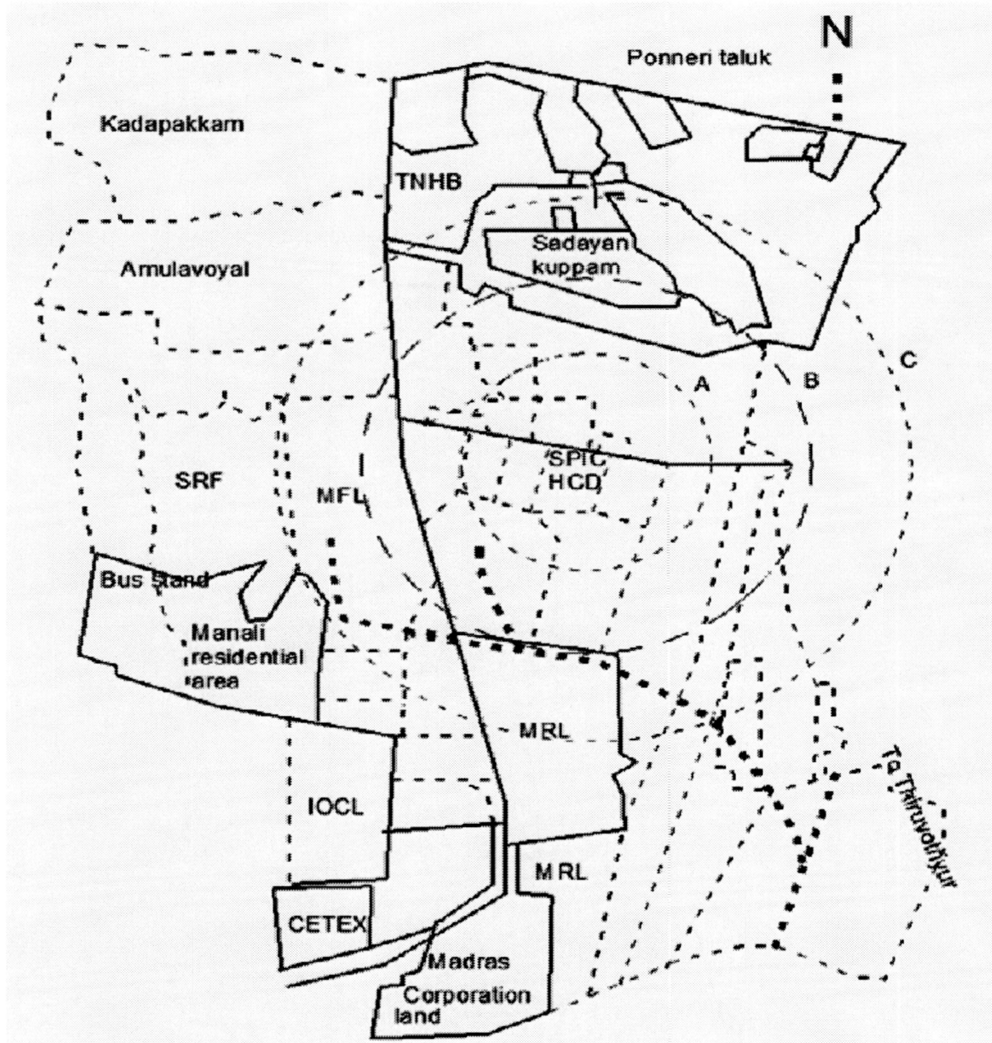

Figure 17.10. Risk contours for scenario 3 showing impact area due to severe (A), high (B) and moderate (C) risk.

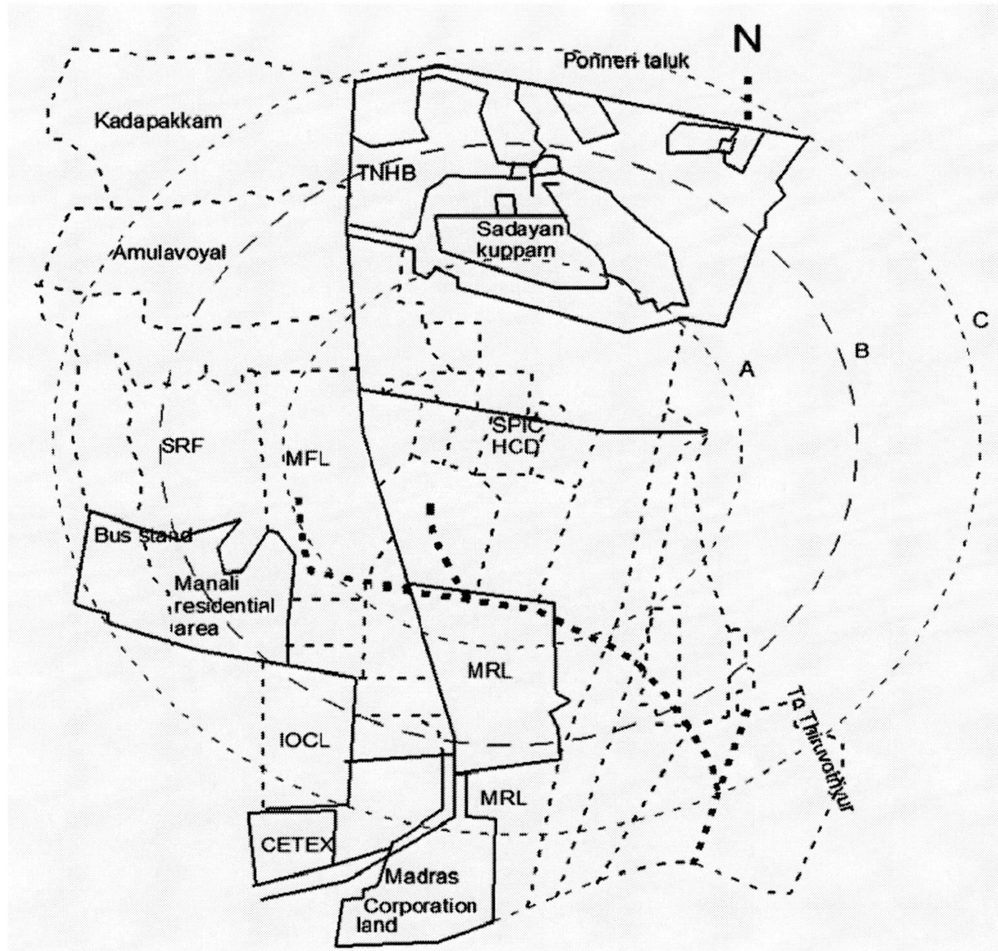

Figure 17.11. Risk contours for scenario 4 showing impact area due to severe (A), high (B) and moderate (C) risk.

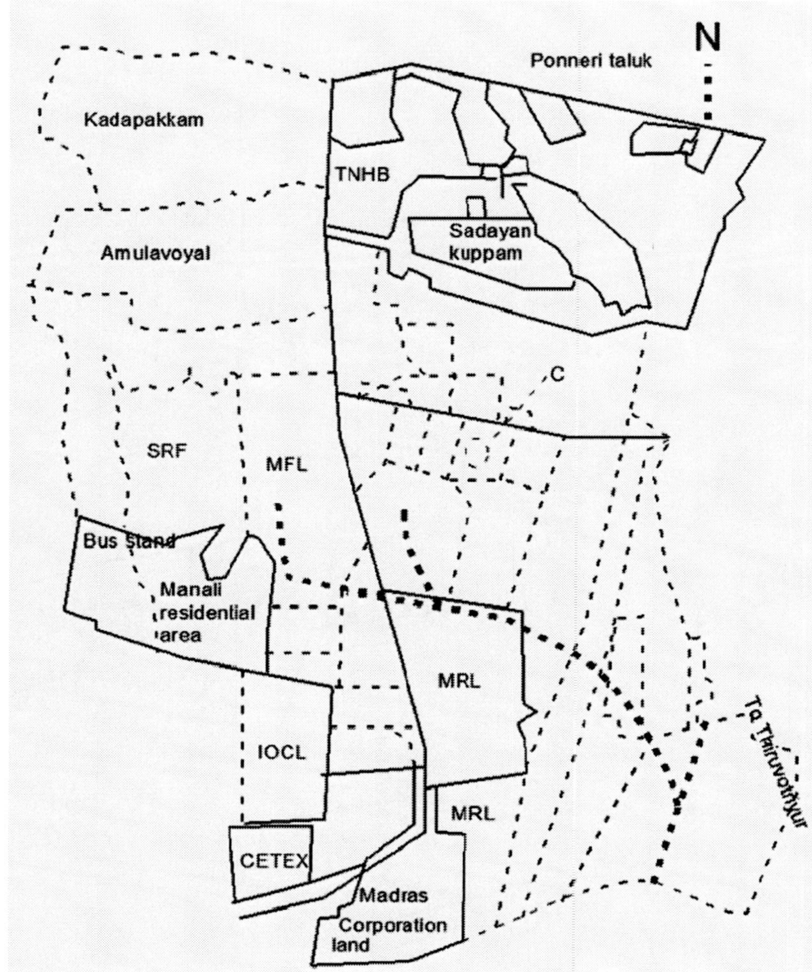

Figure 17.12. Risk contours for scenario 5 showing impact area due to high and moderate (C) risk.

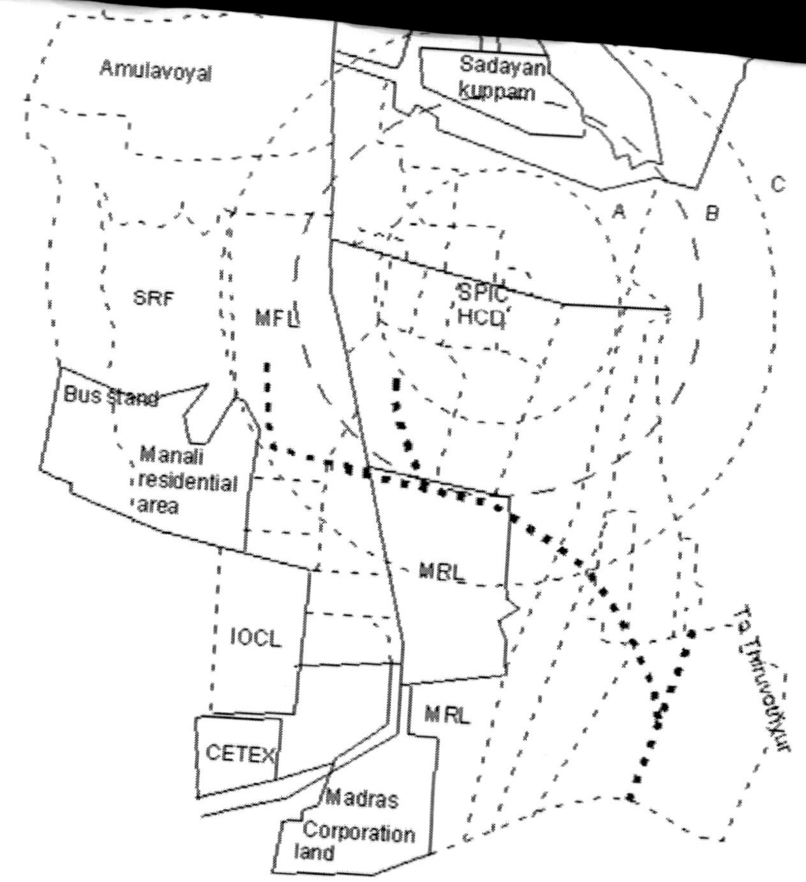

Figure 17.16. Risk countours for scenario 9 showing impact area due to severe (A), high (B) and moderate (C) risk.

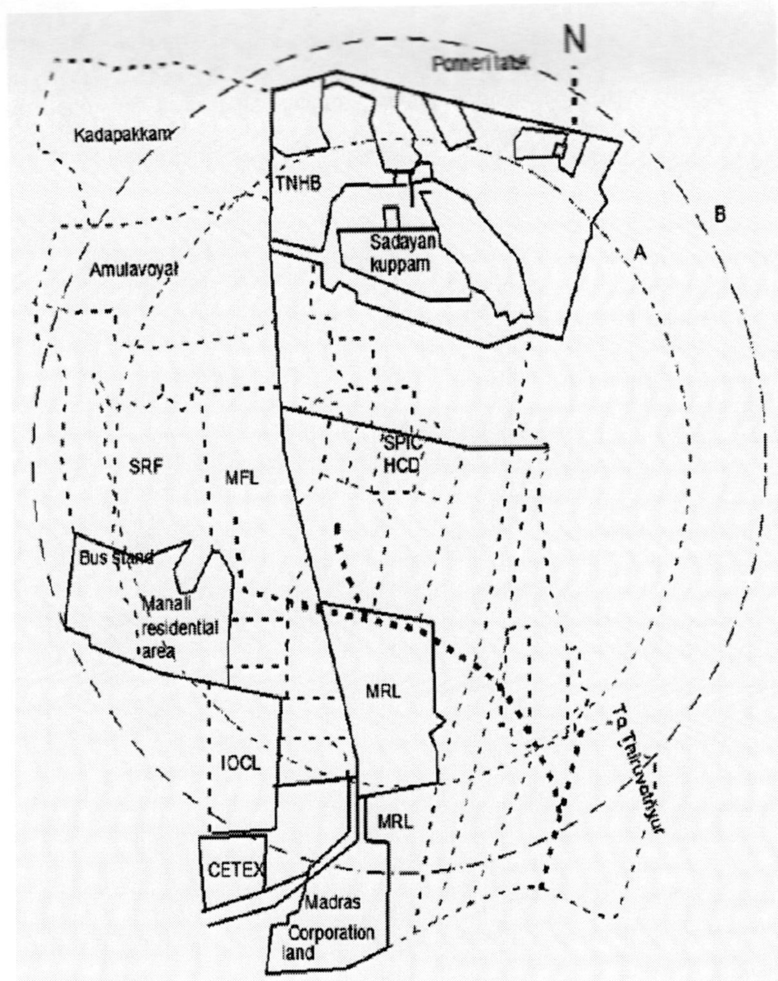

Figure 17.13. Risk contours for scenario 6 showing impact area due to severe (A), and high (B) risk.

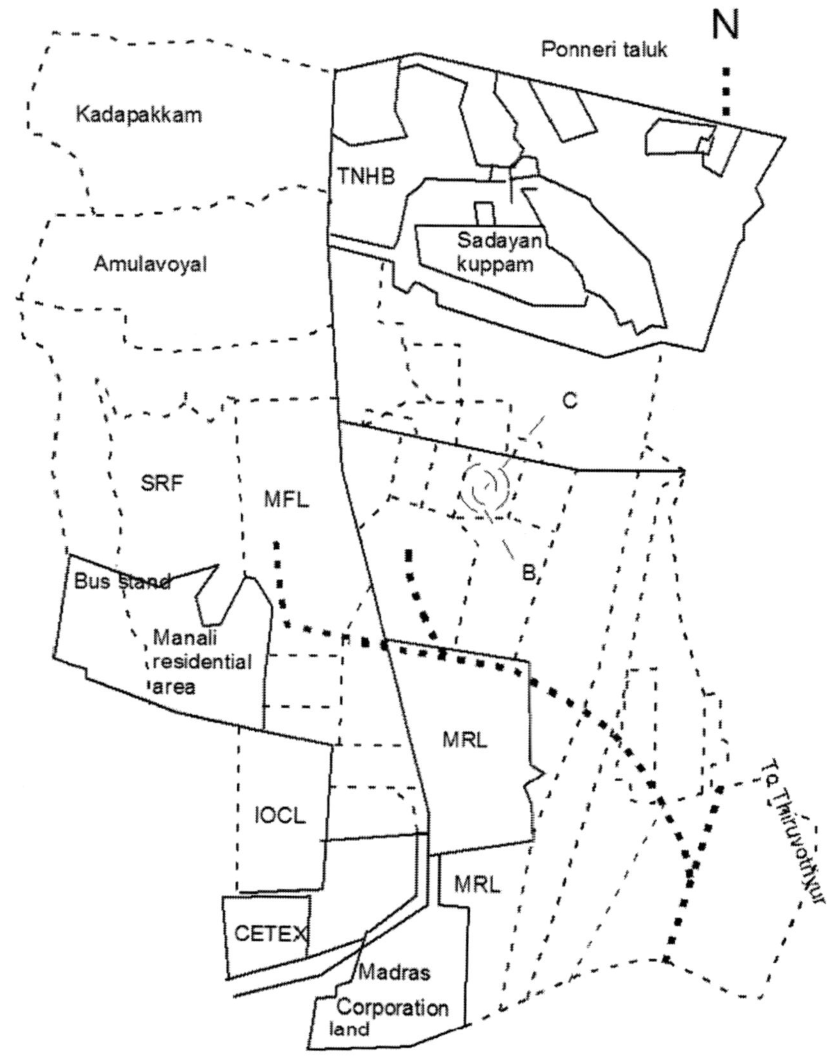

Figure 17.14. Risk contours for scenario 7 showing impact area due to high and moderate (C) risk.

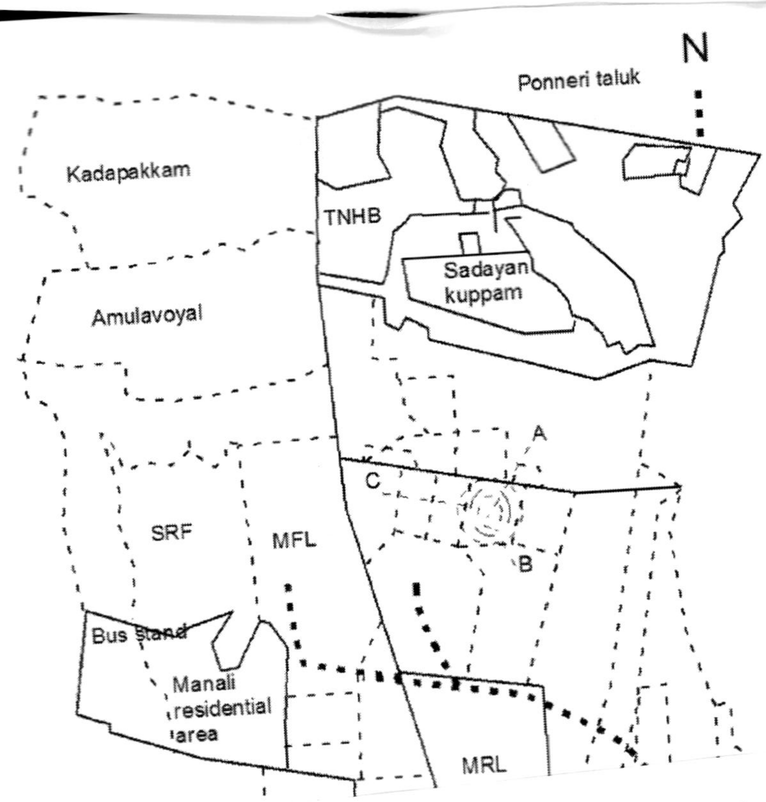

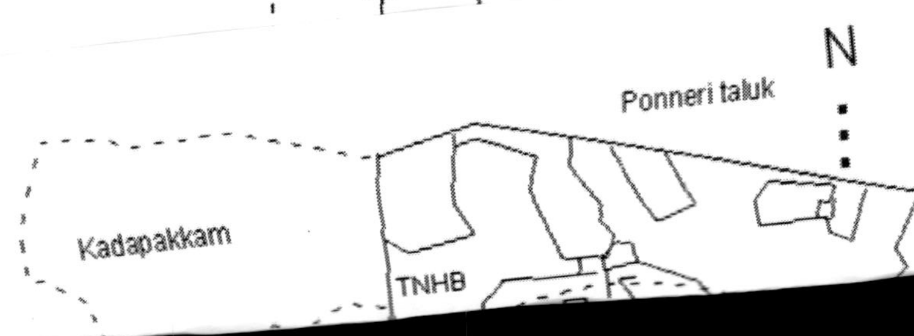

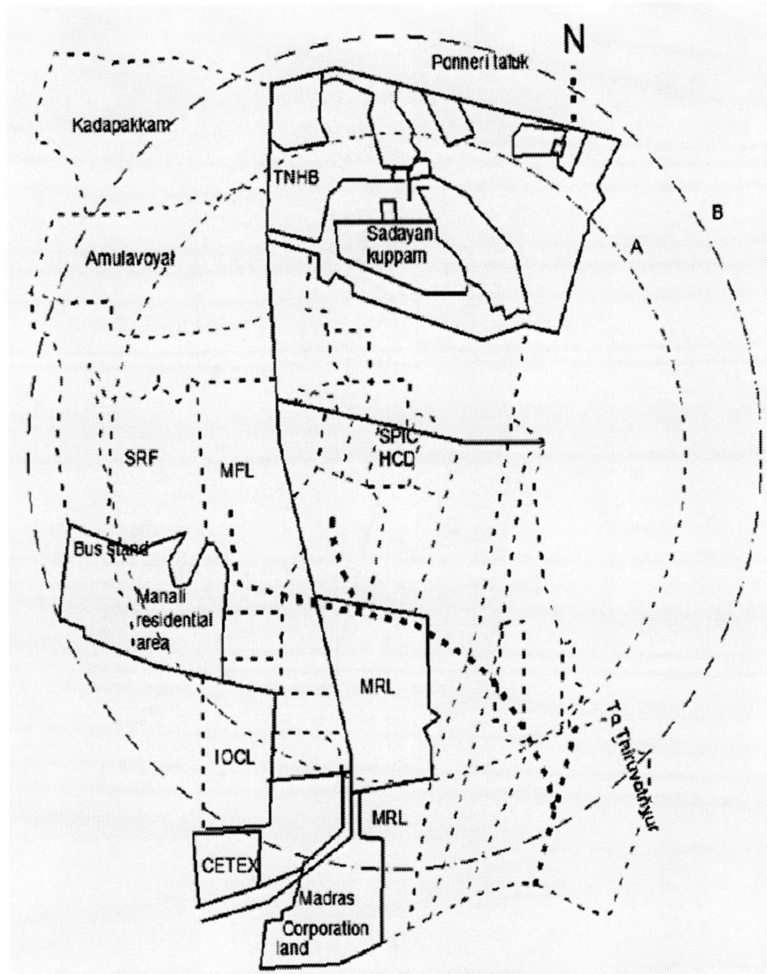

Figure 17.13. Risk contours for scenario 6 showing impact area due to severe (A), and high (B) risk.

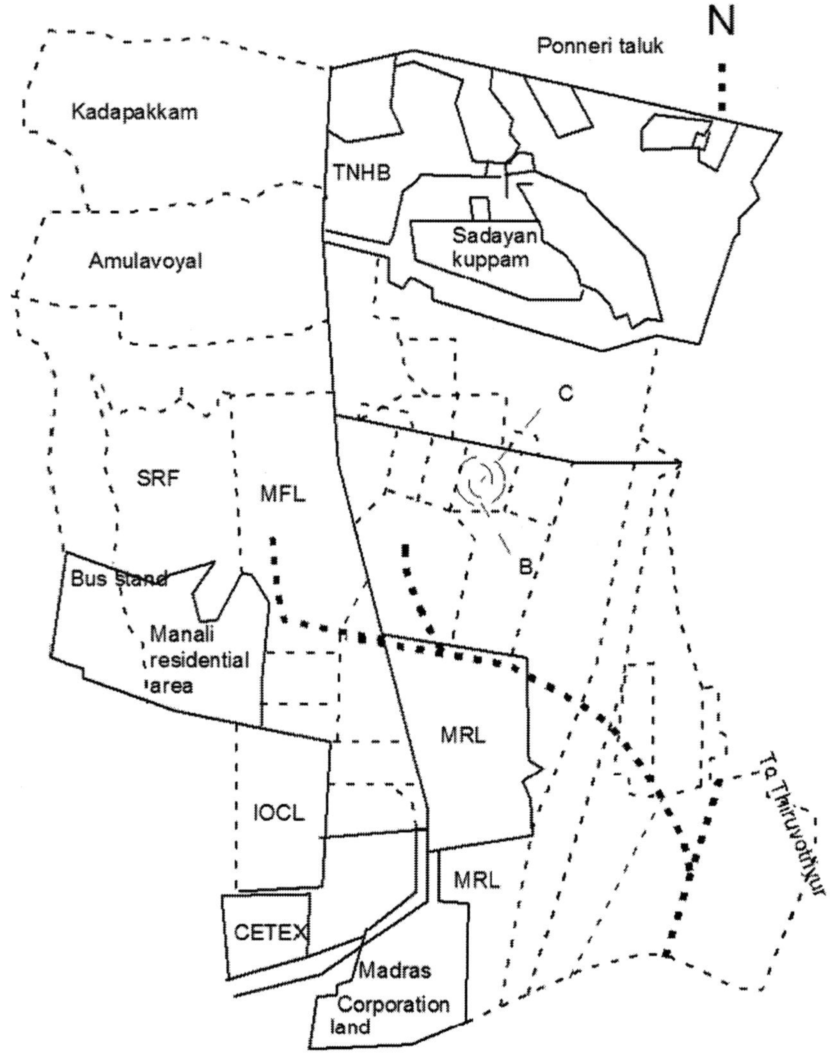

Figure 17.14. Risk contours for scenario 7 showing impact area due to high and moderate (C) risk.

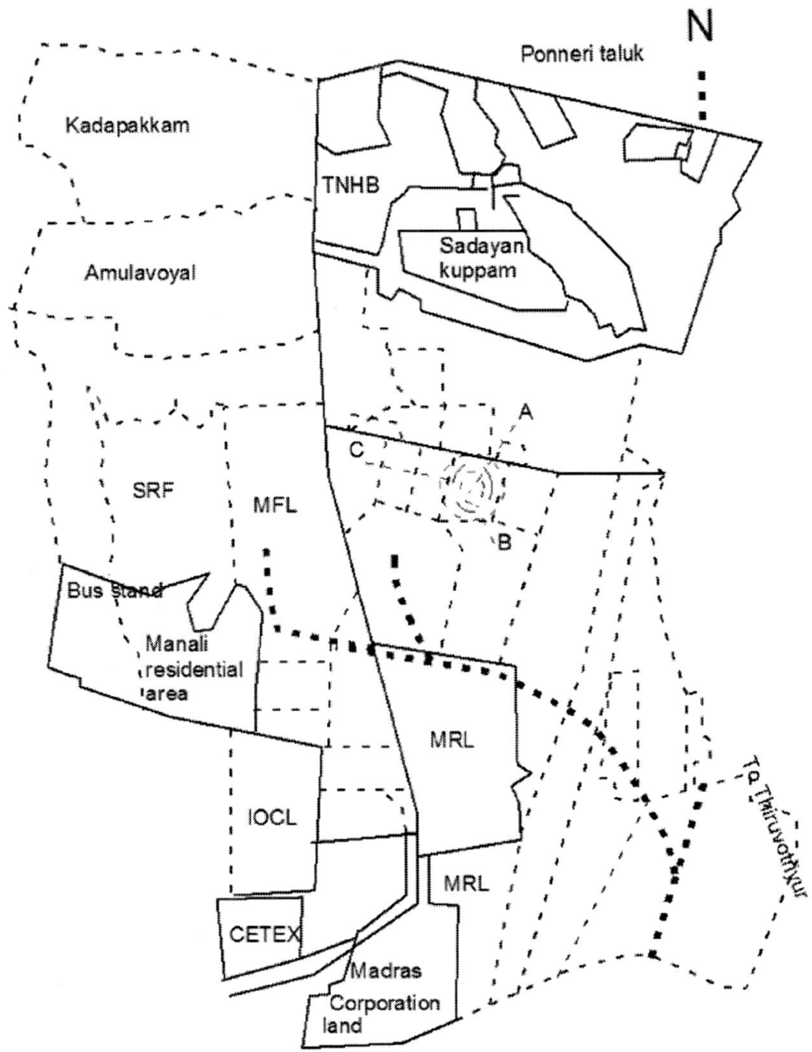

Figure 17.15. Risk contours for scenario 8 showing impact area due to severe (A), high (B) and moderate (C) risk.

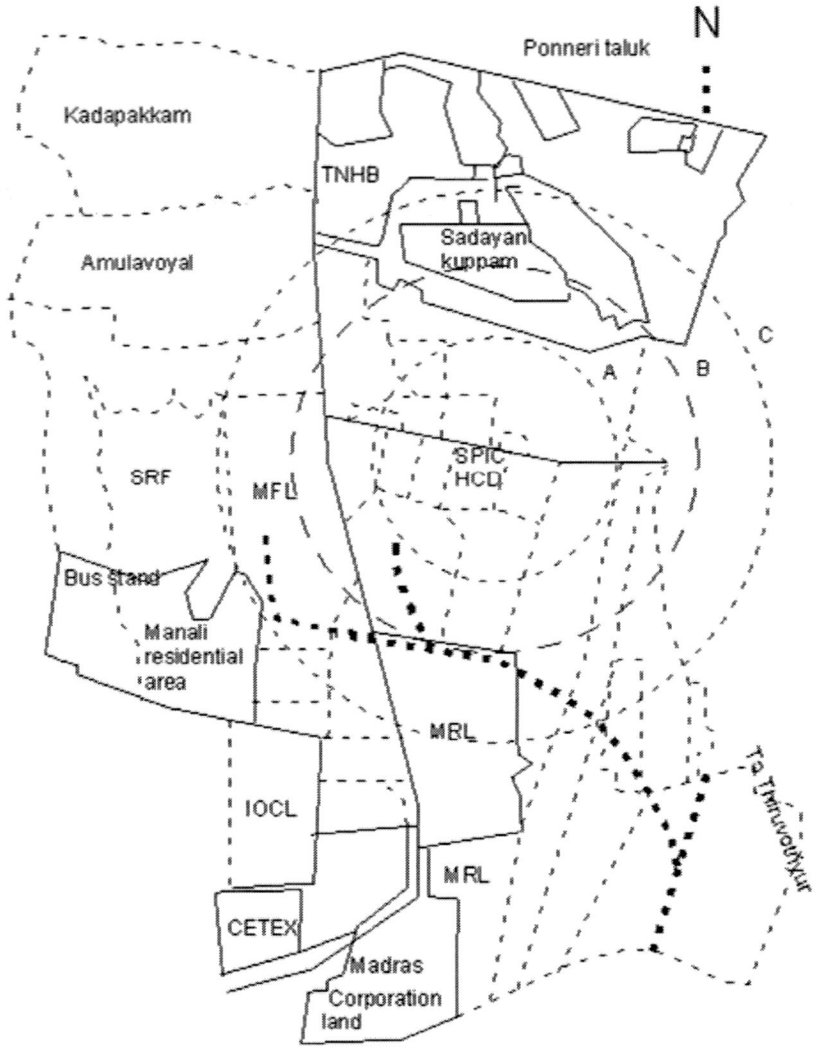

Figure 17.16. Risk countours for scenario 9 showing impact area due to severe (A), high (B) and moderate (C) risk.

Chapter 18

RAPID RISK ASSESSMENT OF SOL

18.1. INTRODUCTION

SOL (SPIC Organic Ltd) produces propylene oxide, propylene glycol, polyol, and DPG (di-propylene glycol). The raw materials for these include chlorine, propylene, pebble line, and limestone. The simplified process block diagram of the process are presented in Figures 18.1 to 18.3

18.2. HAZARD IDENTIFICATION AND RANKING

All the units of the plant (Table 18.1) were screened for their risk potential. The hazards associated with each unit were estimated by applying FEDI (Fire and Explosion Damage Index) and TDI (Toxic Damage Index). The FEDI values are presented in Figure 18.4. It is seen that 13 of the 18 units carry hazard reaching or exceeding the threshold level of 100.

18.3. DEVELOPMENT OF ACCIDENT SCENARIOS

Accident scenarios were developed for all the units found hazardous as per the ranking described above. They are presented below.

Propylene Chlorohydrin Storage: Scenario 1

An instantaneous release of propylene chlorohydrin stored under high pressure leads to boiling liquid expanding vapour explosion (BLEVE). The vapor cloud generated by BLEVE on ignition forms a fireball.

Propylene Oxide Reactor: Scenario 2

An excessive pressure development in the storage vessel causes confined vapor cloud explosion (CVCE). The vapor cloud generated by CVCE on ignition turns into a fireball and consequently damages the other storage vessels.

Reactor Holding Tank: Scenario 3

A continuous release of the contents (propylene oxide, ethylene oxide etc) under pressure (11 Kg/cm^2) leads to vapour cloud explosion and the remaining chemical in dike/vessel burns as pool fire.

Propylene Storage: Scenario 4

An instantaneous release of propylene either through vent valve or through any other accidental opening causes the vessel to explode in the manner of a BLEVE. As the released chemical is highly flammable it turns into a fire ball on meeting an ignition source.

Propylene Oxide Tank: Scenario 5

Pressure build-up in the storage vessel of propylene oxide (either due to absorption of external heat overpressurization or some runaway reactions) leads to the failure of the vessel, causing confined vapor cloud explosion. The released vapor cloud becomes a fireball after catching fire.

MPG Storage: Scenario 6

Instantaneous release is followed by pool fire in the monopropylene glycol (MPG) storage unit.

Chlorine Storage Vessel: Scenario 7

Continuous release of chlorine either from vent valve, pressure relief valve, or un-intended opening results in toxic dispersion.

Ethylene Oxide Storage: Scenario 8

An instantaneous release of ethylene oxide accompanies a BLEVE. The released chemical becomes a fireball on meeting an ignition source. In-spite of highly reactive chemical (ethylene oxide) the probability of vapor cloud or confined vapor cloud explosion is less because the quantity stored is not sufficient to cause that event. The vessel can fail only as BLEVE.

DPG Storage: Scenario 9

Formation of vapour cloud due to release of DPG from the vessel. The released chemical on ignition would cause either vapor cloud explosion (VCE) or flash fire As the quantity of chemical handled in the unit is low and the chemical is not highly flammable/reactive, there is greater probability of flash fire than VCE.

Solvent Recovery Unit: Scenario 10

There is a pool fire in the solvent recovery unit, which causes release and dispersion of chlorine and chlorinated organic.

PO Neutralization Unit: Scenario 11

Instantaneous release of chemical from the unit is followed by pool fire on ignition.

Propylene Glycol Reactor: Scenario 12

Leak of chemical cause formation of flammable vapor cloud which then gets ignited to a fire ball.

Polyol Reactor: Scenario 13

Excessive pressure development in the reactor causes release of chemical through vent valve or pressure relief valve (PRV). The released chemical on meeting an ignition source causes pool fire.

These scenarios have been processed for damage estimation through MAXCRED. A brief note on the damage-effect calculation models used for the detailed study is presented as follows.

18.4. Impact Forecasting: Consequence Assessment

The forecasts for *scenario* 1 are presented in Table 18.2. The shock wave and missiles generated by BLEVE may hit secondary targets and can lead to secondary explosions or toxic releases. The vapor cloud generated by BLEVE on ignition may cause a fireball, which would generate heat radiation effect. The shock wave and heat radiation can cause fatal injury as well as second order accidents by seriously damaging other vessels. It is clear from Table 18.2 that an area of ~25m radius has 50% probability of being damaged due to shock waves whereas an area of 100m radius may be affected due to severe heat load.

The confined vapor cloud explosion in propylene oxide reactor as per *scenario 2* would generate shock waves and missiles (Table 18.3) which may effect a wider area than likely as per *scenario 1*. The released chemicals propylene with other chemicals from the reactor would burn and a lethal heat load may be observed over an area of ~50m radius, due to the resultant fire ball.

The forecasts based on detailed calculations for *scenario 3* are presented in Table 18.4. This scenario (VCE followed by pool fire associated with propylene oxide release) would impact an area of ~100m radius with the propensity to cause lethality at 50% probability.

As per *scenario 4* BLEVE would generate shock waves. In addition, these would be secondary impact of the released material getting ignited and forming a fireball thereby generating additional heat load (Table 18.5). An area of ~125 m radius is likely to suffer damage.

The forecast for *scenario 5* (CVCE followed by fireball in propylene oxide storage tank) are presented in Table 18.6. Intense shock waves with 50% probability of causing damage would influence an area of more than 150m radius. Missiles of same severity may hit objects upto a distance of ~ 1990m. Heat load due to the fireball (50% probability of damage) would encompass an area of ~330m radius. This accident shall also have the probability of causing secondary accidents by damaging nearby storage/processing units.

The study of the consequences of *scenario 6* (release of MPG) reveals that the likely damage due to this event in terms of heat load would be less intense than forecast by *scenario 5* but higher than *scenarios 1-4*. However, it is evident from the forecasts based on detailed analysis (Table 18.7) that even at a distance ~235m away from the accident epicenter the intensity of heat load would be severe enough to cause lethal damage.

The forecasts for *scenario 7* (continuous release of chlorine followed by dispersions) are presented in Table 18.8. Toxic load having the potential to cause fatality at 50% probability would envelope an area of ~ 7300m radius.

Scenarios 8 for ethylene oxide storage vessel envisage BLEVE followed by fireball. The forecasts for this scenario are summarized in Table 18.9. The shock waves generated due to BLEVE (50% probability of causing damage) would be felt upto ~50m from the point of origin would hit objects with sufficient momentum to cause damage over larger distances. The vapor cloud generated due to release of ethylene as BLEVE on ignition would form a fireball. The resultant heat load shall extend over an area of 140m radius. As this area houses other storage and processing units, secondary and higher order accidents may also be caused.

The forecasts for *scenario 9* (dichloro propylene glycol) are presented in Table 18.10. Lethal heat lead would be observed over an area of ~20m radius.

The solvent recovery unit involves flammable as well as toxic chemicals. The release of the chemicals would cause toxicity as well as fire. The accident scenario developed for this unit is toxic release followed by pool fire. The impact of the pool fire would be observed over an area of ~60m radius while damage consequences for toxic effect (50% probability of causing lethality) would envelope an area of ~2400 m radius. The forecasts of this scenario are depicted in Table 18.11. The neutralization unit may involve instantaneous release of propylene oxide, forming a vapor cloud, as well as pool of liquid in the dike. The released chemicals on ignition would burn as pool fire. The forecasts of the analysis for this scenario are presented in Table 18.12. It is revealed that damage causing heat load would axially reach upto 90m from the accident point. The impacts of this scenario, too, may reach other units, which may fail and cause domino effect.

The accident scenario for propylene glycol is also developed as instantaneous flammable release of chemical, which on ignition turns into fireball. The forecasts based on detailed calculations of scenario are presented in Table 18.13. The accident may cause lethal heat load over an area of 90m radius.

The polyol reactor handles large quantities of flammable chemicals at extreme conditions of temperature and pressure. The accident scenario envisaged in this unit is pool fire (*scenario 13*). The results of damage calculation reveal that heat load would envelop an area of 64m radius (Table 18.14).

18.5. Risk Estimation

On the basis of the results presented above, and probability of occurrence of the various accident scenarios, risk factors have been computed. In order to enable these have been visualization of the accidents risk contours have been drown over the industry layout (Figures 18.5 to 18.17).

A total of thirteen accident scenarios in different units of SOL have been developed and studied. It is revealed that storage of propylene (*scenario* 4), propylene oxide (*Scenario* 5) and ethylene oxide (*scenario* 8) are extremely hazardous. The damaging impact of these scenarios would cover an area of more than 250 m radius. This area encompasses other storage units of SOL as well as neighboring industries. Thus, these scenarios are likely to trigger domino effect (chain of accidents) with more serious consequences.

Storage of chlorine (*Scenario* 7) and the solvent recovery unit poses substantial toxicity hazard. Lethal concentration sufficient to cause fatality would be observed over wider area (more than 1500m). However there is a low probability of causing secondary accidents due to these events.

It is concluded that as per the damage potential of a single event is concerned; *scenario* 7 (accident in the storage of Chlorine) is the worst whereas if one considers the likelihood of domino effect, then *scenario* 4, 5 and 8 forecast the most disastrous consequences.

18.6. Recommendations

To reduce the hazards associated with the storage and process units, proper hazard minimization/mitigation measures should be taken. A few suggestions are made in this context:

I. Instead of one or two large-capacity vessels several vessels of smaller capacity should be used for storage.

II. Adequate space should be kept between the storage vessels and buffers provided between them so that adverse consequences of failure in one of them do not cause second or higher order accidents.

III. Sensitive gas-detecting devices for flammable and toxic gases should be installed in the premises of storage area and other units.

IV. There should be regular and thorough inspections of electronic control equipment followed by meticulous maintenance.

V. Sufficient quantities of inert gases should be readily available to dilute the concentrations of toxic/flammable gases if they escape to the atmosphere, and to control fire.

A thorough emergency preparedness strategy should always be kept in position, fortified by periodic drills or 'dry runs' so that the damage is contained if an accident does occur.

Table 18.1. Different plants of UB Petroleum Products industry

Plant Process units	Legends used in Figure 18.4
Propylene chlorohydrin storage	A
Propylene oxide reactor	B
Reactor holding tank	C
Propylene storage tank	D
Propylene oxide storage tank	E
MPG storage tank	F
Chlorine storage tank	G
Ethylene Oxide storage tank	H
Nitrogen storage tank	I
DPG storage	J
Solvent recovery unit	K
PO neturalization tank	L
Dehydrochlorination tank	M
PG distillation unit	N
Propylene glycol reactor	O
Polyol reactor	P
Stripping unit	Q
PO distillation unit	R

Table 18.2. Result of consequence analysis for scenario 1; accident in Propylene chlorohydrin storage unit

Parameters	Values
Explosion : BLEVE	
Total energy released	(kJ) :4.1E+06
Peak over pressure	(kPa) :150.00
Variation of over pressure in air	(kPa/s) :99.00
Shock velocity of air	(m/s) :472.00
Duration of shock wave	(ms) :34.0
Missile characteristics	
Initial velocity	(m/s) :19.00
Kinetic energy of fragment	(kJ) :8.66E+02
Parameters	Values
Fragment velocity at study point	(m/s) :19.00
Penetration ability at study point (based on empirical models)	

Concrete structure	(m) :0.0027
Brick structure	(m) :0.0034
Steel structure	(m) :0.0019
Damage radii (DR) for various degree of damage due to over pressure	
DR for 100% complete damage	(m) :17
DR for 100% fatality or 50% complete damage	(m) :25
DR for 50% fatality or 25% complete damage	(m) :37
Damage radii (DR) for various degree of damage due to missile	
DR for 100% damage or 100% fatality	(m) :1288
DR for 50% damage or 100% fatality	(m) :1404
DR for 100% fatality or 10% damage	(m) :1507
Fire: Fire Ball Radius of the fire ball	(m) :76.00
Duration of the fire ball	(s) :31.00
Energy released by fire ball	(kJ) :2.33E+08
Radiation heat flux	(kJ/sq.m) :88867.00
Damage Radii (DR) due to thermal load	(m) :91
DR for 100% fatality/damage	
DR for 50% fatality/damage	(m) :114
DR for 100% third degree of burn	(m) :131
DR for 50% third degree of burn	(m) :169

Table 18.3. Result of consequence analysis for scenario 2; accident in Propylene Oxide reactor

Parameters	Values
Explosion : CVCE	
Total energy released	(kJ) :8.12E+07
Peak over pressure	(kPa) :403.00
Variation of over pressure in air	(kPa/s) :293.00
Shock velocity of air	(m/s) :645.00
Duration of shock wave	(ms) :44.0
Missile characteristics	
Initial velocity of fragment	(m/s) :58.00
Kinetic energy of fragment	(kJ) :8.36E+03
Fragment velocity at study point	(m/s) :57.00
Penetration ability at study point (based on empirical models)	
Concrete structure	(m):0.0146
Brick structure	(m):0.0187
Steel structure	(m):0.0058
Damage radii (DR) for various degree of damage due to over pressure	
DR for 100% complete damage	(m) :30
DR for 100% fatality or 50% complete damage	(m) :46
DR for 50% fatality or 25% complete damage	(m) :67
Damage radii (DR) for various degree of damage due to missile	
DR for 100% damage or 100% fatality	(m) :1832
DR for 50% damage or 100% fatality	(m) :1948
DR for 100% fatality or 10% damage	(m) :2052

Table 18.3. (Continued)

Parameters	Values
Fire: Fire Ball	
Radius of the fire ball	(m):44.00
Duration of the fire ball	(s):18.00
Energy released by fire ball	(kJ):5.06E+07
Radiation heat flux	(kJ/sq.m):19329.00
Damage Radii (DR) due to thermal load	
DR for 100% fatality/damage	(m) :49
DR for 50% fatality/damage	(m) :63
DR for 100% third degree of burn	(m) :75
DR for 50% third degree of burn	(m) :89

Table 18.4. Result of consequence analysis for scenario 3; accident in Reactor holding tank

Parameters	Values
Explosion : VCE	
Total energy released by explosion	(kJ) :1.28E+05
Peak over pressure	(kPa) :7.00
Variation of over pressure in air	(kPa/s) :2.00
Shock velocity of air	(m/s) :338.00
Duration of shock wave	(ms) :11.0
No Missile effect:	
Damage radii (DR) for various degree of damage due to over pressure	
DR for 100% complete damage	(m) :30
DR for 100% fatality or 50% complete damage	(m) :45
DR for 50 % fatality or 25% complete damage	(m) :75
FIRE: POOL FIRE	
Continuous Model	
Burning area	(sq.m) :100448.00
Burning rate	(kg/s) :7.00
Heat flux	(kJ/sq.m) :87.00
Damage Radii (DR) due to thermal load	
DR for 100% fatality/damage	(m) :65
DR for 50% fatality/damage	(m) :97
DR for 100% third degree of burn	(m) :105
DR for 50% third degree of burn	(m) :145

Table 18.5. Result of consequence analysis for scenario 4; accident in Propylene storage vessel

Parameters	Values
Explosion : BLEVE	
Total energy released	(kJ) :1.32E+07
Peak over pressure	(kPa) :1745.4
Variation of over pressure in air	(kPa/s) :814.1
Shock velocity of air	(m/s) :1245.2
Duration of shock wave	(ms) :21.4

Parameters	Values
Missile characteristics	
Initial velocity	(m/s) :445.3
Kinetic energy of fragment	(kJ) :1.66E+06
Fragment velocity at study point	(m/s) :345.4
Penetration ability at study point (based on empirical models Concrete structure	(m) :0.50
Brick structure	(m) :0.68
Steel structure	(m) :0.10
Damage radii (DR) for various degree of damage due to over pressure	
DR for 100% complete damage	(m) :245
DR for 100% fatality or 50% complete damage	(m) :325
DR for 50% fatality or 25% complete damage	(m) :510
Damage radii (DR) for various degree of damage due to missile	
DR for 100% damage or 100% fatality	(m) :2029
DR for 50% damage or 100% fatality	(m) :2144
DR for 100% fatality or 10% damage	(m) :2248
Fire: Fire Ball	
Radius of the fire ball	(m) :97.3
Duration of the fire ball	(s) :39.7
Energy released by fire ball	(KJ) :9.06E+07
Radiation heat flux	(KJ/sq.m) :11384.7
Damage Radii (DR) due to thermal load	
DR for 100% fatality/damage	(m) :182
DR for 50% fatality/damage	(m) : 228
DR for 100% third degree of burn	(m) :263
DR for 50 % third degree of burn	(m) :338

Table 18.6. Result of consequence analysis for scenario 5; accident in Propylene oxide storage tank

Parameters	Values
Explosion ; CVCE	
Energy released during explosion	(kJ) :5.84E+06
Peak over pressure	(kPa) :1295.3
Variation of over pressure in air	(kPa/s) :540.2
Shock velocity of air	(m/s) :1082.6
Duration of shock wave	(ms) :14.6
Missile characteristics	
Initial velocity	(m/s) :214.9
Kinetic energy of fragment	(kJ) :115458.6
Fragment velocity at study point	(m/s) :135.7
Penetration ability at study point (based on empirical models	
Concrete structure	(m) :0.09
Brick structure	(m) :0.012
Steel structure	(m) :0.02
Damage radii (DR) for various degree of damage due to over pressure	
DR for 100% complete damage	(m) :102
DR for 100% fatality or 50% complete damage	(m) :155
DR for 50% fatality or 25% complete damage	(m) :229

Table 18.6. (Continued)

Parameters	Values
Damage radii (DR) for various degree of damage due to missile	
DR for 100% damage or 100% fatality	(m) :1832
DR for 50% damage or 100% fatality	(m) :1948
DR for 100% fatality or 10% damage	(m) :2052
Fire: Fire Ball	
Radius of the fire ball	(m) :145.00
Duration of the fire ball	(s) :59.00
Energy released by fire ball	(kJ) :2.02E+09
Radiation heat flux	(kJ/sq.m) :77314.00
Damage Radii (DR) due to thermal load	
DR for 100% fatality/damage	(m) :268
DR for 50% fatality/damage	(m) :335
DR for 100% third degree of burn	(m) :387
DR for 50 % third degree of burn	(m) :497

Table 18.7. Result of consequence analysis for scenario 6; accident in MPG storage vessel

Parameters	Values
FIRE: POOL FIRE	
Instantaneous Model	
Radius of the pool fire	(m) :5.00
Burning area	(sq.m) :79.00
Burning rate	(kg/s) :7.00
Heat flux	(kJ/sq.m) :383583.00
Damage Radii (DR) due to thermal load	
DR for 100% fatality/damage	(m) :189
DR for 50% fatality/damage	(m) :236
DR for 100% third degree of burn	(m) :273
DR for 50% third degree of burn	(m) :350

Table 18.8. Result of consequence analysis for scenario 7; accident in chlorine Storage vessel

Parameters Values	
Toxic Release & Dispersion	
PG Continuous : Model	
Ground level source concentration	(kg/cu.m) :0.0001
Damage radii (DR) for various degree of damage	
DR for 100% lethality	(m) :5151
DR for 50% lethality	(m) :7309
DR for 10% lethality	(m) :16477
Plume characteristics	
Ground level concentration of plume on axis	(kg/cu.m) : 0.0146
Width of plume	(m) :60.9
Cloud concentration	(kg/cu.m) : 0.0129
Maximum ground level concentration	(kg/cu.m) : 0.0418

Table 18.9. Result of consequence analysis for Scenario 8; accident in Ethylene Oxide storage unit

Parameters	Values
Explosion ; BLEVE	
Total energy released	(kJ) :3.33E+07
Peak over pressure	(kPa) :1904.7
Variation of over pressure in air	(kPa/s) :1010.5
Shock velocity of air	(m/s) :1705.2
Duration of shock wave	(ms) :21.5
Missile characteristics	
Initial velocity	(m/s) :665.4
Kinetic energy of fragment	(kJ) :5.90E+06
Fragment velocity at study point	(m/s) :1415.2
Penetration ability at study point (based on empirical models	
Concrete structure	(m) :0.67
Brick structure	(m) :0.80
Steel structure	(m) :0.15
Damage radii (DR) for various degree of damage due to over pressure	
DR for 100% complete damage	(m) :267
DR for 100% fatality or 50% complete damage	(m) :394
DR for 50% fatality or 25% complete damage	(m) :502
Damage radii (DR) for various degree of damage due to missile	
DR for 100% damage or 100% fatality	(m) :1855
DR for 50% damage or 100% fatality	(m) :1971
DR for 100% fatality or 10% damage	(m) :2074
Fire: Fire Ball	
Radius of the fire ball	(m) :92.00
Duration of the fire ball	(s) :38.00
Energy released by fire ball	(KJ) :3.82E+08
Radiation heat flux	(KJ/sq.m) :14538.00
Damage Radii (DR) due to thermal load	
DR for 100% fatality/damage	(m) :117
DR for 50% fatality/damage	(m) :146
DR for 100% third degree of burn	(m) : 168
DR for 50 % third degree of burn	(m) :216

Table 18.10. Result of consequence analysis for scenario 9; accident in DPG storage vessel

Parameters	Values
Fire: Flash Fire	
Volume of vapour cloud	(cub.m) :134.00
Effective time of fire	(sec) :12942.00
Effective thermal load	(kJ/sq.m) :3463.00
Damage Radii (DR) due to thermal load	
DR for 100% fatality/damage	(m) :18
DR for 50% fatality/damage	(m) :23
DR for 100% third degree of burn	(m) :26
DR for 50 % third degree of burn	(m) : 34

Table 18.11. Result of consequence analysis of scenario 10; accident in Solvent recovery unit

Parameters	Values
Toxic Release & Dispersion	
Heavy Gas Dispersion characteristics	
Box Instantaneous: Model	
Concentration at cloud axis	(kg/cu.m) :4.33E+01
Value of source height	(m) :15.0000
Damage radii (DR) for various degree of damage	
DR for 100% lethality	(m) :1871
DR for 50% lethality	(m) :2405
DR for 10% lethality	(m) :4308
Puff characteristics	
Puff concentration at centre of cloud	(kg/cu.m) :1.16E+00
Concentration at cloud edges	(kg/cu.m) :2.41E+00
Distance along downwind	(m) :10.0000
FIRE: POOL FIRE	
Instantaneous Model	
Radios of the pool fire	(m) :5.00
Burning area	(sq.m) :79.00
Burning rate	(kg/s) :6.00
Heat flux	(kJ/sq.m) :25650.00
Damage Radii (DR) due to thermal load	
DR for 100% fatality/damage	(m) :49
DR for 50% fatality/damage	(m) :61
DR for 100% third degree of burn	(m) :71
DR for 50% third degree of burn	(m) :91

Table 18.12. Result of consequence analysis for scenario 11; accident in PO neuterilization unit

Parameters	Values
FIRE: POOL FIRE	
Instantaneous Model	
Radius of the pool fire	(m) :5.00
Burning area	(sq.m) :79.00
Burning rate	(kg/s) :8.00
Heat flux	(kJ/sq.m) :66155.00
Damage Radii (DR) due to thermal load	
DR for 100% fatality/damage	(m) :79
DR for 50% fatality/damage	(m) :98
DR for 100% third degree of burn DR for 50% third degree of burn	(m) :113 (m) :146

Table 18.13. Result of consequence analysis for scenario 12; accident in Propylene glycol reactor

Parameters	Values
Fire: Fire Ball	
Radius of the fire ball	(m) :53.00
Duration of the fire ball	(s) :22.00
Energy released by fire ball	(KJ) :8.23E+07
Radiation heat flux KJ/sq.m) :	31421.00

Parameters	Values
Damage Radii (DR) due to thermal load	
DR for 100% fatality/damage	(m) :54
DR for 50% fatality/damage	(m) :68
DR for 100% third degree of burn	(m) :78
DR for 50 % third degree of burn	(m) :101

Table 18.14. Result of consequence analysis for scenario 13; accident in Polyol reactor

Parameters Values	
FIRE: POOL FIRE	
Continuous Model	
Burning area	(sq.m) :98980.00
Burning rate	(kg/s) :8.00
Heat flux	(kJ/sq.m) :607.00
Damage Radii (DR) due to thermal load	
DR for 100% fatality/damage	(m) :34
DR for 50% fatality/damage	(m) : 64
DR for 100% third degree of burn DR for 50% third degree of burn	(m) :95 (m) :106

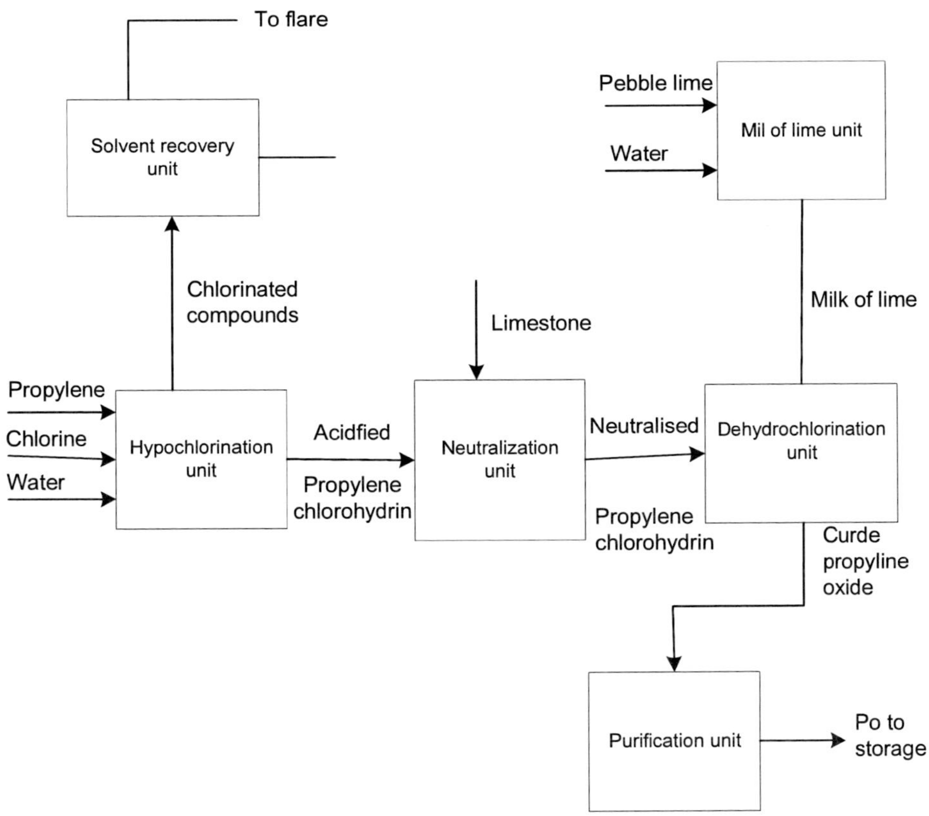

Figure 18.1. Block diagram for propylene oxide plant.

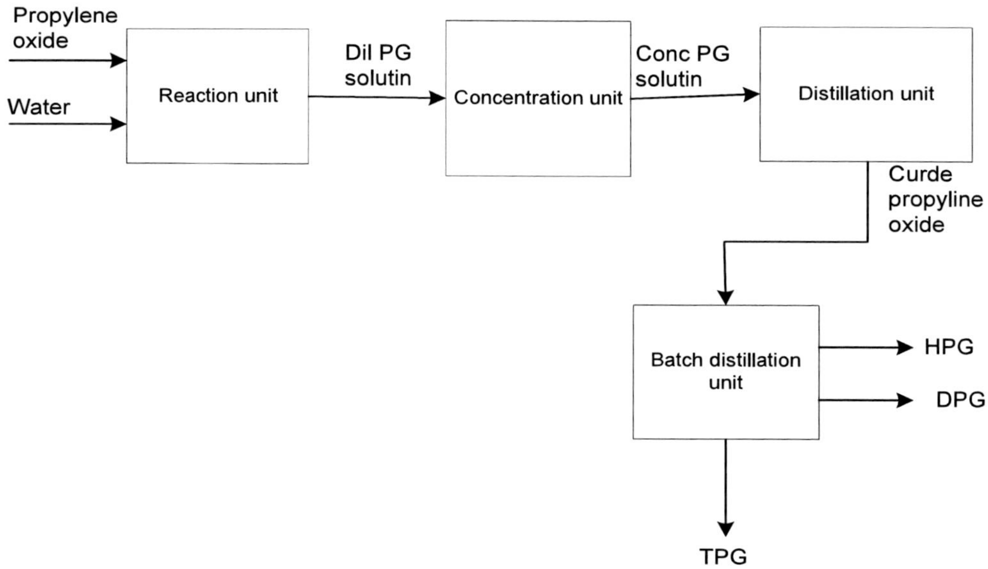

Figure 18.2. Block diagram for propylene glycol plant.

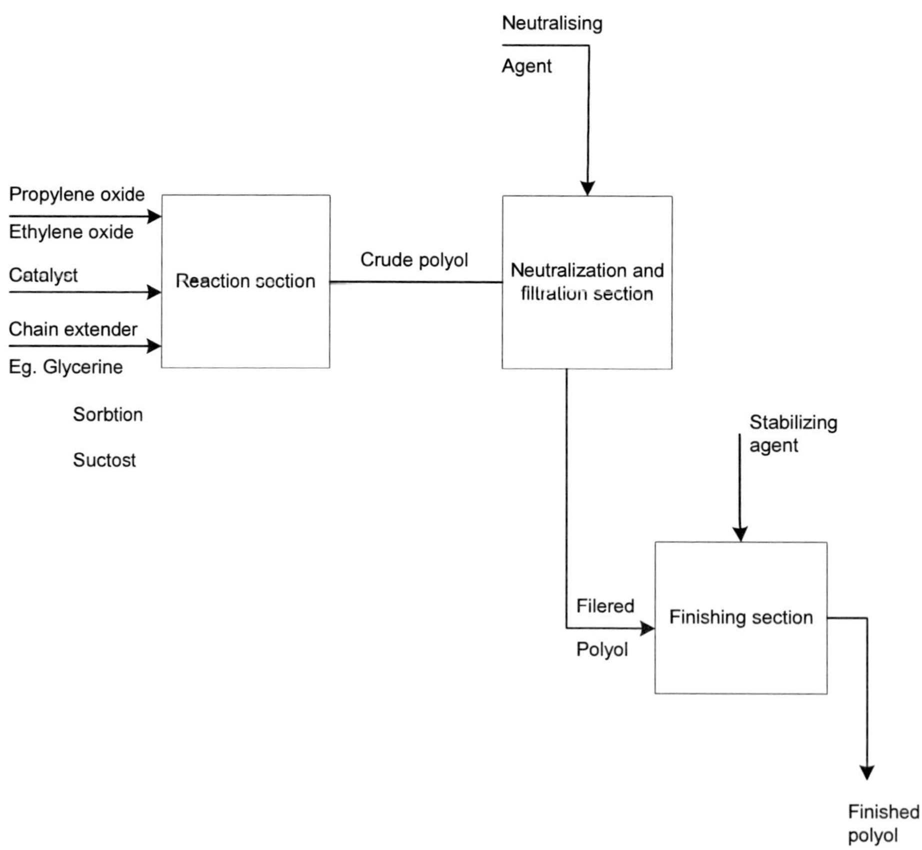

Figure 18.3. Block diagram for polyol plant.

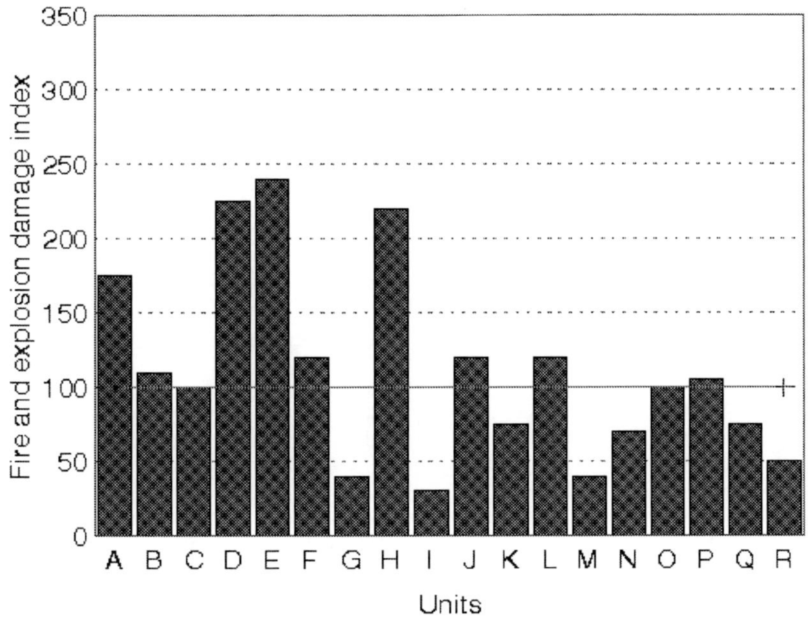

Figure 18.4. Fire and explosion damage index values for various units in UBP (cf Table 14).

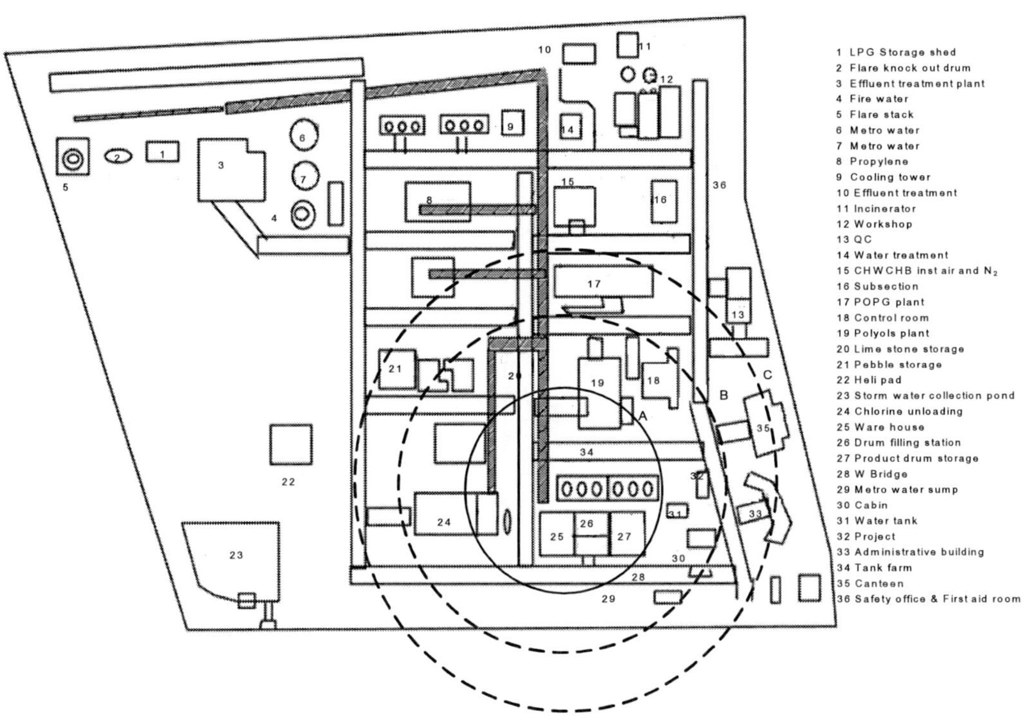

Figure 18.5. Risk contours for scenario 1 showing impact area due to severe (A), high (B), moderate (C) risk.

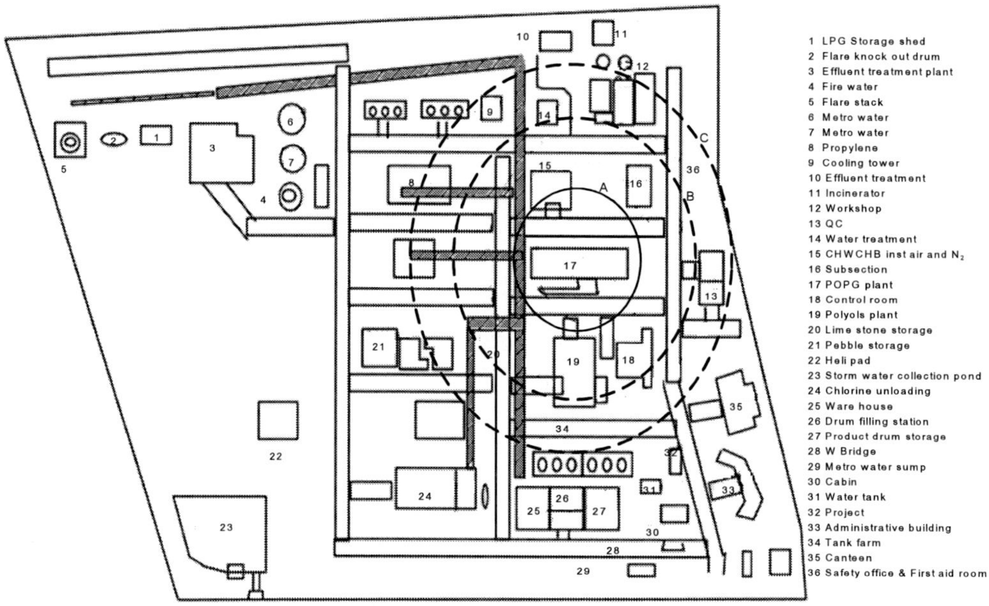

Figure 18.6. Risk contours for scenario 2 showing impact area due to severe (A), high (B), and moderate (C) risk.

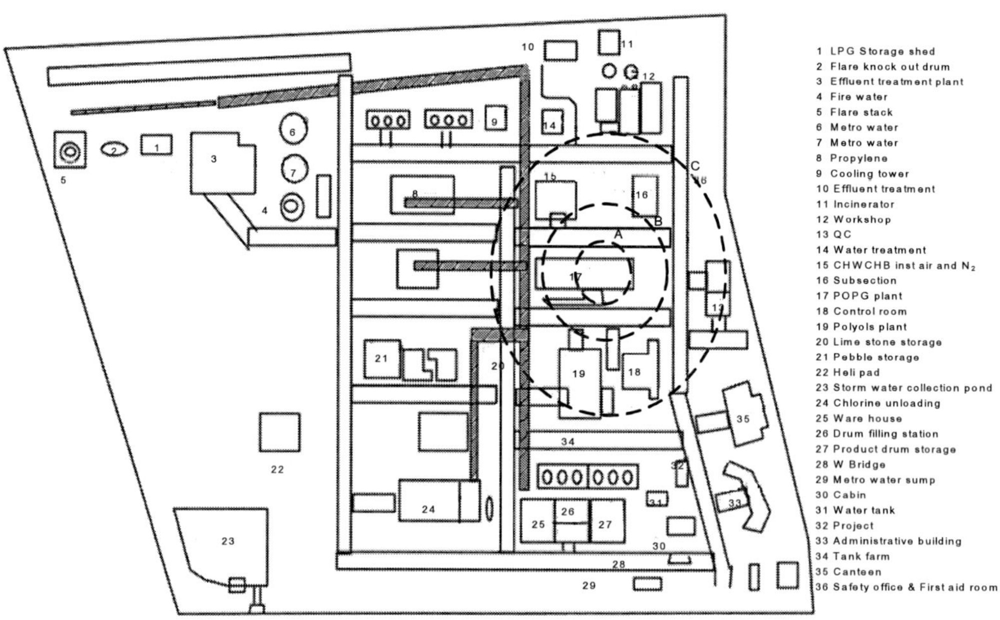

Figure 18.7. Risk contours for scenario 3 showing impact area due to severe (A), high (B), and moderate (C) risk.

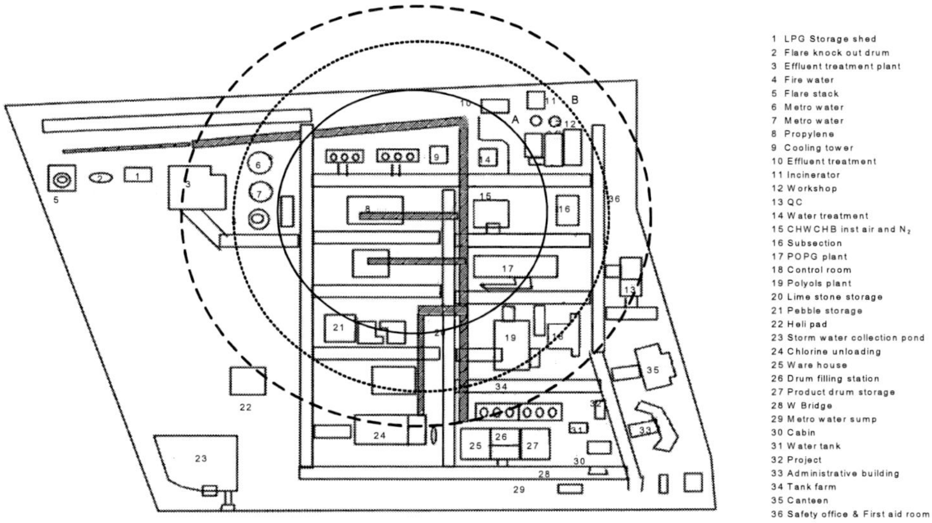

Figure 18.8. Risk contours for scenario 4 showing impact area due to severe (A), high (B), and moderate (C) risk.

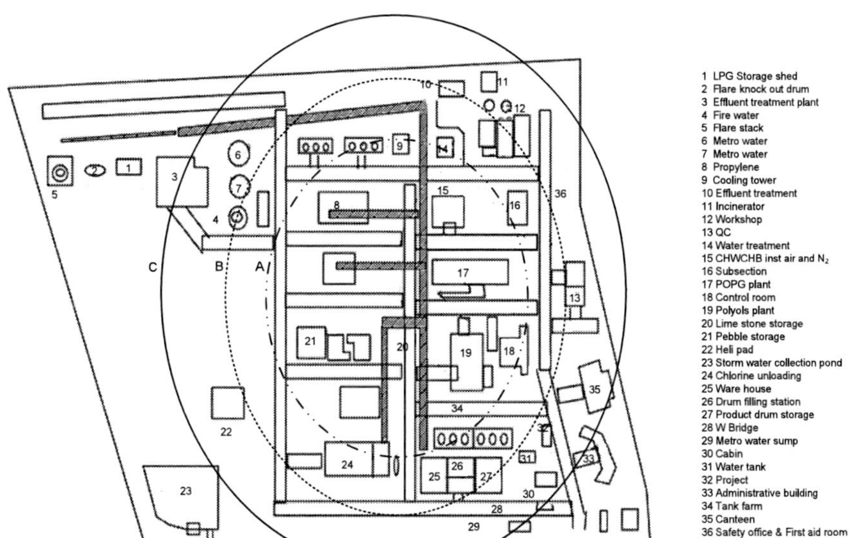

Figure 18.9. Risk contours for scenario 5 showing impact area due to severe (A), high (B), and moderate (C) risk.

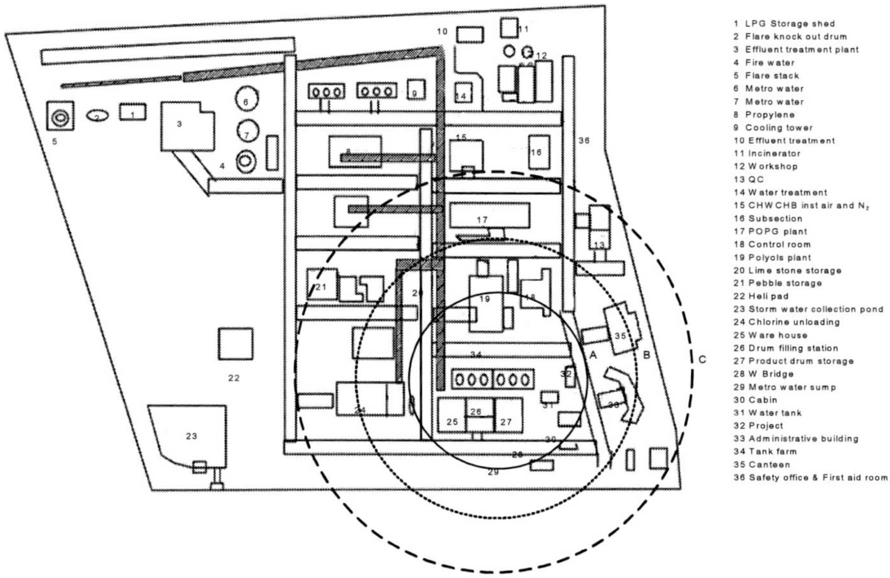

Figure 18.10. Risk contours for scenario 6 showing impact area due to severe (A), high (B), and moderate (C) risk.

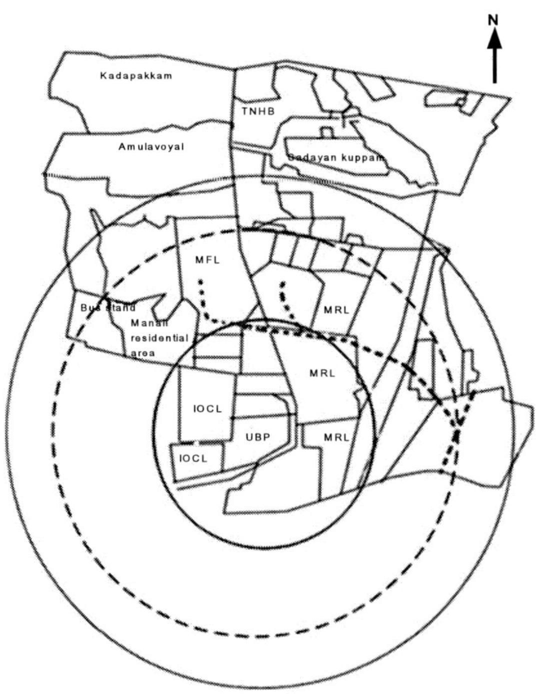

Figure 18.11. Risk contours for scenario 7 showing impact area due to severe (A), high (B), and moderate (C) risk.

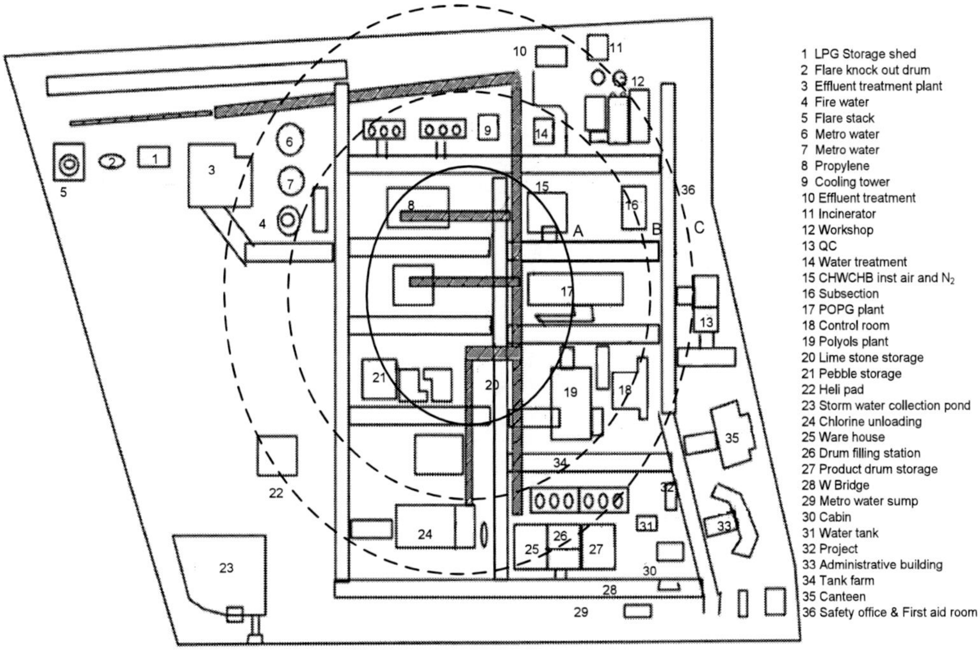

Figure 18.12. Risk contours for scenario 8 showing impact area due to severe (A), high (B), and moderate (C) risk.

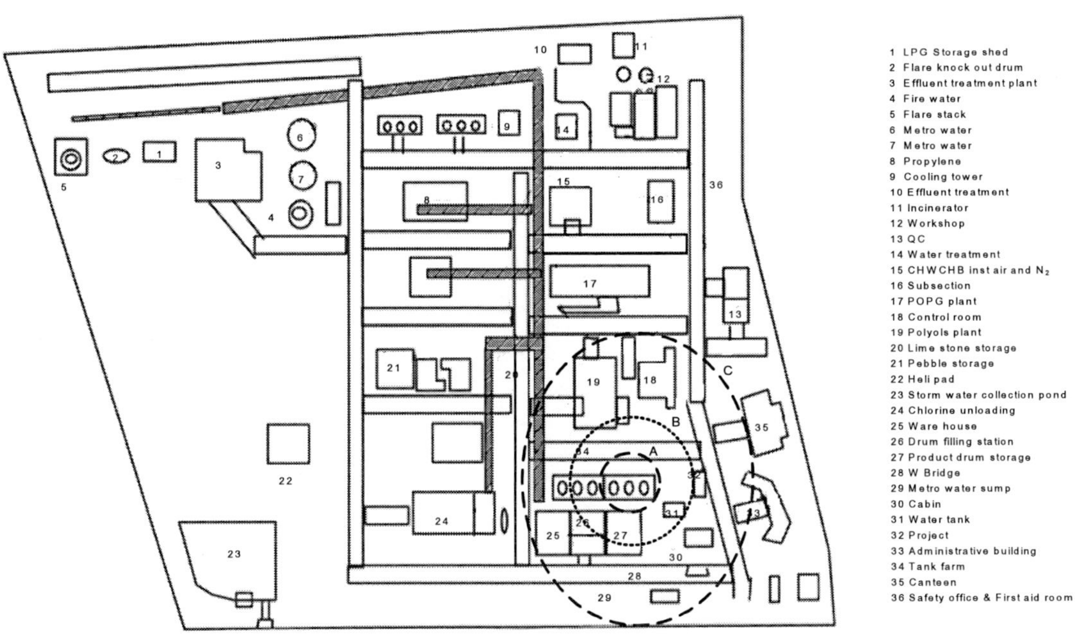

Figure 18.13. Risk contours for scenario 9 showing impact area due to severe (A), high (B), and moderate (C) risk.

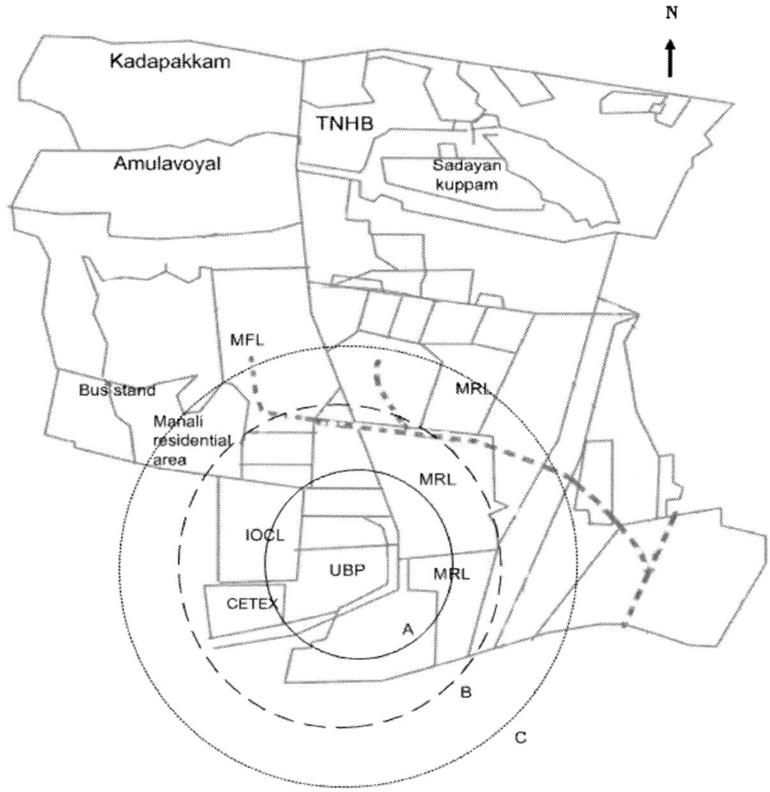

Figure 18.14. Risk contours for scenario 10 showing impact area due to severe (A), high (B), and moderate (C) risk.

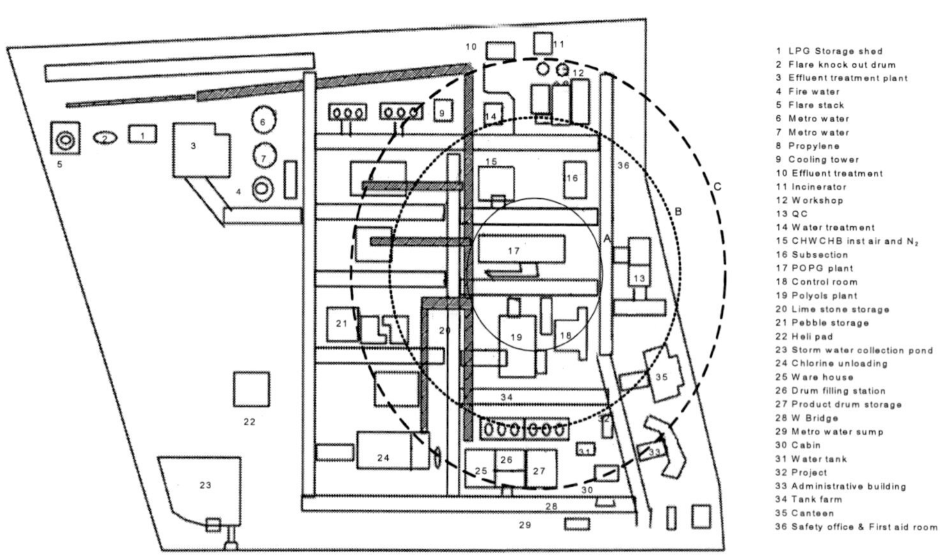

Figure 18.15. Risk contours for scenario 11 showing impact area due to severe (A), high (B), and moderate (C) risk.

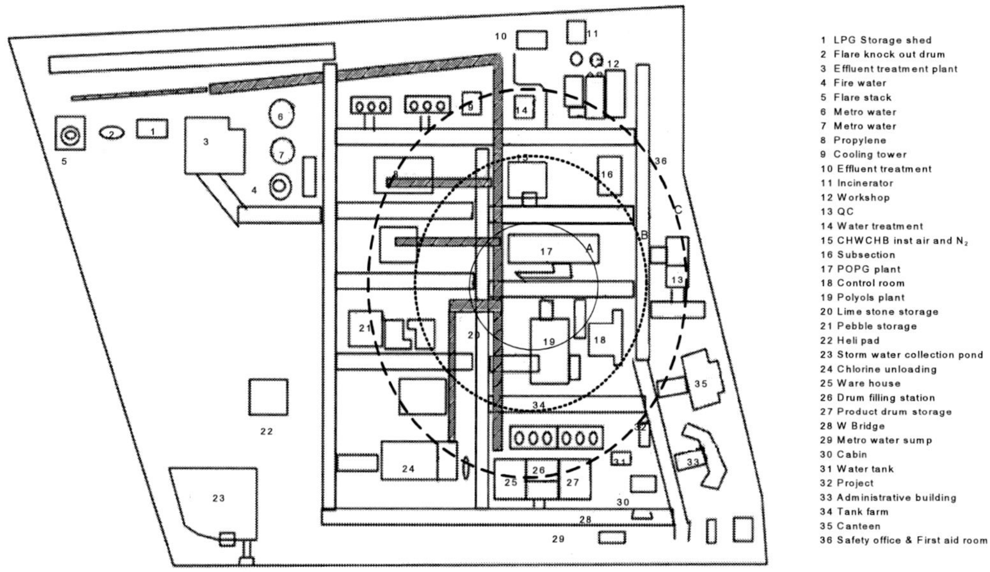

Figure 18.16. Risk contours for scenario 12 showing impact area due to severe (A), high (B), and moderate (C) risk.

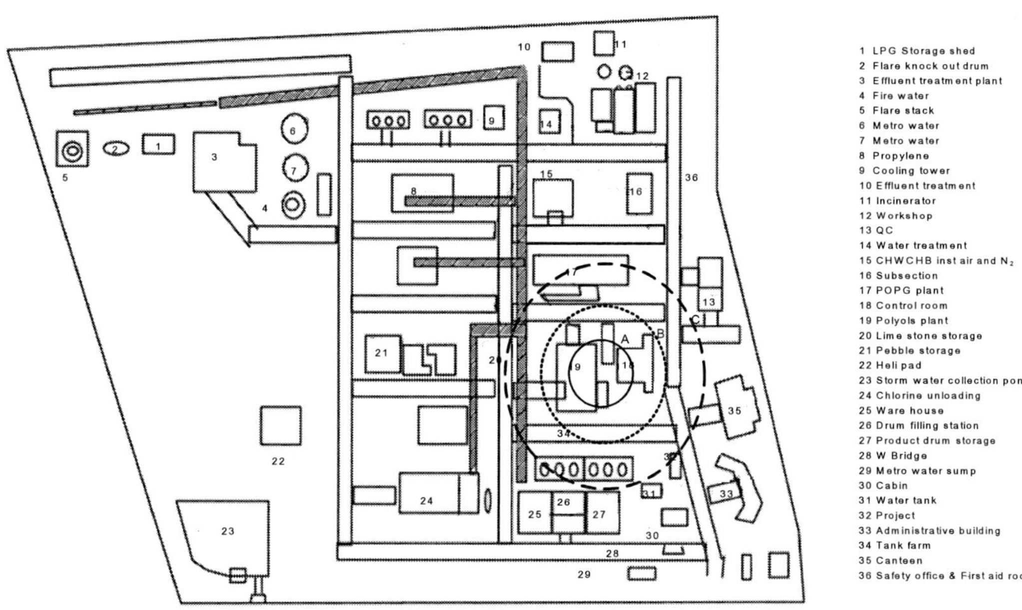

Figure 18.17. Risk contours for scenario 13 showing impact area due to severe (A), high (B), and moderate (C) risk.

Chapter 19

RAPID RISK ASSESSMENT OF TPL

19.1. INTRODUCTION

Tamilnadu Petroproducts Limited (TPL) is a joint sector company of the Southern Petrochemical Industries Corporation Limited (SPIC). TPL's linear alkyl benzene (LAB) plant has an above-average track record. It has gone on stream ahead of its schedule, has recorded profits right from the first year of production, has won 1991 National Safety Council (UK) award for its excellent safety management and is identified by its technical collaborators as their "referral" plant.

The LAB Plant comprises of five major units, namely

1) PREFRACTIONATION
2) HYDROTREATOR
3) MOLEX
4) PACOL
5) DETERGENT ALKYLATION

During 1990s TPL has added an epichlorohydrin manufacturing plant to its LAB plant. The study described have has been focussed on the epichlorohydrin (EPI) plant. The main units of the plant are presented in Table 19.1. Simplified process flow diagram is presented in Figure 19.1.

19.2. HAZARD IDENTIFICATION AND RANKING

Various units of the plant were processed through HIRA system and the FEDI and the TDI index values were computed. Figures 19.2 and 19.3 present the results of this analysis (Legends of the figures are illustrated in Table 19.1.The units L (chlorinated organic tanks) and M (chlorine storage) are above the threshold limit of 100 and hence warrant further scrutiny. Similarly from the FEDI values it is inferred that units C (allyl chloride purification), I (propylene bullets), J (EPI tanks), K (fuel oil tanks) and M (chlorine storage) exceed of the threshold 100.

19.3. Development of Accident Scenarios

Accident scenarios were developed for all the units found hazardous as per the ranking described above.

Propylene Storage: Scenario 1

An excessive pressure development in the storage vessel of propylene may lead to confined vapor cloud explosion (CVCE). The vapor cloud generated by CVCE on ignition turns into a fire ball and consequently damages the other storage vessels (chlorine, allylchloride, epichlorohydrin, fuel oil).

Chlorine Storage: Scenario 2

A sudden release of pressurized chlorine triggers a boiling liquid expanding vapor explosion (BLEVE) and dispersion of toxic vapor cloud; in other words explosive release followed by toxic dispersion.

Epichlorohydrin Storage: Scenario 3

Vapor cloud explosion VCE occurs in the epichlorohydrin unit accompanied by a pool fire. This shall happen if a small leak in the storage vessel releases material at moderate flow rate and forms a vapor cloud in the atmosphere which on ignition leads to VCE. Due to the heat generated by VCE or by external source of ignition the remaining material in the vessel or dike may catch fire causing a pool fire. A VCE followed by pool fire can damage neighboring vessels, due to excess shock waves and heat load. These may also trigger secondary and higher order accidents.

Allylchloride Storage: Scenario 4

Instantaneous release followed by pool fire is envisaged in the allylchloride storage unit.

Fuel Oil Storage: Scenario 5

Pool fire in the fuel oil storage unit.

Chlorinated Organic Storage: Scenarios 6

Release of toxic chlorinated organic followed by evaporation and dispersion.

19.4. Impact Forecasting: Consequence Assessment

Table 19.2 presents the summary of forecasts for *scenario 1*. The missiles generated by CVCE may hit nearby targets and can lead to secondary explosions or toxic release. The vapor cloud generated by CVCE on ignition may cause a fire ball and hence severe heat radiation effect. The shock wave generated due to CVCE can cause injury as well as second order accidents by seriously damaging other vessels. It has been estimated that shock waves with 50% probability of causing injury would be observed over an area of 500 m radius. The heat radiation effect with 100% probability of lethality would be observed over an area of 300 m radius and missile effects with 50% chances of damage would be observed across 750 m radius.

The sudden release of chlorine and the consequent BLEVE as per *scenario 2* would lead to severe shock waves, and toxic release (Table 19.3). In this scenario the damage potential of the shock waves is estimated to be comparatively lesser than *scenario* 1. 100% damage - causing shock wave would be operative over an area of 200m radius. As chlorine is non-combustible, no heat radiation effects would be observed, but a build-up of lethal concentration would take place over an area of 1500 m radius.

A VCE as per *scenario 3* would give rise to heat radiation effects, shock waves, and missile effects. In addition there would be secondary impact of burning of released material in the dike/vessel which would again lead to additional heat load. In *scenario 3*, missile effect is not significant (Table 19.4). However, shock waves of intensity high enough to damage all objects coming in their way would impact in an area of 200m radius. The combined effect of heat radiation, VCE, and pool fire would be lethal over an area of 600 m radius. Due to intense heat load some of the chemical may evaporate and disperse causing its build up to a lethal concentration over an area of 400m radius.

The model forecasts for *scenario 4* are presented in Table 19.5. Lethal heat load would be observed over the study area; the maximum damage distance would envelop an area of 300m radius.

The forecasts based on detailed calculation for *scenario 5* is presented in Table 19.6. Lethal heat load would be observed over an area of 400m radius.

The forecasts for *scenarios 6* have been presented in Table 19.7. In both cases the lethal concentration (100% probability of fatality) is confined to an area of 200m radius.

19.5. Risk Estimation

On the basis of the results presented above, and probability of occurrence of the various accident scenarios, risk factors have been computed. In order to enable visualization of the accidents, risk contours have been drown over a map of the study area (Figures 19.4 to 19.9).

It may be seen that the risk contours for *scenarios 1-3* (Figures 19.4-19.6) are extending beyond the boundaries of the industry and enveloping other industries and nearby populated areas. The risk contours for other scenarios are confined to the campus of the TPL plant but the risk contours for *scenario*s 4 and 5 extend upto other storage vessels process units in the factory (Figures 19.7 and 19.8). This may cause secondary accidents, the impacts of which may go beyond the factory confines.

In summary, *scenario 2* represents the worst likely disaster within the realm of credibility. It has the largest area-of-lethal-impact (shock wave over an area of radius 200 m and lethal concentration across an area of radius 1500 m). Further, the most thickly populated areas lie within its range. If one considers the cumulative effects, then *scenario 1* would come out as the worst, as more intense impacts (in terms of heat radiation, shock waves, missiles) may be observed per unit area in this scenario. Of the eight scenarios, *scenario 1* is the most likely to cause 'domino' or cascading effects as missiles, shock waves, and radiation effects would be generated simultaneously and other industries or units dealing with flammable and toxic materials are situated within the striking distance of the primary accident of *scenario 1. Scenarios 3, 4* and *5* also have the potential to lead to secondary accidents, as severe heat load generated in these would encompass other storage vessels. All in all, *scenario 2* is the worst as far as primary effects are concerned whereas *scenario 1* is the worst in terms of its potentiality of causing cascading (domino) effects.

19.6. Recommendations

To reduce the hazards associated with the storage of chlorine, propylene and epichlorohydrin, proper hazard minimization/mitigation measures should be taken. A few suggestions are made in this context:

I. Instead of one or two large-capacity vessels several vessels of smaller capacity should be used for storage.
II. Adequate space should be kept between the storage vessels and buffers provided between them so that adverse consequences of failure in one of them do not cause second or higher order accidents.
III. Sensitive gas detecting devices for flammable and toxic gases should be installed in the premises of storage area and other units.
IV. There should be regular and thorough inspections of electronic control equipment followed by meticulous maintenance.
V. Sufficient quantities of inert gases should be readily available to dilute the concentrations of toxic/flammable gases if they escape to the atmosphere, and to control fire.
VI. A thorough emergency preparedness strategy should always be kept in position, fortified by periodic drills or 'dry runs' so that the damage is contained if an accident does occur.

Table 19.1. The main units of epichlorohydrin (EPI) plant

Units	Reference used in Figures 19.2 and 19.3	Operation
Propylene purification	A	Fractionation
Propylene chlorination	B	Reaction
AllylChloride purification	C	Fractionation
Chlorine absorption in water	D	Absorption
Chlorohydrination	E	Reaction
Hydrolysis and EPI purification	F	Reaction & Distillation

Units	Reference used in Figures 19.2 and 19.3	Operation
Quenching	G	Heat transfer
Vent scrubbing	H	Separation
Propylene bullets	I	Storage
EPI tanks	J	Storage
Fuel oil tanks	K	Storage
Chlorinated organic tanks	L	Storage
Chlorine storage	M	Storage

Table 19.2. Consequence analysis for *scenario 1*

Parameters	Values
Distance from accident epicenter	(m) : 200
Explosion : CVCE	
Energy released during explosion	(kJ) : 3.25E+07
Peak over pressure	(kPa) : 1541.2
Variation of over pressure in air	(kPa/s) : 801.4
Shock velocity of air	(m/s) : 875.6
Duration of shock wave	(s) : 24.0
Missile characteristics	
Initial velocity	(m/s) : 495.4
Kinetic energy of fragment	(kJ) : 2.25E+06
Fragment velocity at study point	(m/s) : 179.3
Penetration ability at study point	
Concrete structure	(m) : 0.28
Brick structure	(m) : 0.40
Steel structure	(m) : 0.07
Fire : Fire Ball	
Radius of the fire ball	(m) : 175.4
Duration of the fire ball	(s) : 850
Energy released by fire ball	(kJ) : 6.15E+09
Radiation heat intensity	(kJ/sq. m) : 64431.2

Table 19.3. Consequence analysis for *scenario 2*

Parameters	Values
Distance from accident epicenter	(m) : 200
Explosion : BLEVE	
Total energy released	(kJ) : 5.7E+06
Peak over pressure	(kPa) : 134.8
Variation of over pressure in air	(kPa/s) : 57.9
Shock velocity of air	(m/s) : 343.9
Duration of shock wave	(s) : 33.4
No missile effect	
Toxic release & dispersion	
Light gas dispersion characteristics	
Gaussian instantaneous: model	
Concentration at distance 200 m	(kg/cu. m) : 1.8E-04
Concentration at cloud axis	(m) : 1.3E-02
Value of source height	(kg/cu. m) : 5.0
Puff characteristics:	
Puff concentration at centre of cloud	(kg/cu. m) : 8.16E-04

Table 19.3. (Continued)

Parameters	Values
Concentration at cloud edges	(kg/cu. m) : 8.13E-04
Distance along downwind	(m) : 200.0
Dosage at study point	(kg/cu. m) : 0.078

Table 19.4. Consequence analysis for *scenario 3*

Parameters	Values
Distance from accident epicenter	(m) : 200
Explosion ; UVCE	
Total energy released	(kJ) : 6.6E+06
Peak over pressure	(kPa) : 115.6
Variation of over pressure in air	(kPa/s) : 42.5
Shock velocity of air	(m/s) : 453.5
Duration of shock wave	(s) : 21.9
Missile characteristics:	
Initial velocity of missile	(m/s) : 876.5
Kinetic energy of associated with missile	(kJ) : 1988426.37
Fragment velocity at study point	(m/s) : 234.74
Penetration ability at study point	
Concrete structure	(m) : 0.21
Brick structure	(m) : 0.24
Steel structure	(m) : 0.002
Fire: Fire Ball	
Radius of the pool fire	(m) : 5.0
Burning area	(sq. m) : 78.5
Burning rate	(kg/s) : 38.4
Heat intensity	(kJ/sq. m) : 1160.1
Toxic release & dispersion	
Light gas dispersion characteristics	
Gaussian instantaneous: model	
Concentration at cloud axis	(kg/cu. m) : 5.3E-05
Puff characteristics:	
Puff concentration at centre of cloud	(kg/cu. m) : 2.93E-05
Concentration at cloud edges	(kg/cu. m) : 2.97E-05
Distance along downwind	(m) : 200.00
Dosage at study point	(kg/cu.m) : 0.00088

Table 19.5. Consequence analysis for *scenario 4*

Parameters	Values
Distance from accident epicenter	(m) : 200
Fire: Pool fire	
Continuous model	
Burning area	(sq. m) : 75.00
Burning rate	(kg/h) : 15042.82
Heat flux	(kj/sq. m) : 6896.1

Table 19.6. Consequence analysis for *scenario 5*

Parameters	Values	
Distance from accident epicenter	(m)	200
Fire: Pool fire		
Instantaneous model		
Radius of the pool fire	(m) :	5.00
Burning area	(sq. m) :	78.53
Burning rate	(kg/s) :	38.42
Heat flux	(kJ/sq. m)	14418.9

Table 19.7. Consequence analysis for *scenario 6*

Parameters	Values
Distance from accident epicenter	(m) : 200
Toxic Release & Dispersion	
Heavy gas dispersion characteristics	
Box Instantaneous: Model	
Concentration at cloud axis	(kg/cu. m) : 1.08E-03
Value of source height	(m) : 4.0
Puff characteristics:	
Puff concentration at centre of cloud	(kg/cu. m) : 2.91E-07
Concentration at cloud edges	(kg/cu. m) : 6.03E-08
Distance along downwind	(m) : 10.0000
Dosage at study point	(kg/cu. m) : 2.3E-03

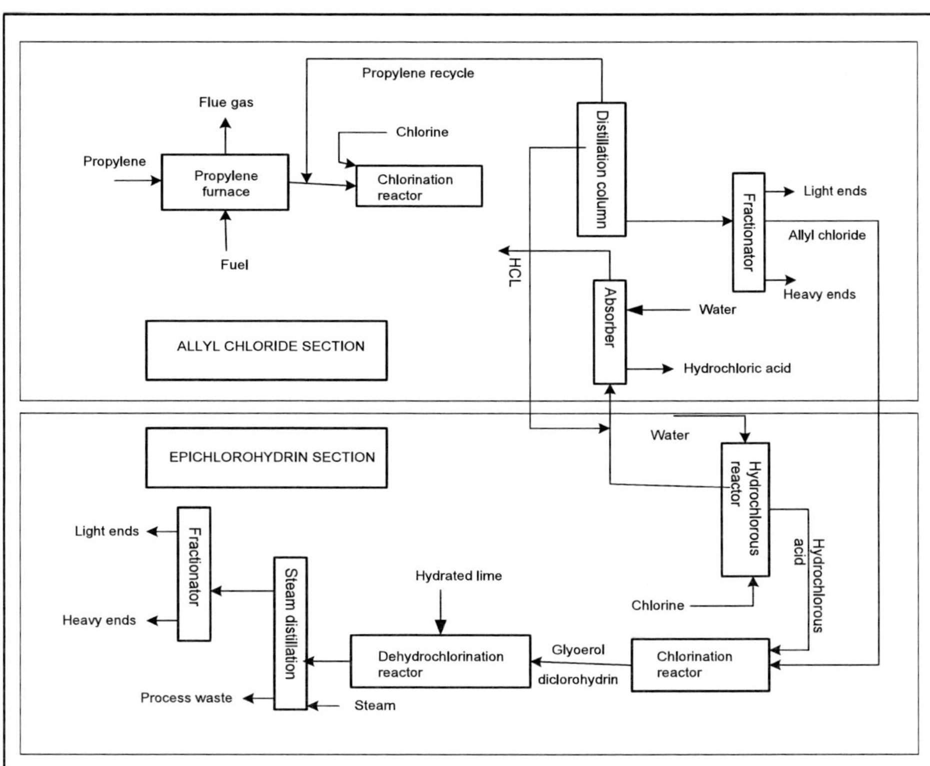

Figure 19.1. Simplified process flow diagram.

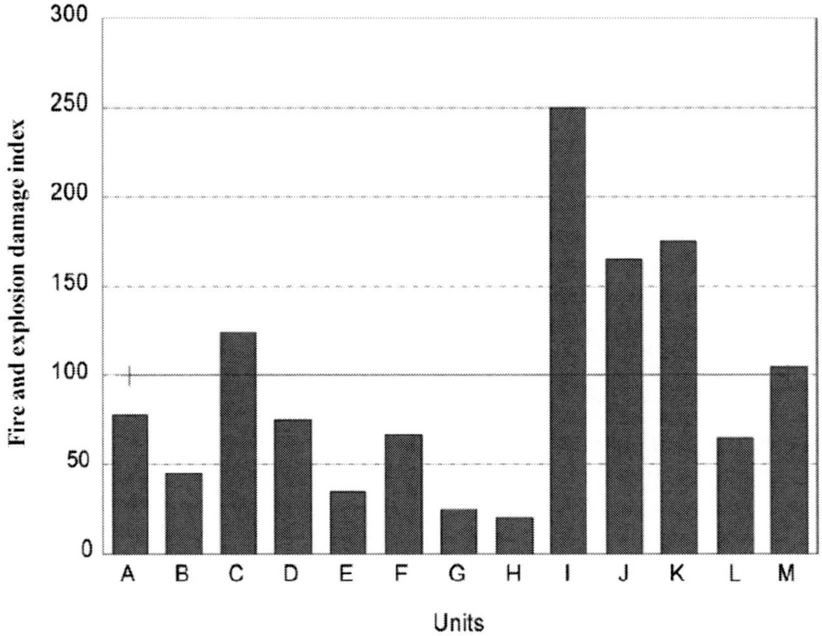

Figure 19.2. Fire and explosion damage index for various units in TPL (Legends A…M is explained in Table 8.1).

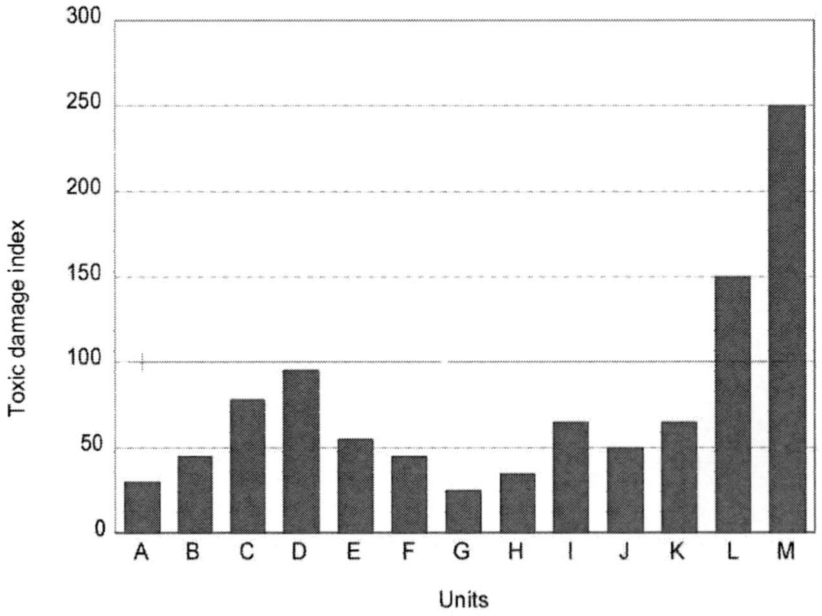

Figure 19.3. Toxic damage index for various units in TPL (Legends A…M is explained in Table 8.1).

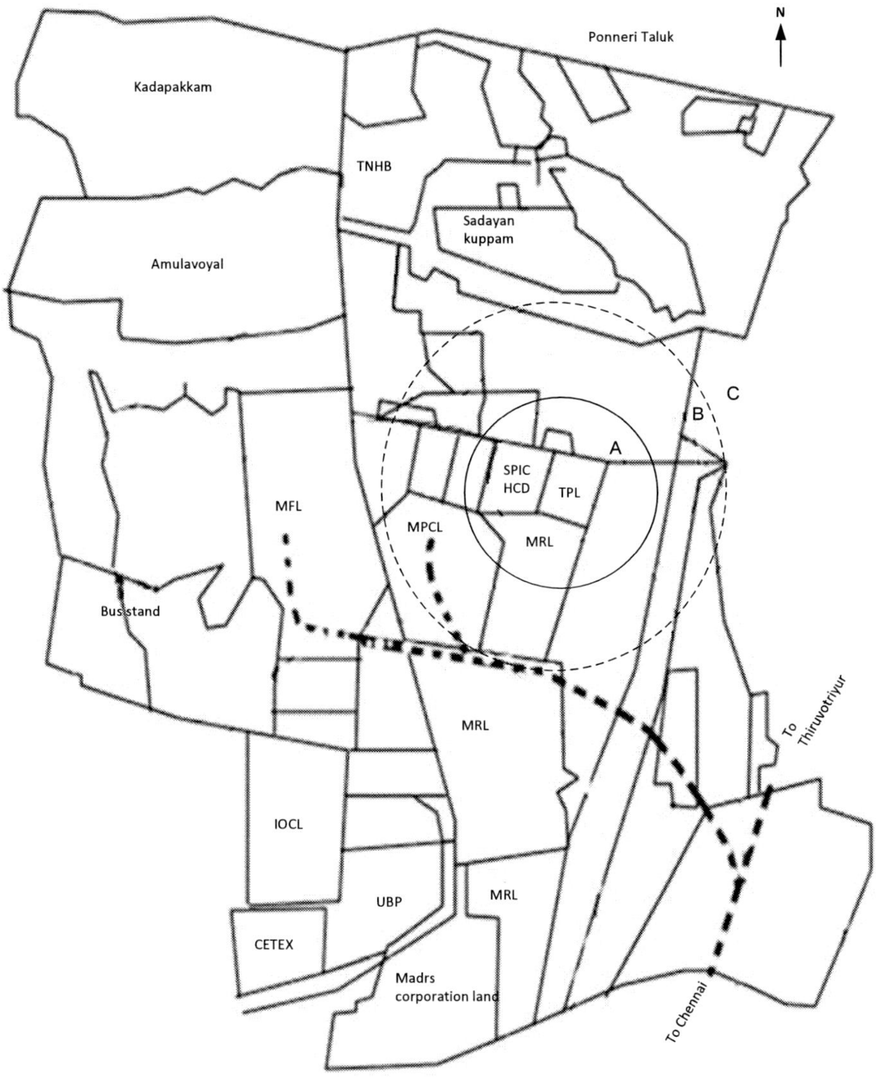

Figure 19.4. Risk contours for scenario 1 showing impact are due to severe (A), high (B) and moderate (C) risk.

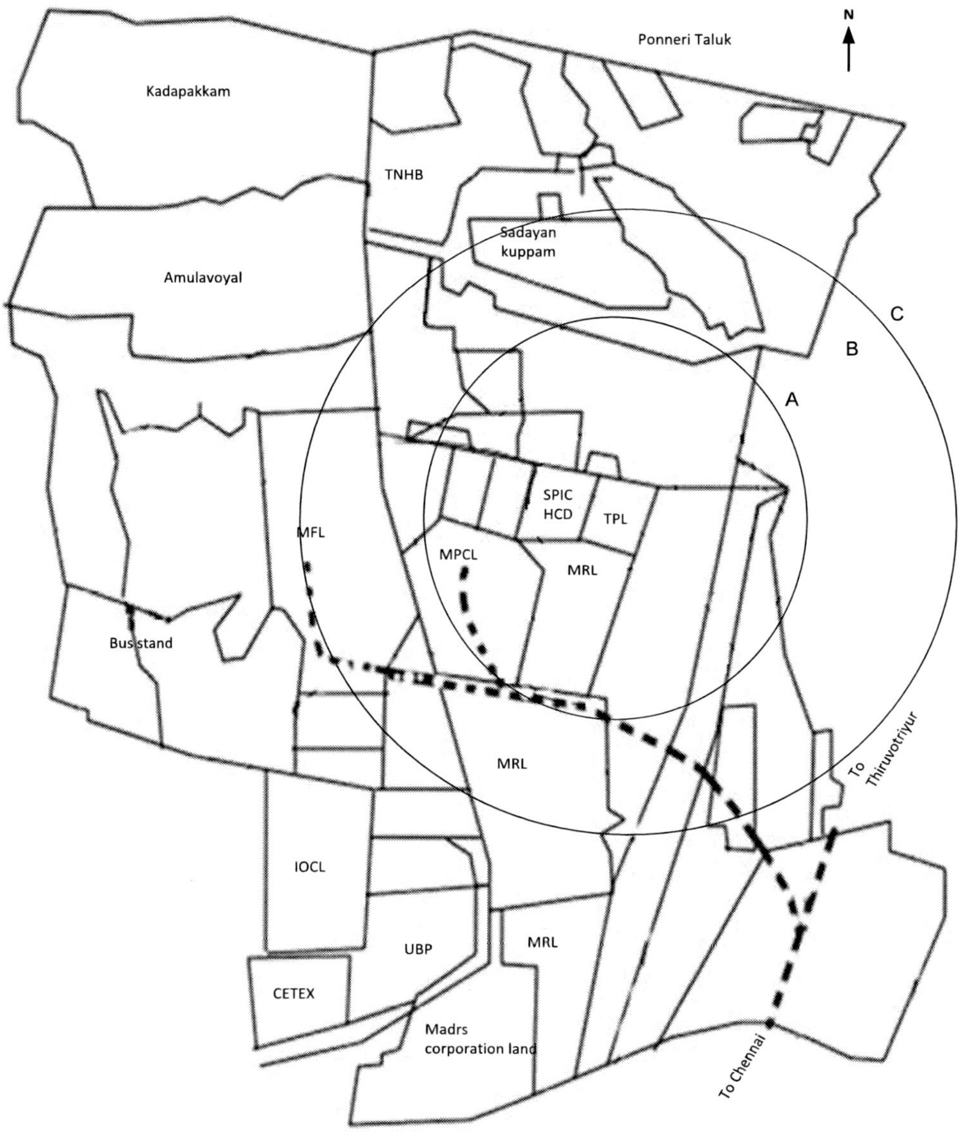

Figure 19.5. Risk contours for scenario 2 showing impact are due to severe (A), high (B) and moderate (C) risk.

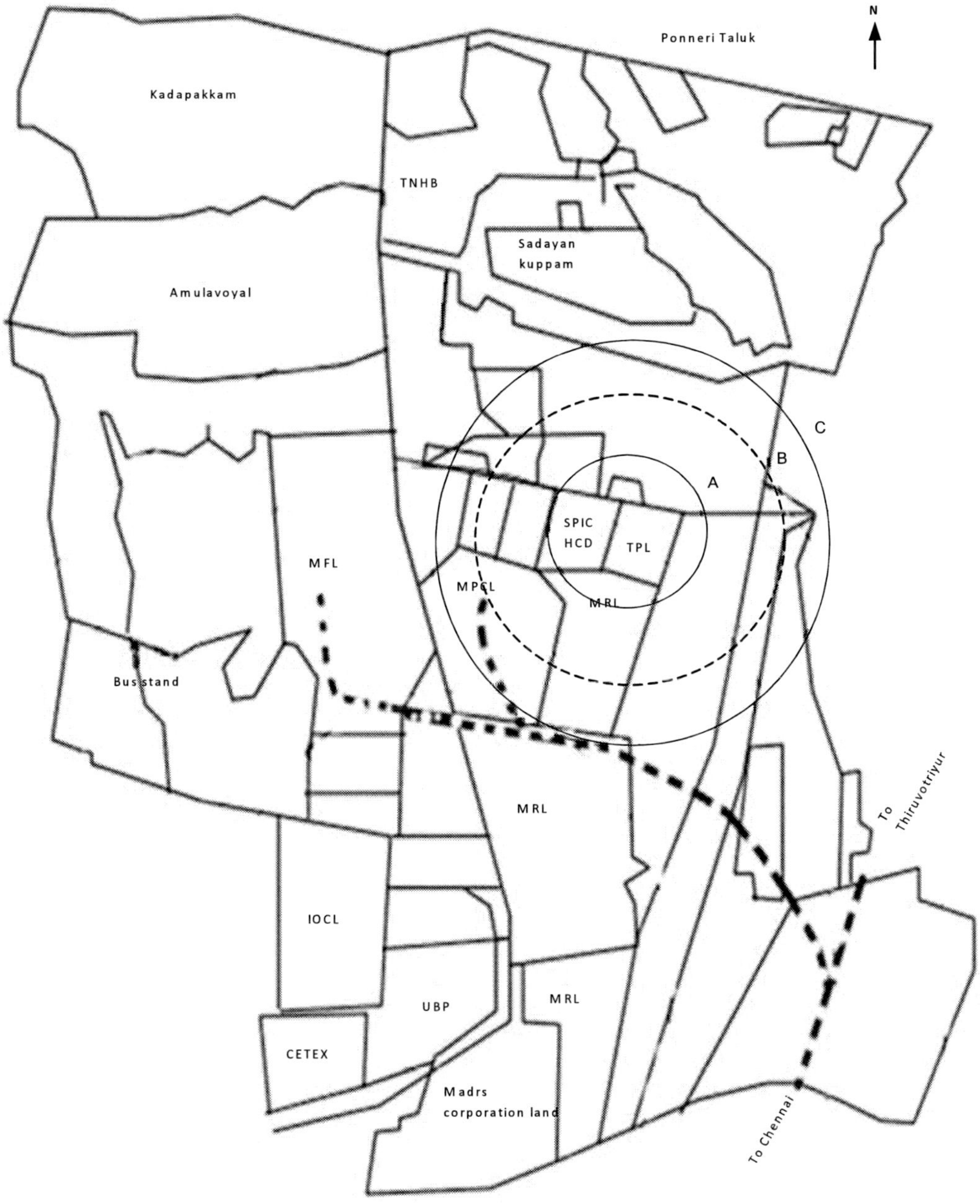

Figure 19.6. Risk contours for scenario 3 showing impact are due to severe (A), high (B) and moderate (C) risk.

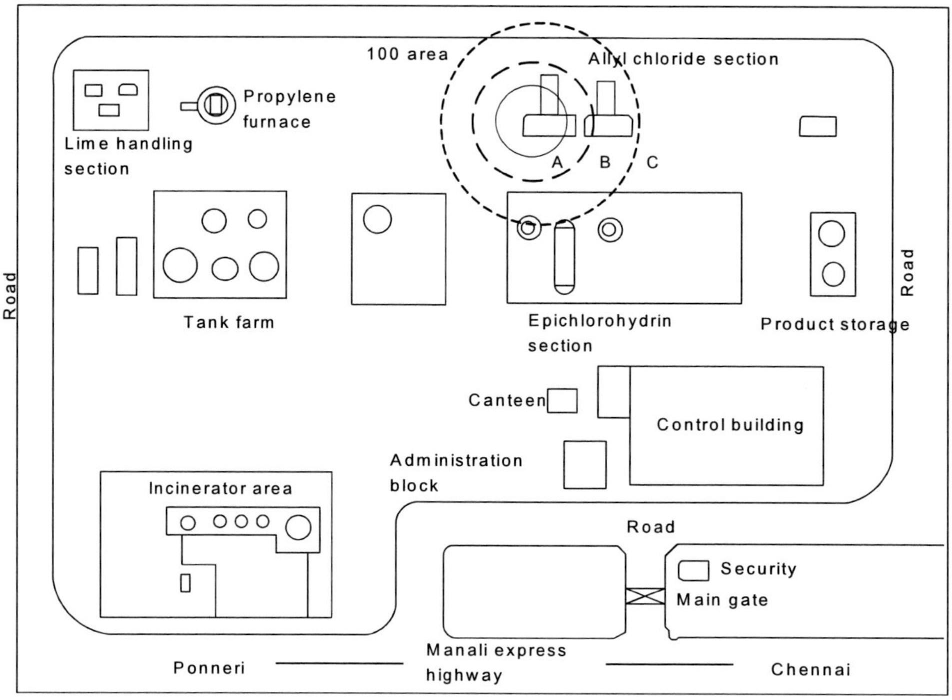

Figure 19.7. Risk contours for scenario 4 showing impact are due to severe (A), high (B) and moderate (C) risk

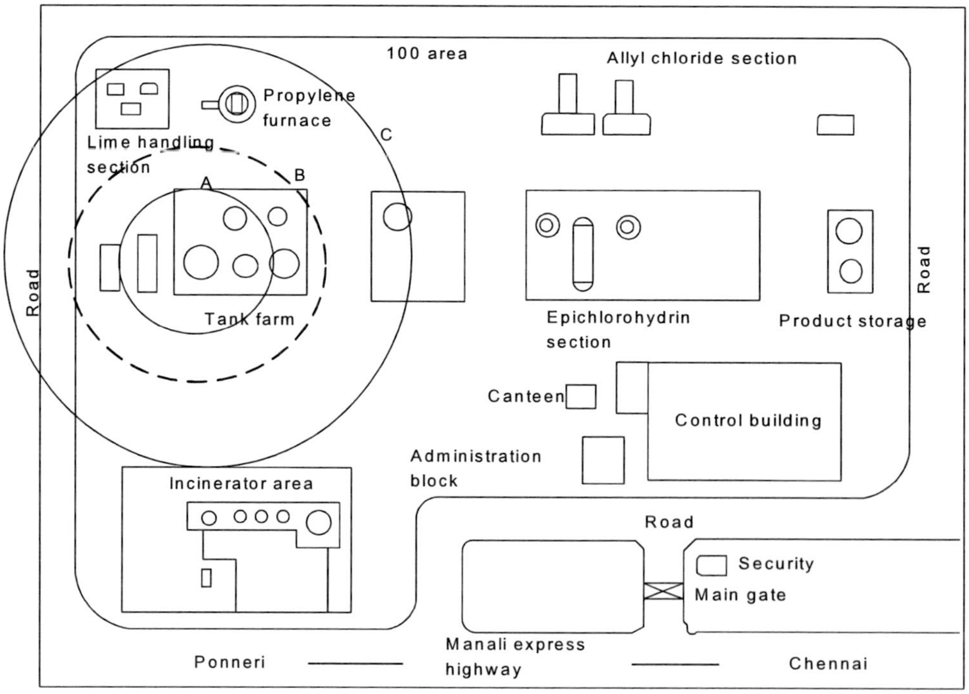

Figure 19.8. Risk contours for scenario 5 showing impact are due to severe (A), high (B) and moderate (C) risk.

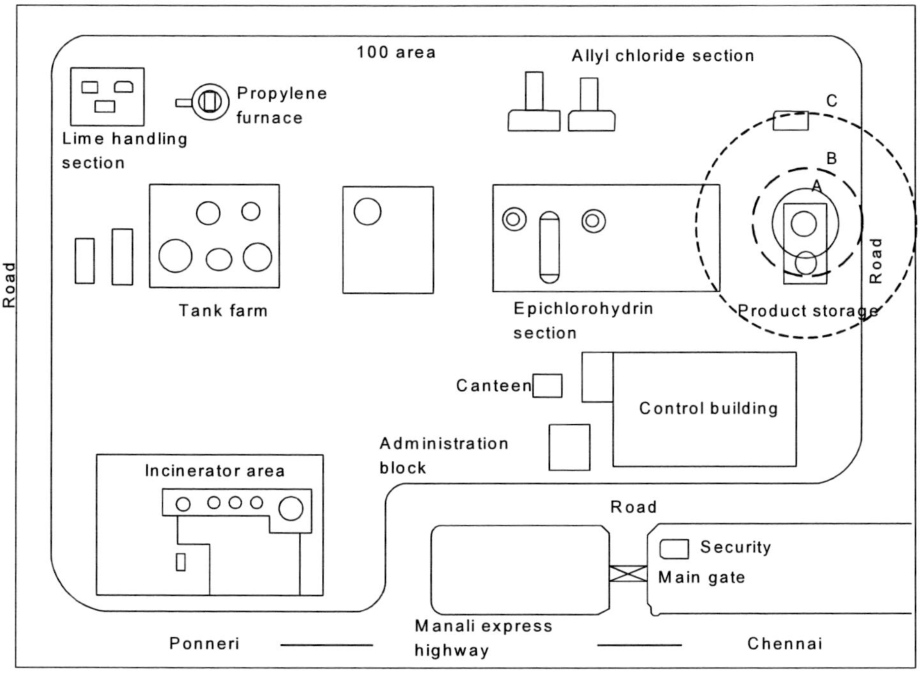

Figure 19.9. Risk contours for scenario 6 showing impact are due to severe (A), high (B) and moderate (C) risk.

Chapter 20

Rapid Risk Assessment of MPL

20.1. Introduction

Manali Petrochemicals Limited (MPL), one of the major industries in Manali Industrial Complex, produces the following:

I. Propylene oxide
II. Propylene glycol
III. Polyol

The simplified block diagrams of the processes are depicted in Figures 20.1 to 20.8.

20.2. Hazard Identification and Ranking

The different components of the MPL process lines: the storage units, fractionation unit, stripping-unit, distillation unit etc were screened for TDI (toxic dispersion index) and FEDI (fire and explosion damage index). The results are summarized in Figures 20.9 and 20.10. As may be seen form Figure 20.9, the unit L (chlorine pipeline) exceeds the threshold TD level of 100. Other units are below this limit. Hence this unit was short-listed for more detailed study. Figure 20.10 reveals that FEDI values for most of the reactor and storage units (A to F, H to K; legends are illustrated in Table 20.1) exceed the threshold, necessitating more detailed assessment.

20.3. Development of Accident Scenarios

Propylene Storage: Scenario 1

An instantaneous release of propylene stored under high-pressure leads to BLEVE. The vapor cloud generated by BLEVE on ignition leads to a fire ball.

Propylene Oxide Storage: Scenario 2

An excessive pressure development in the storage vessel of propylene oxide (highly reactive and flammable chemical stored under pressure) causes a confined vapor cloud explosion (CVCE). The vapor cloud generated by CVCE on ignition turns into a fireball and consequently damages the other storage vessels.

Ethylene Oxide Storage: Scenario 3

An instantaneous release of ethylene oxide under high pressure, leads to BLEVE and, as the chemical is highly flammable, the released cloud on meeting an ignition source turns to a fire ball.

LPG Storage: Scenario 4

An instantaneous release of LPG (liquefied petroleum) either through vent valve or through any other accidental opening causes the vessel to explode in the manner of a BLEVE. As the released chemical is highly flammable it turns into a fire ball on meeting an ignition source.

MPG Storage: Scenario 5

Instantaneous release is followed by pool fire in the monopropylene glycol (MPG) storage unit.

Fuel Oil Storage: Scenario 6

There is a pool fire in the fuel oil storage unit.

DCP Storage: Scenario 7

Instantaneous release is followed by pool fire in the di-chloropropylene (DCP) storage unit.

Propylene Oxide Reactor: Scenario 8

A leak causes formation of pool of liquid chlorohydrin, which on meeting ignition source causes pool fire.

Propylene Oxide Reactor: Scenario 9

Excessive pressure development in the reactor causes release of chemical (propylene) through vent valve or pressure relief valve (PRV). The resultant inflammable cloud gets ignited to a flash fire.

Propylene Glycol Reactor: Scenario 10

There is formation of vapour cloud due to the release of propylene oxide from the propylene glycol reactor. The cloud, on meeting a source of ignition either leads to a vapor cloud explosion, or a flash fire, depending upon the strength of ignition source and the quantity of chemical involved in the vapour cloud.

Propylene Glycol Reactor: Scenario 11

Formation of fire ball due to the release and ignition of the vapour cloud of propylene oxide.

Polyol Reactor: Scenario 12

Instantaneous release of the contents of polyol reactor in explosive manner generate shock waves. The vapour cloud on meeting a source of ignition turns to a fire ball.

Polyol Reactor: Scenario 13

Release of chemical (Propylene oxide) to form a vapour cloud, and subsequently a fire ball.

Transportation Line: Scenario 14

Chlorine is a non-flammable toxic gas, which is normally transported through pipelines in liquefied state. The accident scenario envisages a continuous release of chlorine followed by dispersion.

20.4. Impact Forecasting: Consequence Assessment

The forecasts for *scenario 1* (BLEVE followed by fire ball) are presented in Table 20.2. The shock wave and missiles generated by BLEVE may hit secondary targets and can lead to secondary explosions or toxic releases. The vapor cloud generated by BLEVE on ignition may cause a fire ball which would generate heat radiation effect. It is clear from Table 20.2 that an area of ~200 m radius faces 100% probability of being damaged due to missiles, shock waves, and heat load. The shock wave and heat radiation can cause fatality as well as second order accidents by seriously damaging other vessels.

The confined vapour cloud explosion in propylene oxide tank as per *scenario* 2 would generate strong shock waves and missiles (Table 20.3). The harmful impacts would cover a wide area (upto 354m from the epicenter). In addition, the fire ball resulting from ethylene would scotch, lethally, every living object over an area of 200 m radius. The shock wave and heat radiation can cause fatality as well as second order accidents by seriously damaging other vessels.

The forecasts based on detailed calculations for *scenario 3* (ethylene oxide release) are presented in Table 20.4. BLEVE, followed by fire ball, would cause intensive damage. It is evident from the table that damage of high degree of severity would be likely over an area of ~100 m radius while moderate damage (50% probability of lethality) would occur over an area of ~150 m radius.

As per *scenario 4,* BLEVE would generate shock waves. The released material may get ignited and form a fire ball thereby generating additional heat load. The forecasts for this scenario, presented in Table 20.5,results that the impact of the shock waves would not be severe but the lethal impact of heat load may go up to and beyond a radius of ~75 m.

The study of the consequences of *scenario 5* (release of MPG) reveals that the likely damage due to this event in terms of heat load would be less intense than forecasted by *scenarios* 1-4. However, it is evident from the forecasts (Table 20.6) that even at a distance of ~125 m from the accident epicenter the intensity of heat load would be severe enough to cause secondary accident.

The forecasts based on detailed calculations for *scenario 6* are presented in Table 20.7. Lethal heat load would be observed over an area of ~150 m radius.

The forecasts for *scenario 7* are tabulated in Table 20.8. Lethal heat load would be observed over an area of ~90 m radius.

Two accident scenarios have been generated for propylene oxide reactor (*scenarios 8 and 9*). The forecasts based on detailed calculations for these scenarios are presented in Tables 20.9 and 20.10. The damage potential of first scenario (pool fire) would be limited to an area of ~15 m radius from the point of accident, while the impact of flash fire may be more severe, damaging objects upto ~40 m from the accident epicentre.

Two accident scenarios (*scenarios 10 and 11*) have also been developed for propylene glycol (PG) reactor. The forecasts based on detailed calculations for these scenarios are presented in Tables 20.11 and 20.12. The first likely accident (flash fire) may cause lethal heat load over an area of ~60 m radius, while consequences of the second scenario (pool fire) may extend over an area of ~15 m radius.

Polyol reactor handles large quantities of flammable chemicals. The results of damage calculation of the first scenario (*scenario 12*) instantaneous release followed by fire ball reveals that shock waves and heat load may envelop an area of more than ~25 m radius (Table 20.13). The results for the second accident scenario (*scenario* 13: fire ball) reveal that the lethal heat load (having 50% probability of causing fatality) may occur over an area of ~25 m (Table 20.14).

Scenario 14 envisages a continuous release of chlorine followed by dispersion (Table 20.15). Toxic level with the potential to cause (50% probability) fatality would occur over an area of ~4000 m radius. For another 4000 m the harmful concentration shall persist, through with lesser propensity to cause fatality.

20.5. Risk Estimation

Using the results obtained as above and the probability of occurrence of the relevant accidents, individual fatal risk factors have been estimated. The risk factor is a direct representation of the threat (taking into consideration both damage potential and probability of occurrence) to an individual in an area.

In order to enable swift visualization of accident scenarios, their risk contours have been drawn over the industrial layout (Figures 20.11 – 20.24). It may be seen that the risk contours for *scenarios* 1-7 and 14 (Figures 20.11 – 20.17 and 20.24) are just extending beyond the boundaries of the industry and enveloping other industries and nearby populated areas. The risk contours for other scenarios are confined to the campus of the MPL plant but the risk contours for *scenarios* 5 through 7 extend upto other storage vessels in the factory (Figures 20.15 and 20.17). This may cause secondary accidents, the impacts of which may go beyond the factory confines.

In summary *scenario* 14 represents the worst likely disaster within the realm of credibility. It has the largest area-of-lethal-impact (4000m radius). Further, several thickly populated villages and neighborhoods lie within its range. If one considers the cumulative effects, *scenario 2* appears to be worst, as more intense impacts (in terms of heat radiation, shock waves, missiles) may be caused per unit impact area in this scenario. *Scenario 2* is also the most likely to cause cascading effects as missiles, shock waves, and radiation effects would be generated simultaneously and other industries or units dealing with flammable and toxic materials are situated within the striking distance of the primary accident of *scenario* 2. *Scenarios* 3, 4 and 5 also have the potential to lead to secondary accidents as severe heat load generated in these would encompass other storage vessels. All in all, *scenario* 14 is the worst as far as primary effects are concerned whereas *scenario* 2 is the worst in terms of its potentiality of causing cascading (domino) effects.

20.6. Recommendations

To reduce the hazards associated with the storage of chlorine, propylene and epichlorohydrin, proper hazard minimization/mitigation measures should be taken. A few suggestions are made in this context:

I. Instead of one or two large-capacity vessels several vessels of smaller capacity should be used for storage.

II. Adequate space should be kept between the storage vessels and buffers provided between them so that adverse consequences of failure in one of them do not cause second or higher order accidents.

III. Sensitive gas-detecting devices for flammable and toxic gases should be installed in the premises of storage area and other units.

IV. There should be regular and thorough inspections of electronic control equipment followed by meticulous maintenance.

V. Sufficient quantities of inert gases should be readily available to dilute the concentrations of toxic/flammable gases if they escape to the atmosphere, and to control fire.

VI. A thorough emergency preparedness strategy should always be kept in position, fortified by periodic drills or 'dry runs' so that the damage is contained if an accident does occur.

Table 20.1. Legends used in figures 20.9 and 20.10

Legend	Specification
A	propylene
B	propylene oxide
C	Ethylene oxide
D	LPG storage
E	MPG storage
F	Fuel oil storage
G	polyol storage
H	DCP storage
I	PO reactor
J	PG reactor
K	polyol reactor
L	chlorine pipeline
M	Mix drum
N	surge drum
O	Glycol fractionation units
P	MPG & DPG intermediate storages
Q	Stripping unit
R	PO distillation
S	Propylene glycol evaporation

Table 20.2. Result of consequence analysis for scenario 1; accident in propylene storage vessel

Parameters	Values
Explosion ; BLEVE	
Total energy released	(kJ) : 5.45E+08
Peak over pressure	(kPa) : 17743.4
Variation of over pressure in air	(kPa/s) :9.42E+03
Shock velocity of air	(m/s) : 3155.3
Duration of shock wave	(ms) : 34.5
Missile characteristics	
Initial velocity	(m/s) : 1894.2
Kinetic energy of fragment	(kJ) : 4.54E+07
Fragment velocity at study point	(m/s) : 975.30
Penetration ability at study point (based on empirical models	
Concrete structure	(m) : 0.35
Brick structure	(m) : 0.52
Steel structure	(m) : : 0.08
Damage radii (DR) for various degree of damage due to over pressure	
DR for 100% complete damage	(m) : 255
DR for 100% fatality or 50% complete damage	(m) : 361

Parameters	Values
DR for 50% fatality or 25% complete damage	(m) : 411
Damage radii (DR) for various degree of damage due to missile	
DR for 100% damage or 100% fatality	(m) : 2775
DR for 50% damage or 100% fatality	(m) : 2891
DR for 100% fatality or 10% damage	(m) : 2994
Fire: Fire Ball	
Radius of the fire ball	(m) : 155.5
Duration of the fire ball	(s) : 50.0
Energy released by fire ball	(kJ) : 1.06E+09
Radiation heat flux	(kJ/sq. m) :112841.5
Damage Radii (DR) due to thermal load	
DR for 100% fatality/damage	(m) : 163
DR for 50% fatality/damage	(m) : 204
DR for 100% third degree of burn	(m) : 235
DR for 50% third degree of burn	(m) : 302

Table 20.3. Result of consequence analysis for scenario 2; accident in propylene oxide storage vessel

Parameters	Values
Explosion : CVCE	
Energy released during explosion	(kJ) : 5.45E+07
Peak over pressure	(kPa) : 17743.4
Variation of over pressure in air	(kPa/s) :9224.3
Shock velocity of air	(m/s) :3155.5
Duration of shock wave	(ms) : 34.5
Missile characteristics	
Initial velocity	(m/s) : 1894.2
Kinetic energy of fragment	(kJ) : 4.45E+07
Fragment velocity at study point	(m/s) : 1001.7
Penetration ability at study point (based on empirical model)	
Concrete structure	(m) : 0.35
Brick structure	(m) : 0.52
Steel structure	(m) : 0.08
Damage radii (DR) for various degree of damage due to over pressure	
DR for 100% complete damage	(m) : 157
DR for 100% fatality or 50% complete damage	(m) : 240
DR for 50% fatality or 25% complete damage	(m) : 354
Damage radii (DR) for various degree of damage due to missile	
DR for 100% damage or 100% fatality	(m) : 3190
DR for 50% damage or 100% fatality	(m) : 3305
DR for 100% fatality or 10% damage	(m) : 3409
Fire: Fire Ball	
Radius of the fire ball	(m) : 155.5
Duration of the fire ball	(s) : 50.0
Energy released by fire ball	(kJ) : 1.06E+09
Radiation heat flux	(kJ/sq. m) :112841.5
Damage Radii (DR) due to thermal load	
DR for 100% fatality/damage	(m) :178
DR for 50% fatality/damage	(m) : 213
DR for 100% third degree of burn	(m) : 264
DR for 50% third degree of burn	(m) : 312

Table 20.4. Result of consequence analysis for scenario 3; accident in ethylene oxide storage vessel

Parameters	Values
Explosion ; BLEVE	
Total energy released	(kJ) :4.5E+05
Peak over pressure	(kPa) :139.00
Variation of over pressure in air	(kPa/s) :132.00
Shock velocity of air	(m/s) :463.00
Duration of shock wave	(ms) :284.0
Missile characteristics	
Initial velocity	(m/s) :195.00
Kinetic energy of fragment	(kJ) :9.41E+04
Fragment velocity at study point	(m/s) :175.00
Penetration ability at study point (based on empirical models	
Concrete structure	(m) :0.07
Brick structure	(m) :0.10
Steel structure	(m) :0.01
Damage radii (DR) for various degree of damage due to over pressure	
DR for 100% complete damage	(m) : 77
DR for 100% fatality or 50% complete damage	(m) :118
DR for 50% fatality or 25% complete damage	(m) :174
Damage radii (DR) for various degree of damage due to missile	
DR for 100% damage or 100% fatality	(m) :2413
DR for 50% damage or 100% fatality	(m) :2529
DR for 100% fatality or 10% damage	(m) :2633
Fire: Fire Ball	
Radius of the fire ball	(m) : 68.00
Duration of the fire ball	(s) : 28.00
Energy released by fire ball	(kJ) : 4.06E+08
Radiation heat flux	(kJ/sq. m) : 6197.00
Damage Radii (DR) due to thermal load	
DR for 100% fatality/damage	(m) : 120
DR for 50% fatality/damage	(m) : 150
DR for 100% third degree of burn	(m) : 173
DR for 50% third degree of burn	(m) : 223

Table 20.5. Result of consequence analysis for scenario 4; accident in LPG storage vessel

Parameters	Values
Explosion : BLEVE	
Total energy released	(kJ) : 6.3E+04
Peak over pressure	(kPa) : 48.00
Variation of over pressure in air	(kPa/s) : 45.00
Shock velocity of air	(m/s) : 382.00
Duration of shock wave	(ms) : 213.0
No Missile characteristics	
Damage radii (DR) for various degree of damage due to over pressure	
DR for 100% complete damage	(m) : 40
DR for 100% fatality or 50% complete damage	(m) : 62
DR for 50% fatality or 25% complete damage	(m) : 91
Fire: Fire Ball	
Radius of the fire ball	(m) : 49.00
Duration of the fire ball	(s) : 20.00
Energy released by fire ball	(kJ) : 8.22E+07

Parameters	Values
Radiation heat flux	(kJ/sq. m) : 1256.00
Damage Radii (DR) due to thermal load	
DR for 100% fatality/damage	(m) : 54
DR for 50% fatality/damage	(m) : 68
DR for 100% third degree of burn	(m) : 78
DR for 50% third degree of burn	(m) : 100

Table 20.6. Result of consequence analysis for scenario 5; accident in MPG storage unit

Parameters	Values
Fire: Pool fire	
Instantaneous Model	
Radius of the pool fire	(m) : 5.00
Burning area	(sq. m) : 78.5
Burning rate	(kg/s) : 21.5
Heat flux	(kJ/sq. m) : 89659.8
Damage Radii (DR) due to thermal load	
DR for 100% fatality/damage	(m) :172
DR for 50% fatality/damage	(m) : 221
DR for 100% third degree of burn	(m) : 361
DR for 50% third degree of burn	(m) : 415

Table 20.7. Result of consequence analysis for scenario 6; accident in fuel oil storage unit

Parameters	Values
Fire: Pool fire	
Instantaneous Model	
Radius of the pool fire	(m) :5.00
Burning area	(sq. m) :78.5
Burning rate	(kg/s) :35.0
Heat flux	(kJ/sq. m) :154446.5
Damage Radii (DR) due to thermal load	
DR for 100% fatality/damage	(m) :221
DR for 50% fatality/damage	(m) :307
DR for 100% third degree of burn	(m) :391
DR for 50% third degree of burn	(m) :435

Table 20.8. Result of consequence analysis for scenario 7; accident in Dichloropropane storage unit

Parameters	*Values*
Fire: Pool fire	
Instantaneous Model	
Radius of the pool fire	(m) : 5.00
Burning area	(sq. m) : 79.0
Burning rate	(kg/s) : 7.00
Heat flux	(kJ/sq. m) : 2455.00
Damage Radii (DR) due to thermal load	
DR for 100% fatality/damage	(m) : 76
DR for 50% fatality/damage	(m) : 95
DR for 100% third degree of burn	(m) : 109
DR for 50% third degree of burn	(m) : 140

Table 20.9. Result of consequence analysis for scenario 8; accident in propylene oxide reactor

Parameters	*Values*
Fire: Pool fire	
Continuous Model	
Burning area	(sq. m) : 28701.00
Burning rate	(kg/s) : 4396.00
Heat flux	(kJ/sq. m) : 59.00
Damage Radii (DR) due to thermal load	
DR for 100% fatality/damage	(m) : 12
DR for 50% fatality/damage	(m) : 15
DR for 100% third degree of burn	(m) : 17
DR for 50% third degree of burn	(m) : 22

Table 20.10. Result of consequence analysis for scenario 9; accident in propylene oxide reactor

Parameters	*Values*
Fire: Flash Fire	
Volume of vapour cloud	(cub.m) : 9819.00
Effective time of fire	(sec) : 15635.00
Effective thermal load	(kJ/sq.m) : 443.00
Damage Radii (DR) due to thermal load	
DR for 100% fatality/damage	(m) : 32
DR for 50% fatality/damage	(m) : 41
DR for 100% third degree of burn	(m) : 47
DR for 50 % third degree of burn	(m) : 60

Table 20.11. Result of consequence analysis for scenario 10; accident in propylene glycol reactor

Parameters	Values
Fire: Flash Fire	
Volume of vapour cloud	(cub.m) : 10289.00
Effective time of fire	(sec) : 348794.00
Effective thermal load	(kJ/sq.m) : 971.00
Damage Radii (DR) due to thermal load	
DR for 100% fatality/damage	(m) : 48
DR for 50% fatality/damage	(m) : 60
DR for 100% third degree of burn	(m) : 69
DR for 50 % third degree of burn	(m) : 88

Table 20.12. Result of consequence analysis for scenario 11; accident in propylene glycol reactor

Parameters	Values
Fire: Fire Ball	
Radius of the fire ball	(m) : 22.00
Duration of the fire ball	(s) : 9.00
Energy released by fire ball	(kJ) : 3.76E+06
Radiation heat flux	(kJ/sq. m) : 58.00
Damage Radii (DR) due to thermal load	

Parameters	Values
DR for 100% fatality/damage	(m) : 12
DR for 50% fatality/damage	(m) : 15
DR for 100% third degree of burn	(m) : 17
DR for 50 % third degree of burn	(m) : 22

Table 20.13. Result of consequence analysis for scenario 12; accident in polyol reactor

Parameters	Values
Explosion : BLEVE	
Total energy released	(kJ) : 5.5E+04
Peak over pressure	(kPa) :4.00
Variation of over pressure in air	(kPa/s) : 3.00
Shock velocity of air	(m/s) : 335.00
Duration of shock wave	(ms) : 67.0
No Missile effect	
Damage radii (DR) for various degree of damage due to over pressure	
DR for 100% complete damage	(m) :9
DR for 100% fatality or 50% complete damage	(m) : 13
DR for 50% fatality or 25% complete damage	(m) : 19
Fire: Fire Ball	
Radius of the fire ball	(m) : 26.00
Duration of the fire ball	(s) : 11.00
Energy released by fire ball	(kJ) : 1.01E+07
Radiation heat flux	(kJ/sq. m) :155.00
Damage Radii (DR) due to thermal load	
DR for 100% fatality/damage	(m) : 34
DR for 50% fatality/damage	(m) : 41
DR for 100% third degree of burn	(m) : 56
DR for 50% third degree of burn	(m) : 69

Table 20.14. Result of consequence analysis for scenario 13; accident in polyol reactor

Parameters	Values
Fire: Fire Ball	
Radius of the fire ball	(m) : 26.00
Duration of the fire ball	(s) : 11.00
Energy released by fire ball	(kJ) : 1.01E+07
Radiation heat flux	(kJ/sq. m) :155.00
Damage Radii (DR) due to thermal load	
DR for 100% fatality/damage	(m) : 19
DR for 50% fatality/damage	(m) : 24
DR for 100% third degree of burn	(m) : 28
DR for 50 % third degree of burn	(m) : 36

Table 20.15. Result of consequence analysis for scenario 14; accident in chlorine pipeline

Parameters	Values
Toxic Release & Dispersion	
Box continuous : Model	
Value of wind speed	(m/s) : 2.5000
Ground level X concentration	(kg/cu.m) : 0.0131
Damage radii (DR) for various degree of damage :	
DR for 100% lethality	(m) : 2838
DR for 50% lethality	(m) : 4028
DR for 10% lethality	(m) : 9079
Plume characteristics	
Ground level concentration of plume	(kg/cu.m) : 0.0001
Ground level concentration of plume on axis(kg/cu.m) : 0.0263	
Width of plume	(m) : 33.1
Cloud concentration	(kg/cu.m) : 0.0145
Maximum ground level concentration	(kg/cu.m) : 0.0163

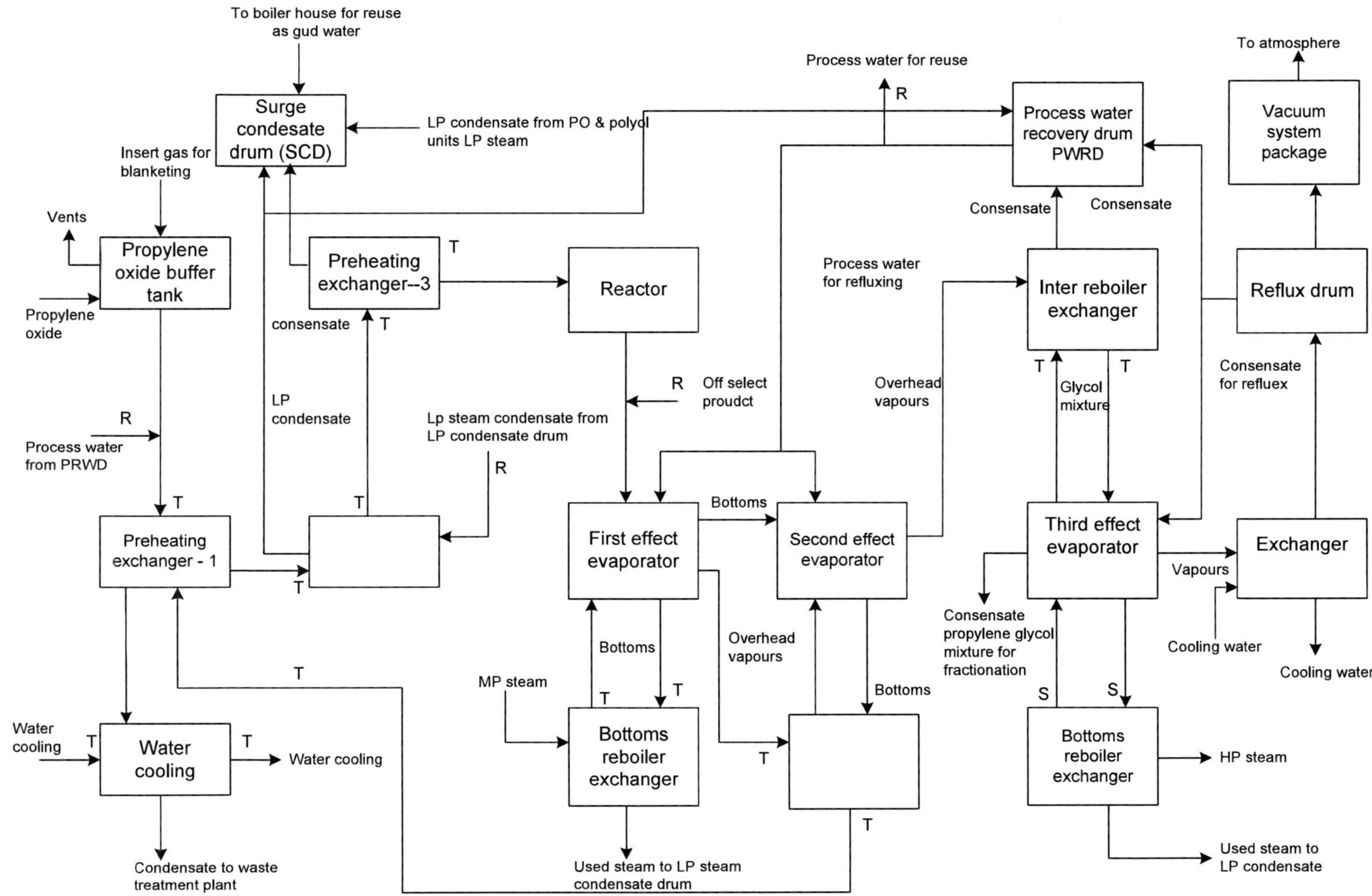

Figure 20.1. Propylene Glycol plant – A.

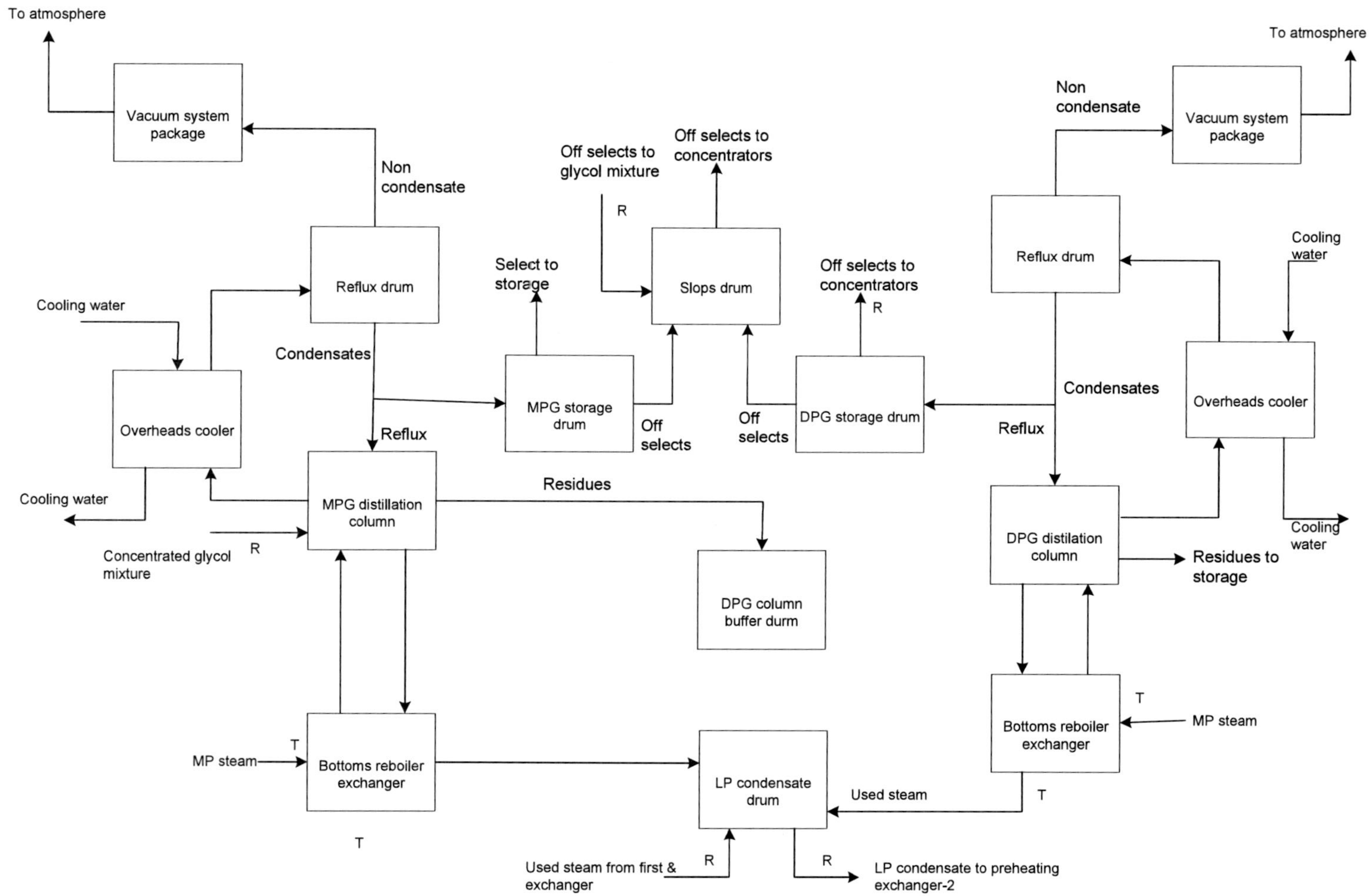

Figure 20.2. Propylene Glycol plant –B.

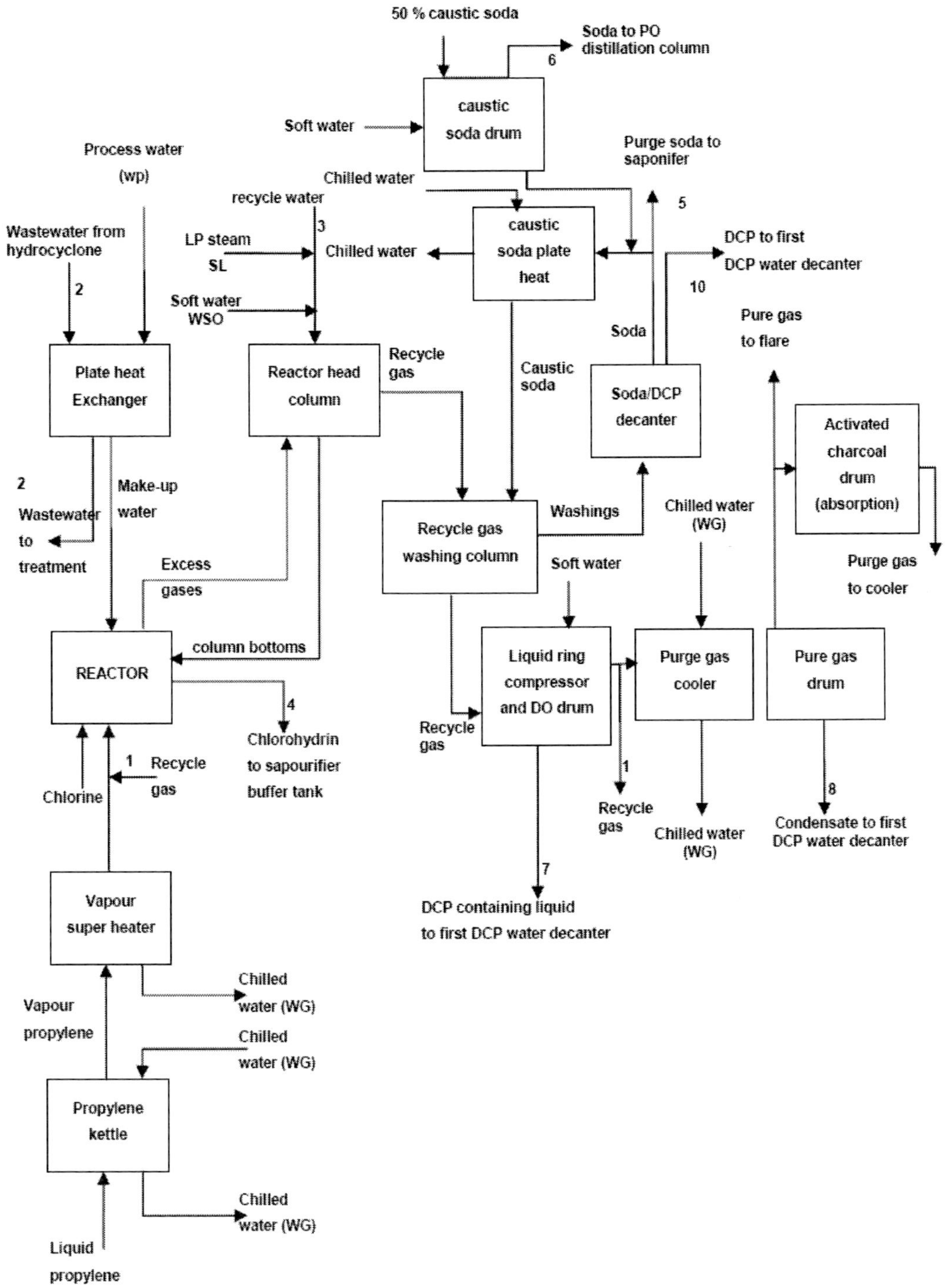

Figure 20.3. Propylene oxide plant A.

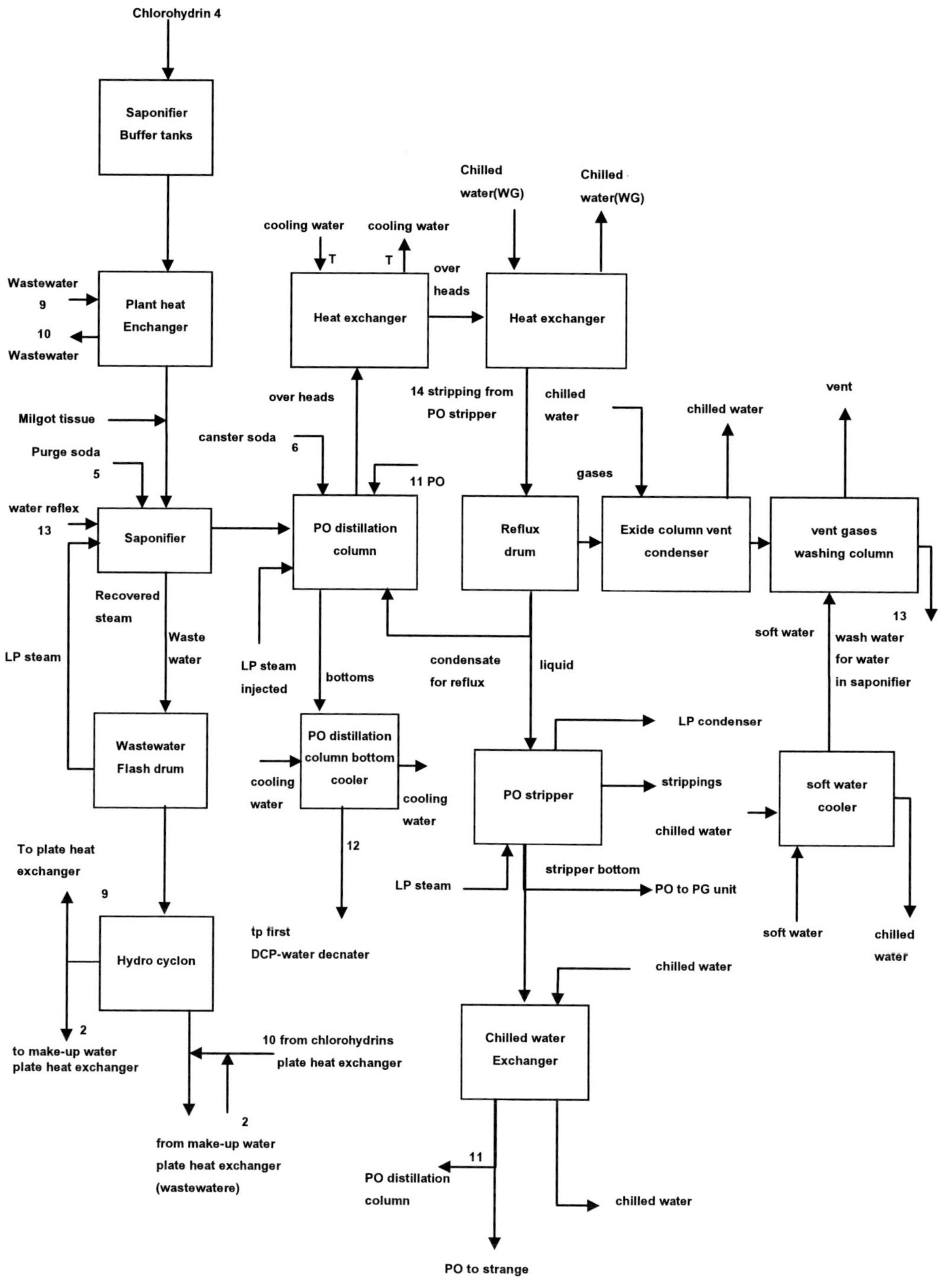

Figure 20.4. Propylene oxide plant – B.

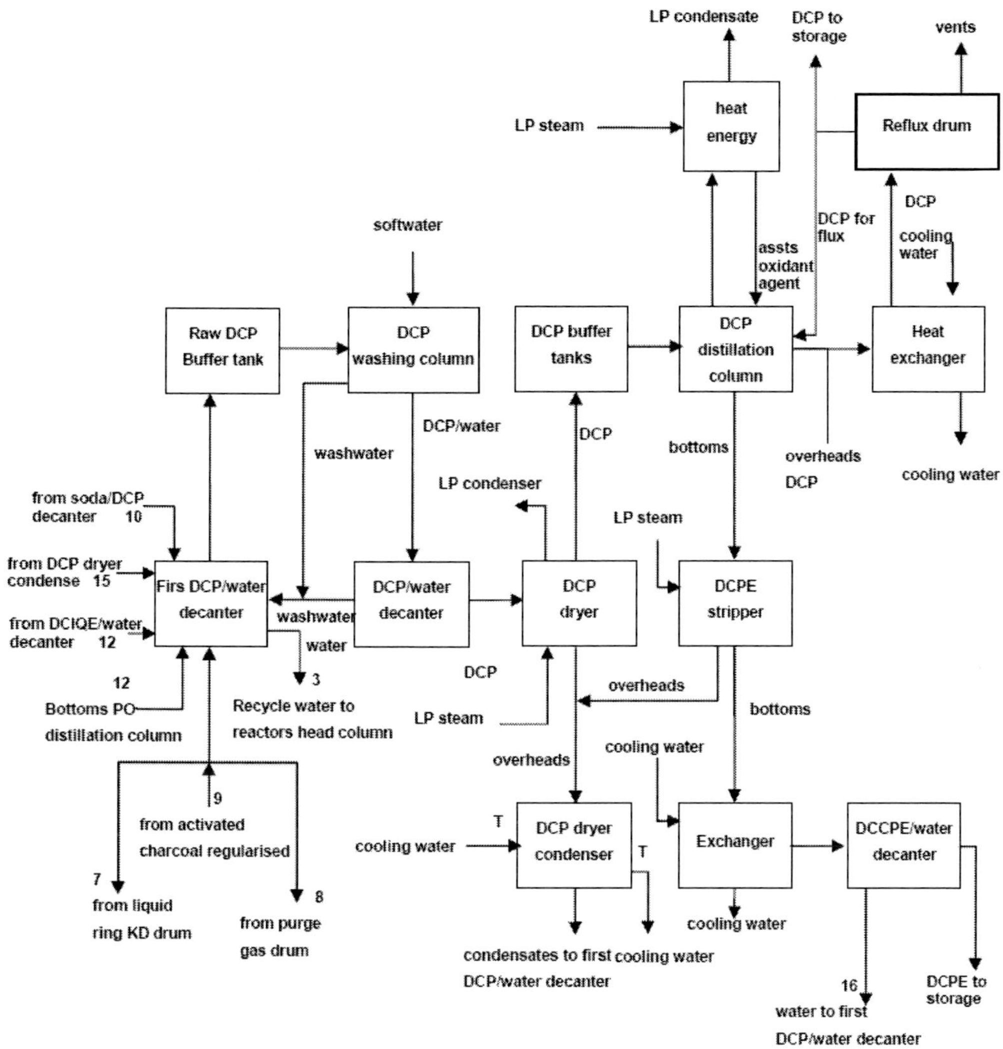

Figure 20.5. Propylene oxide plant – C.

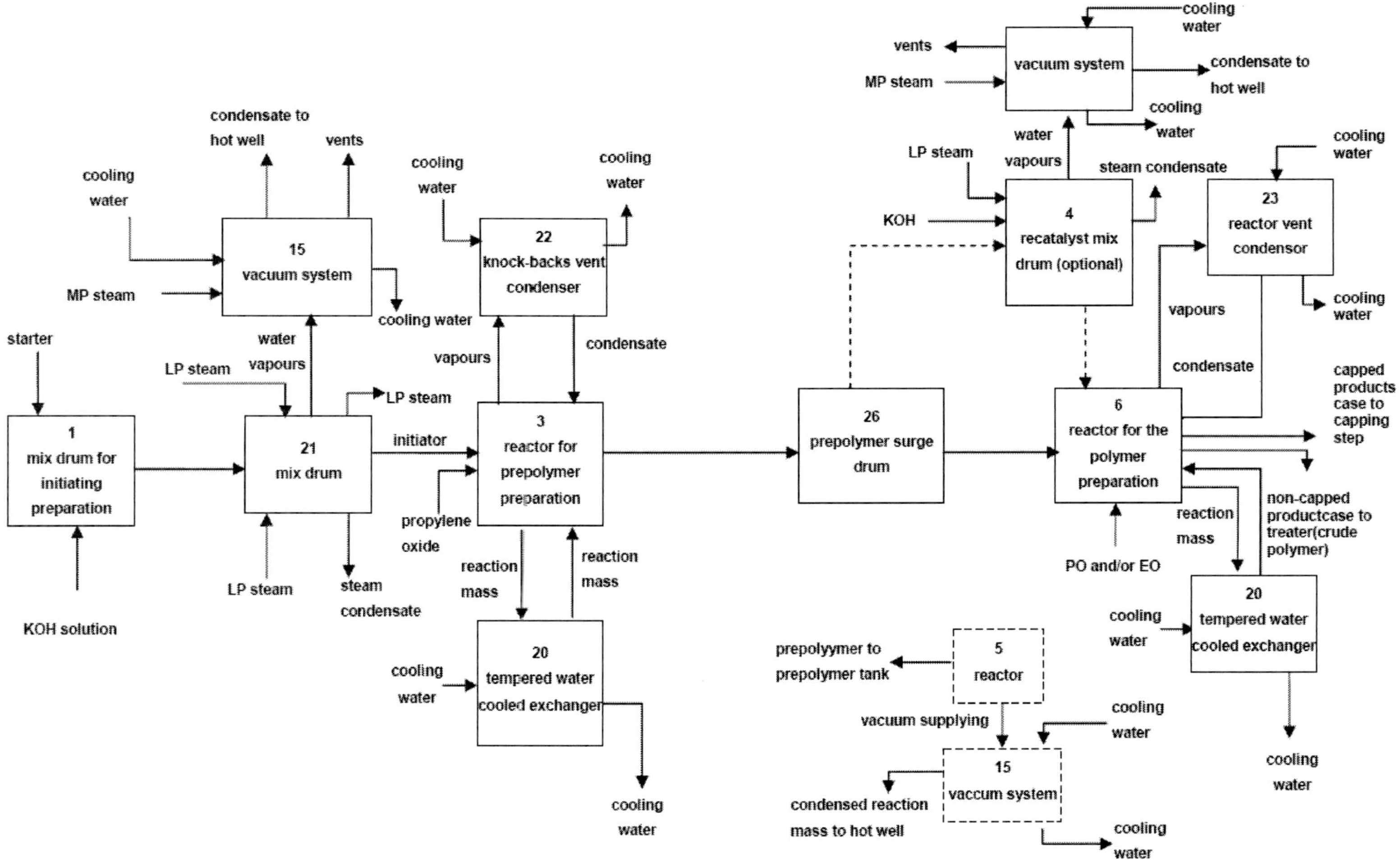

Figure 20.6. Polyols unit – A.

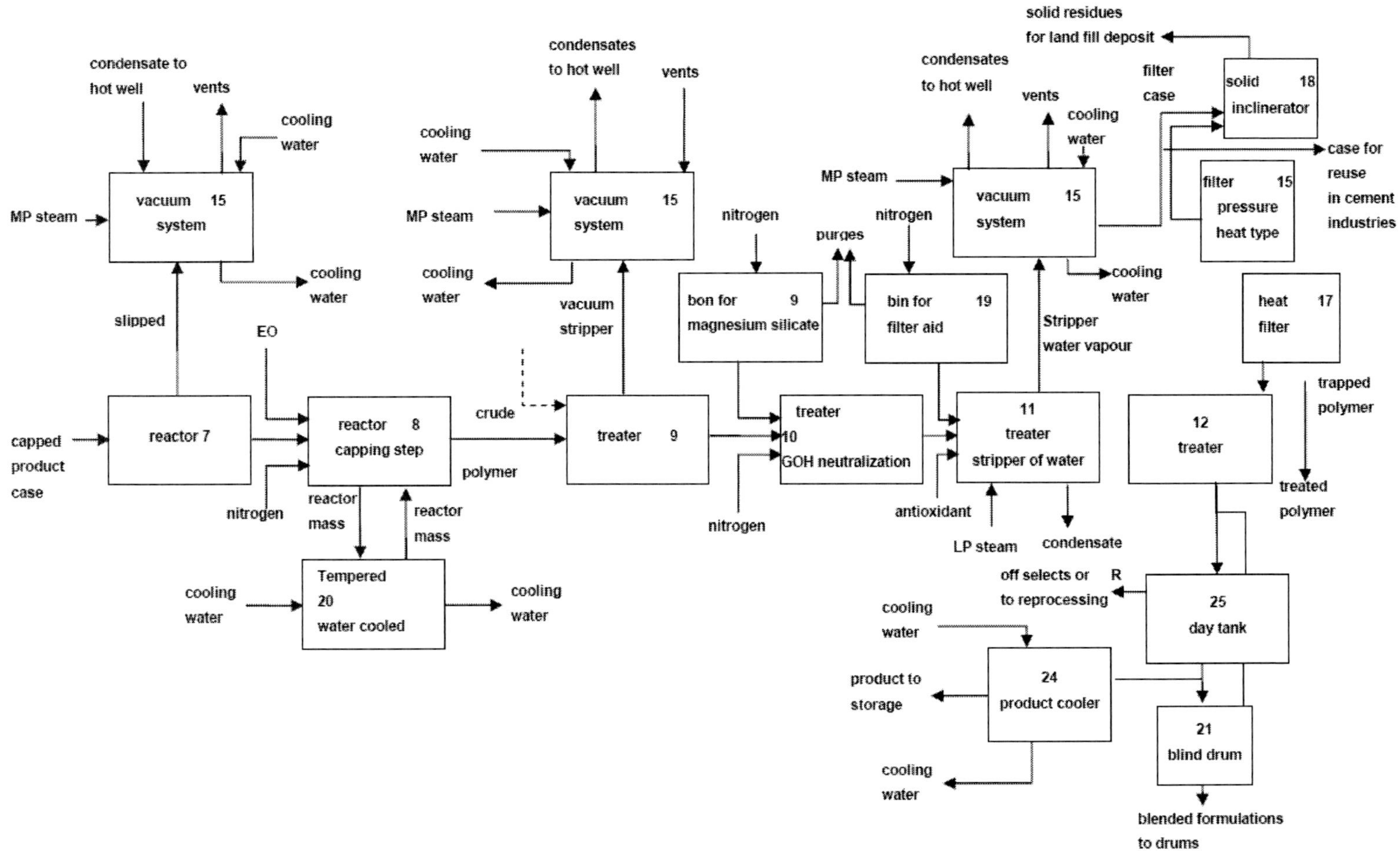

Figure 20.7. Polyols unit – B.

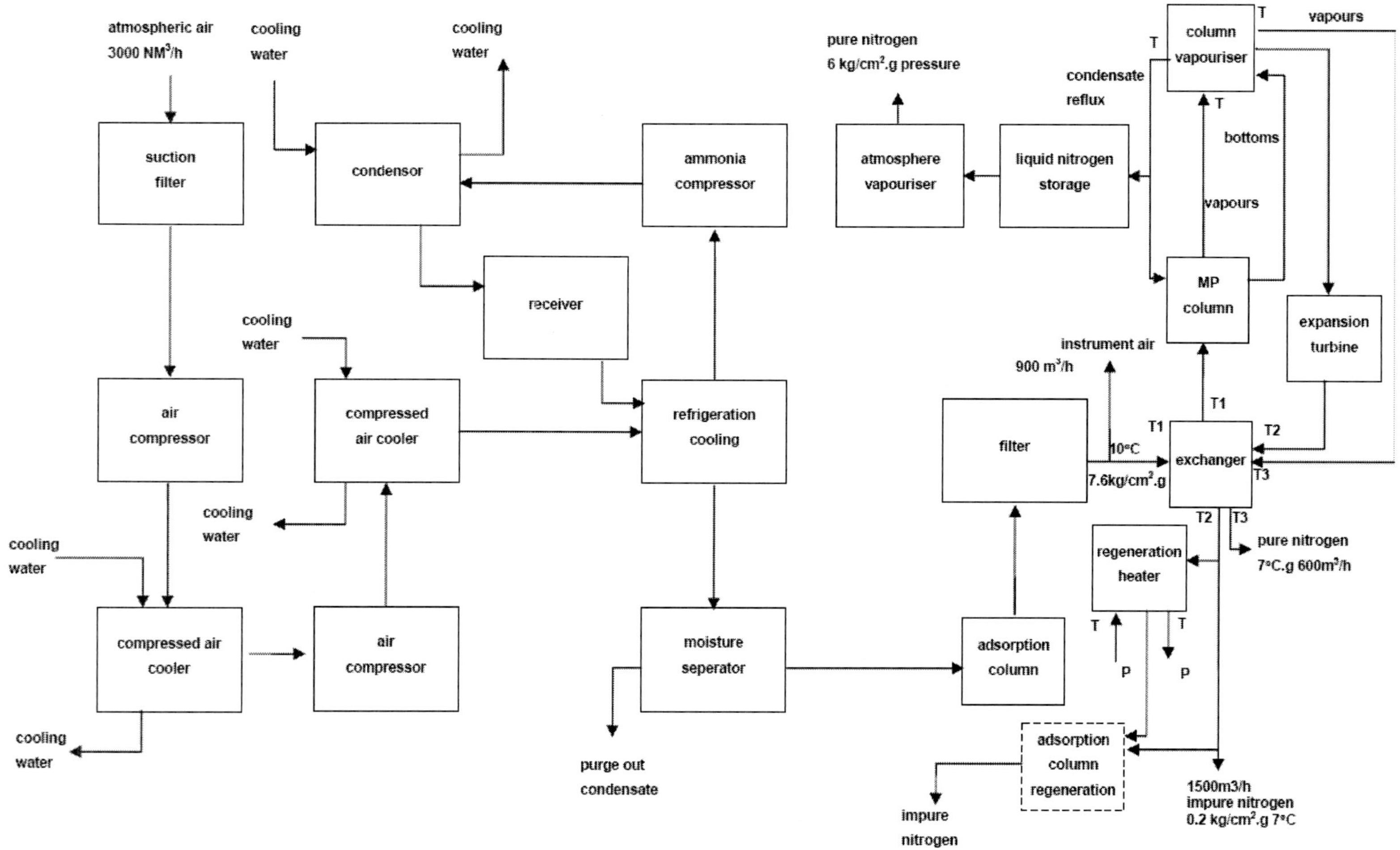

Figure 20.8. Nitrogen plant.

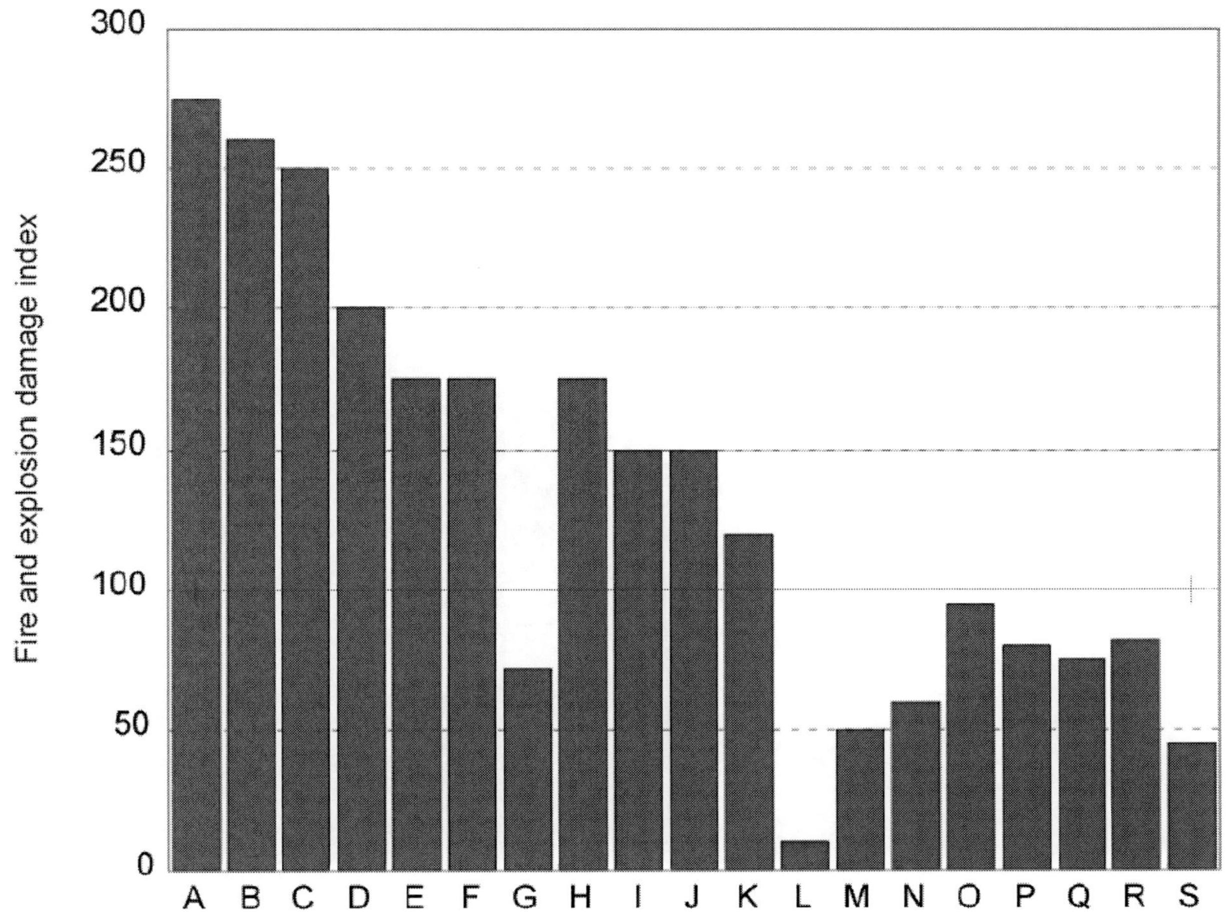

Figure 20.9. TDI (Toxic damage index) values for various units in MPL.

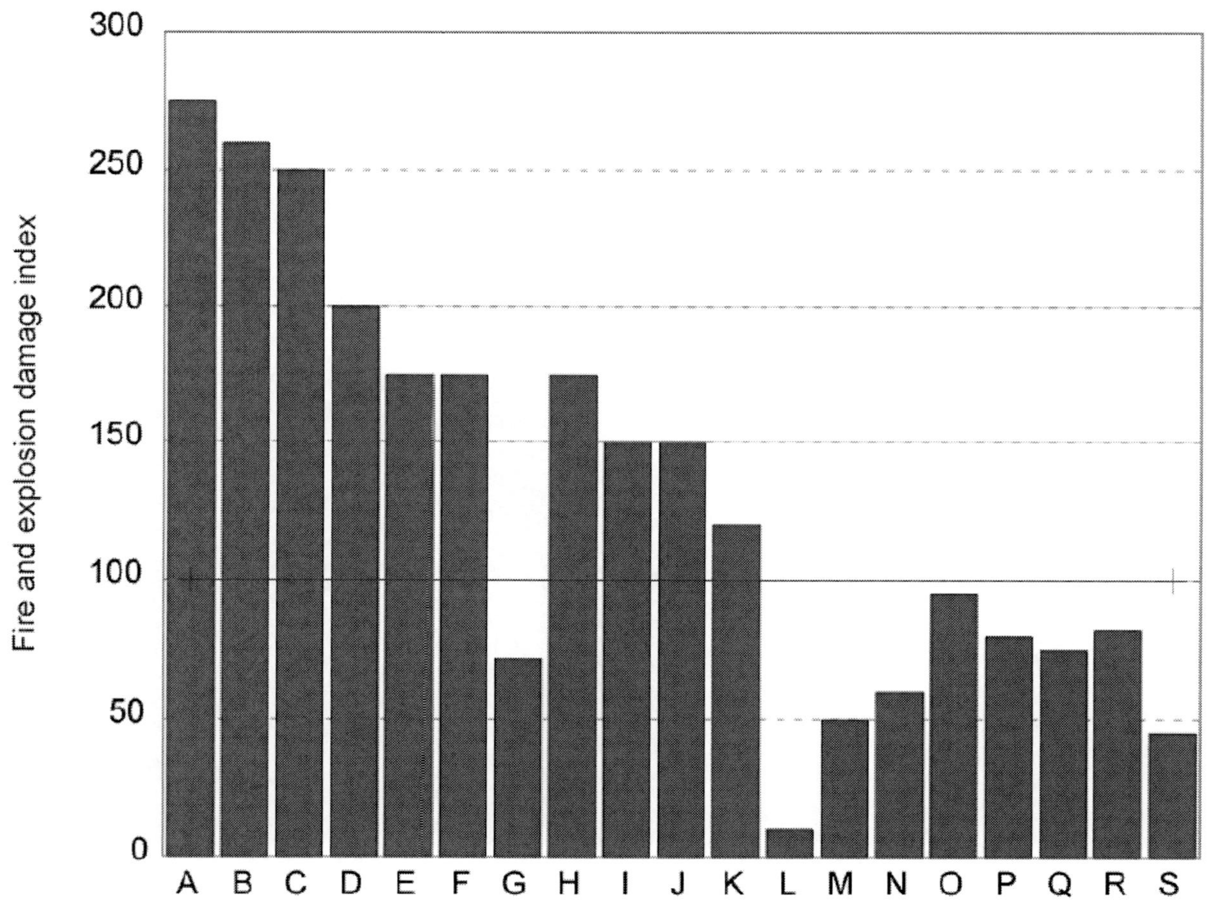

Figure 20.10. FEDI (Fire and explosion damage index) values for various units in MPL.

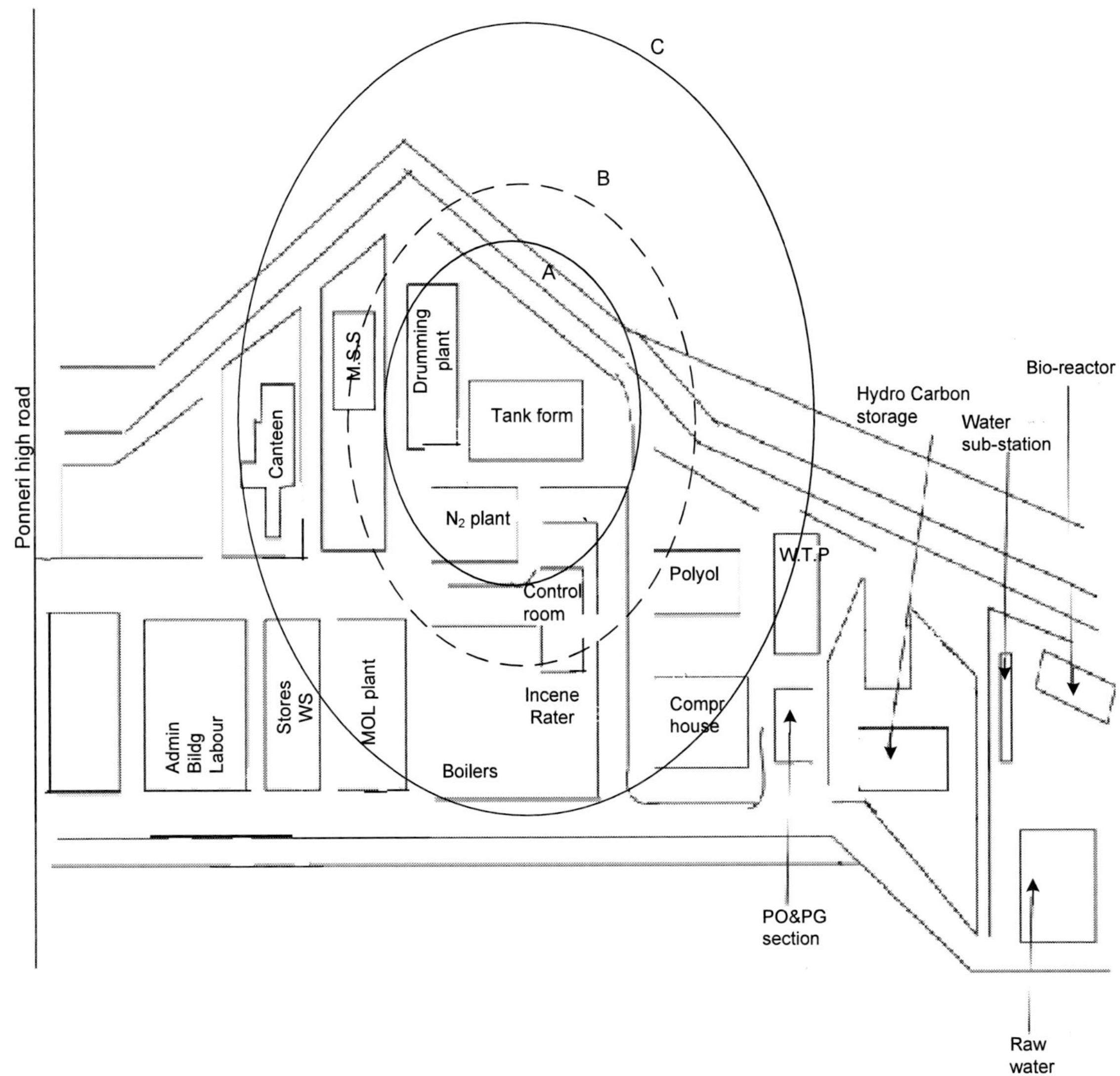

Figure 20.11. Risk contours for scenario 1 indicating the impact area due to severe risk (A), high risk (B), and moderate risk (C).

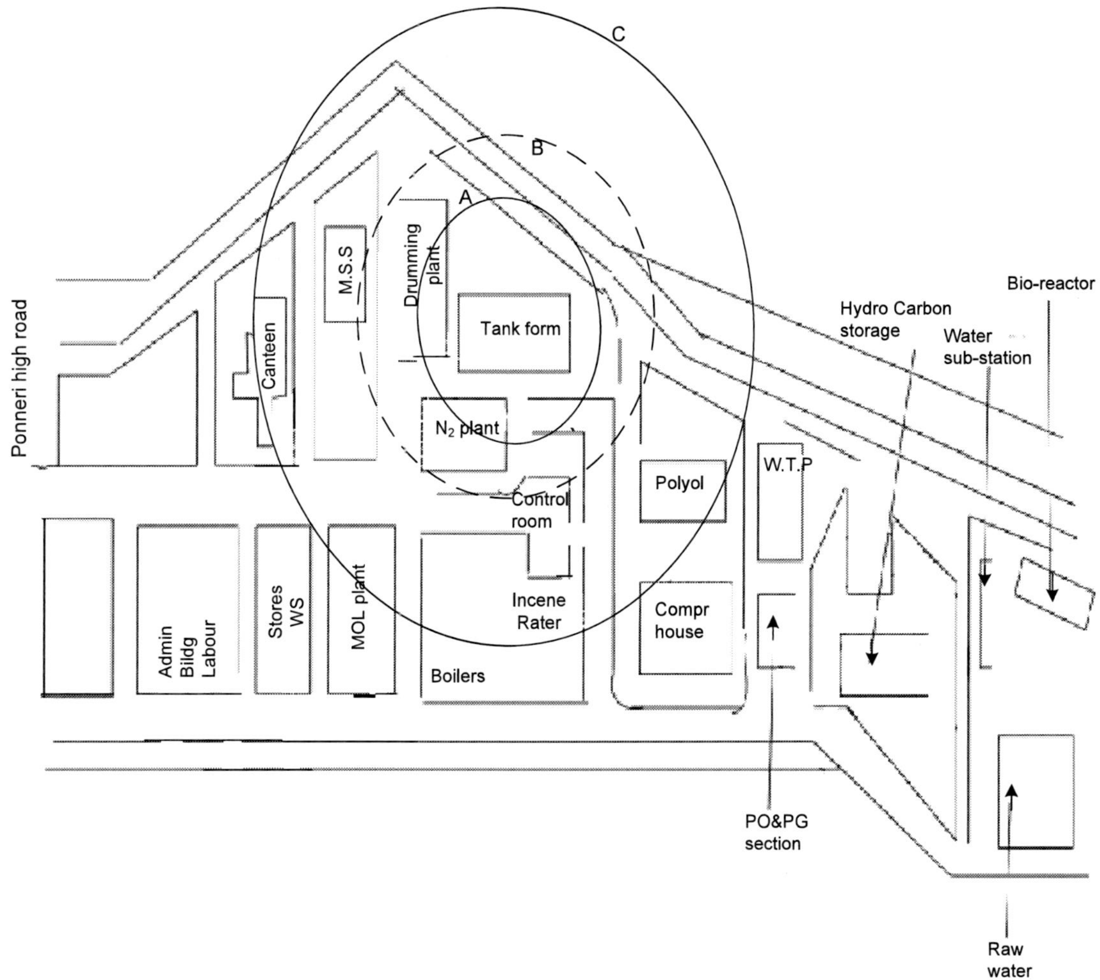

Figure 20.12. Risk contours for scenario 2 indicating the impact area due to severe risk (A), high risk (B), and moderate risk (C).

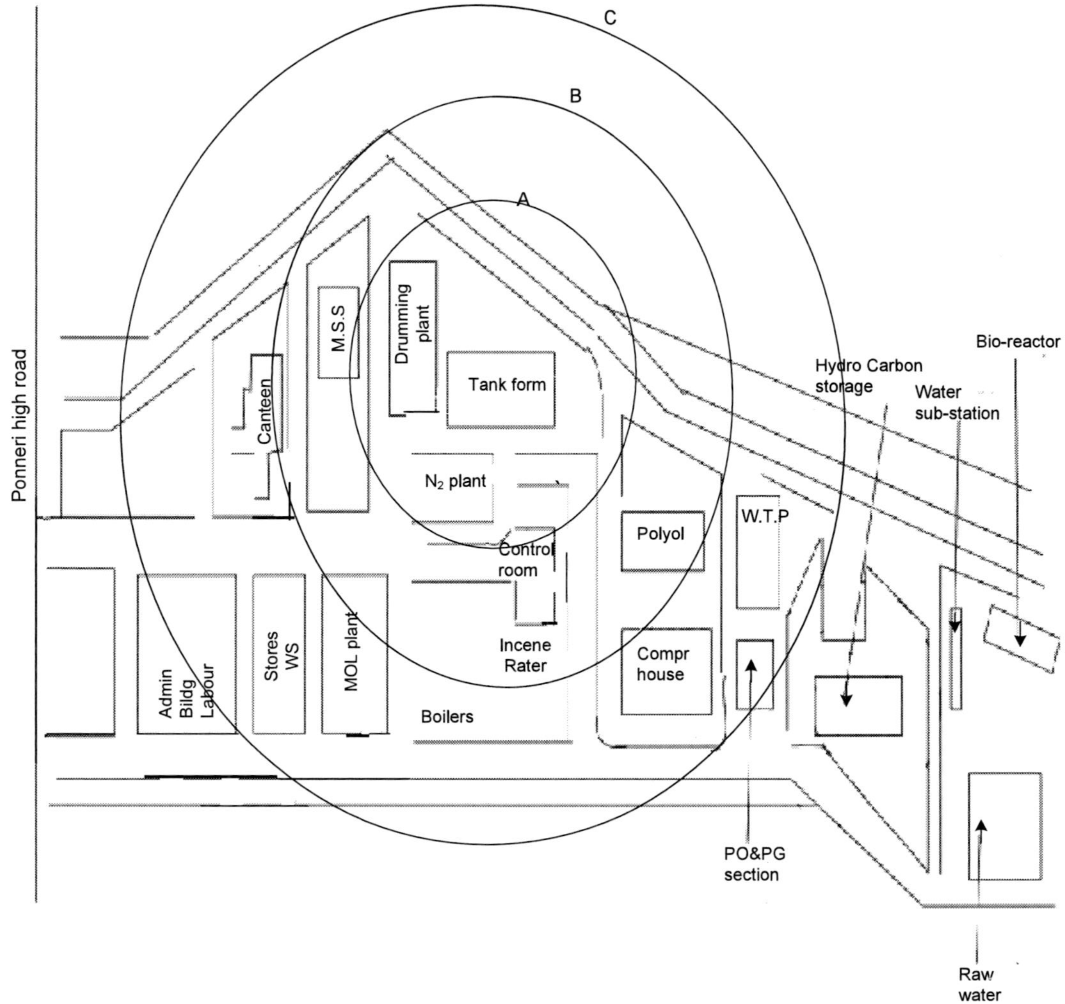

Figure 20.13. Risk contours for scenario 3 indicating the impact area due to severe risk (A), high risk (B), and moderate risk (C).

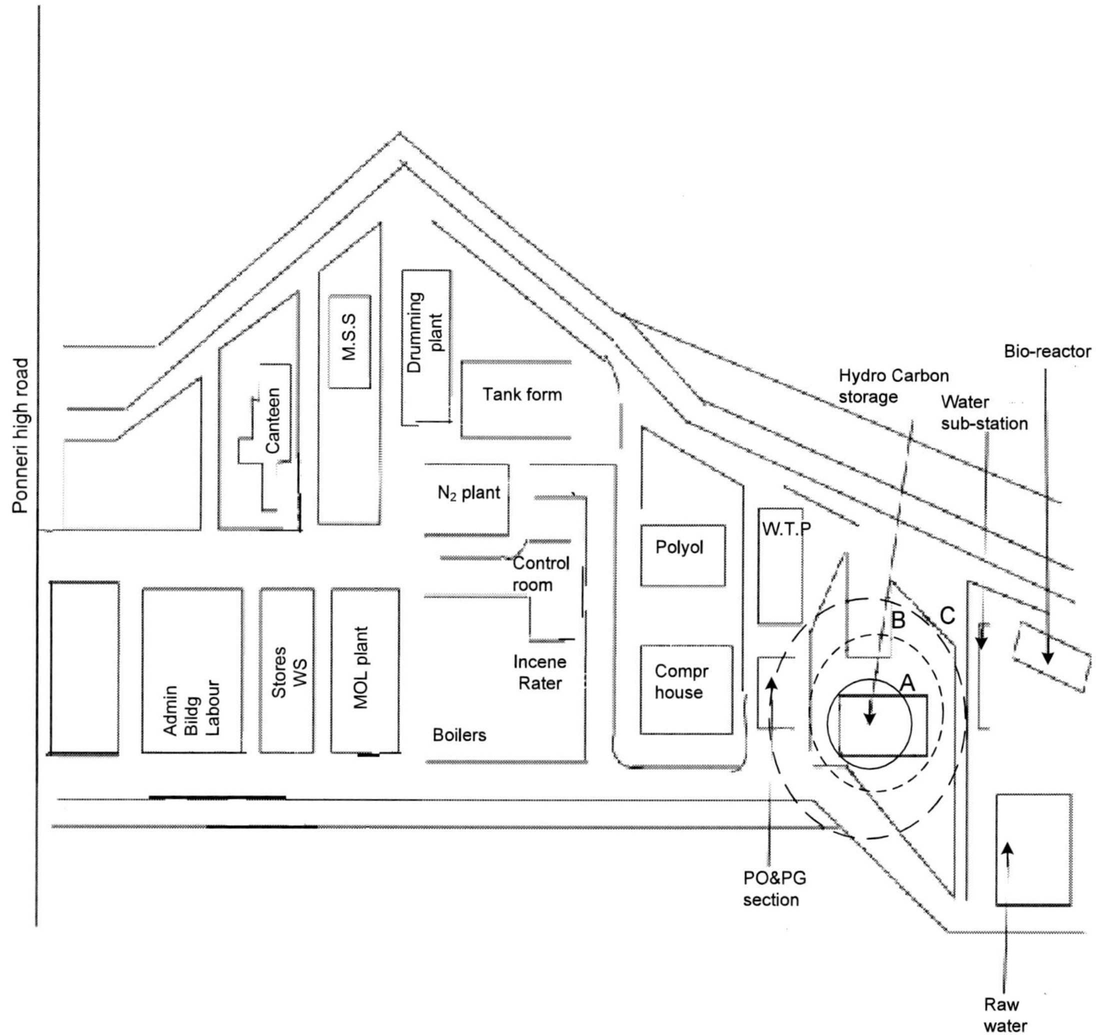

Figure 20.14. Risk contours for scenario 4 indicating the impact area due to severe risk (A), high risk (B), and moderate risk (C).

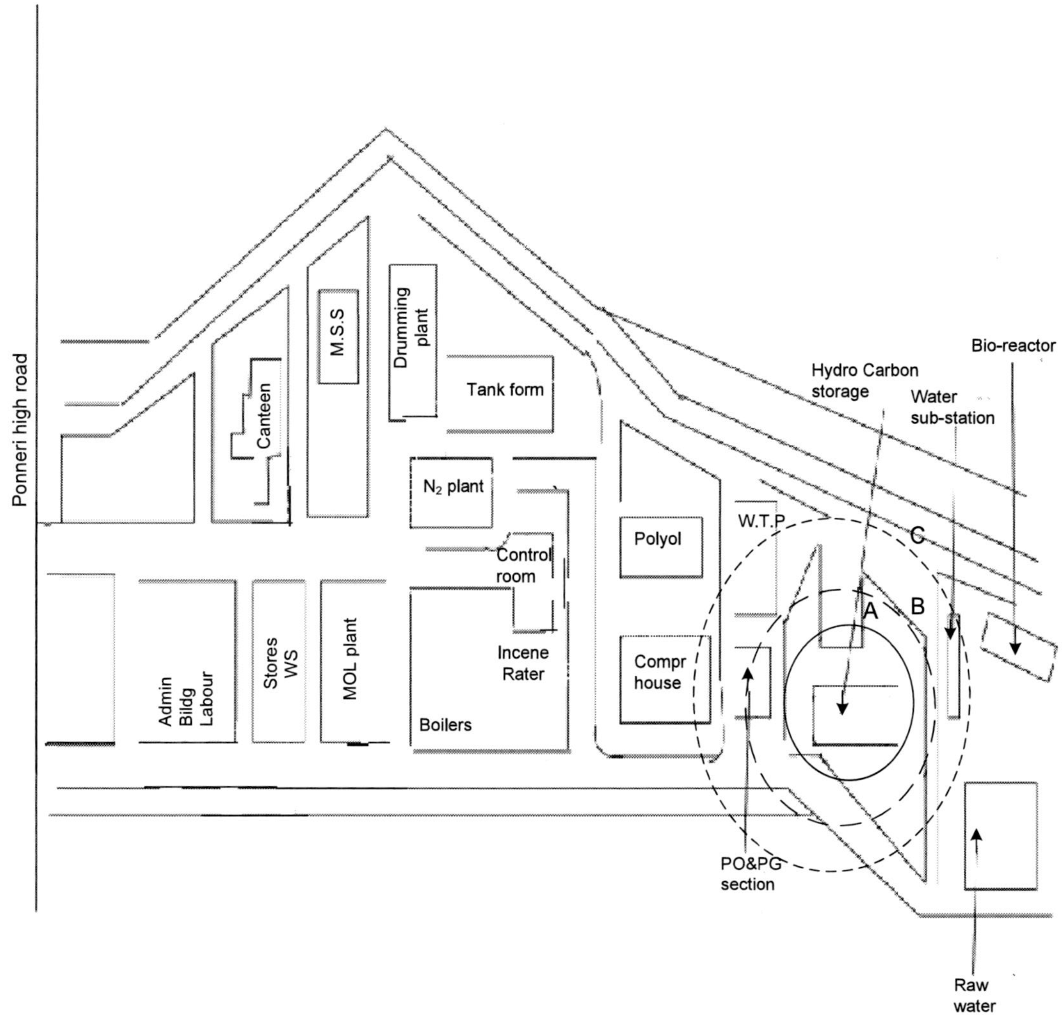

Figure 20.15. Risk contours for scenario 5 indicating the impact area facing very high (A), high (B), and moderate (C) risk.

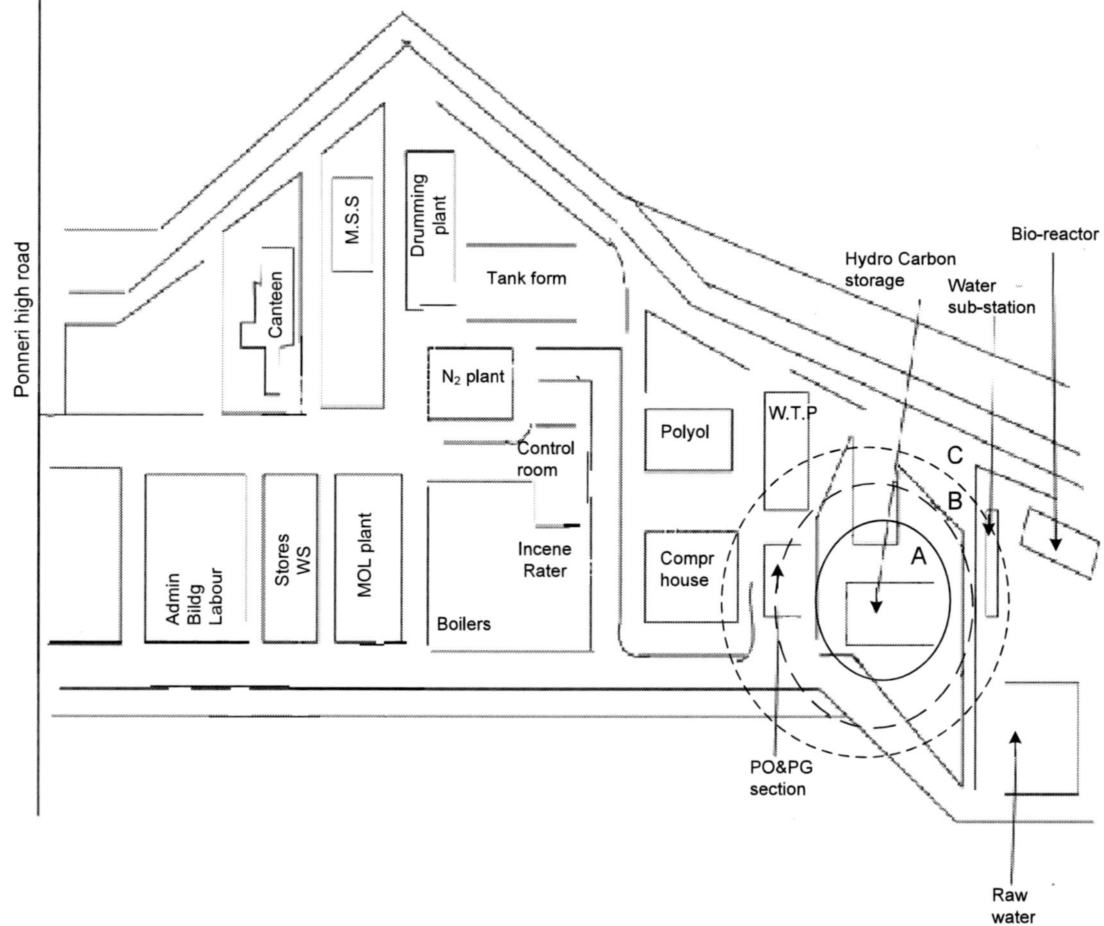

Figure 20.16. Risk contours for scenario 6 indicating the impact area due to severe risk (A), high risk (B), and moderate risk (C).

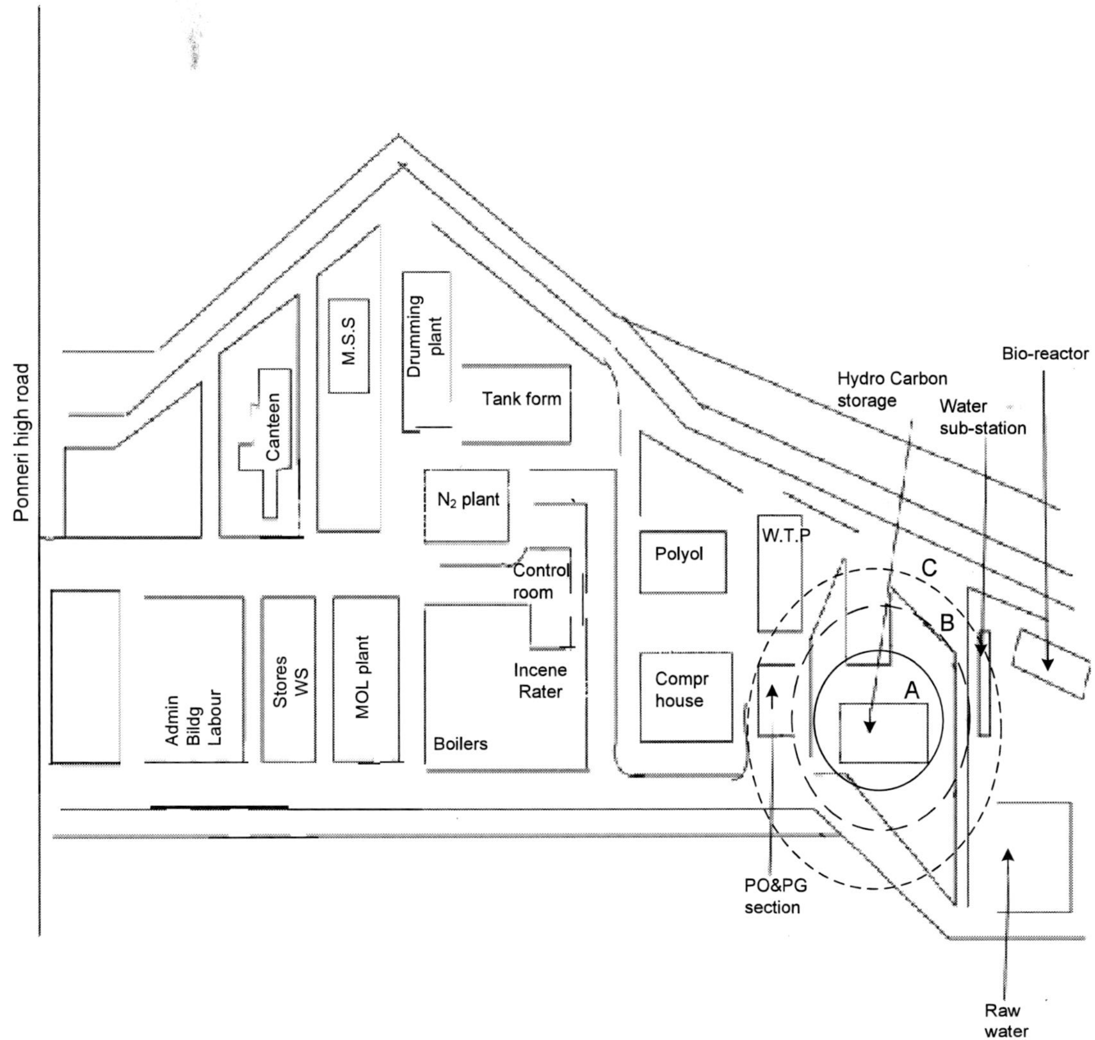

Figure 20.17. Risk contours for scenario 7 indicating the impact area due to severe risk (A), high risk (B), and moderate risk (C).

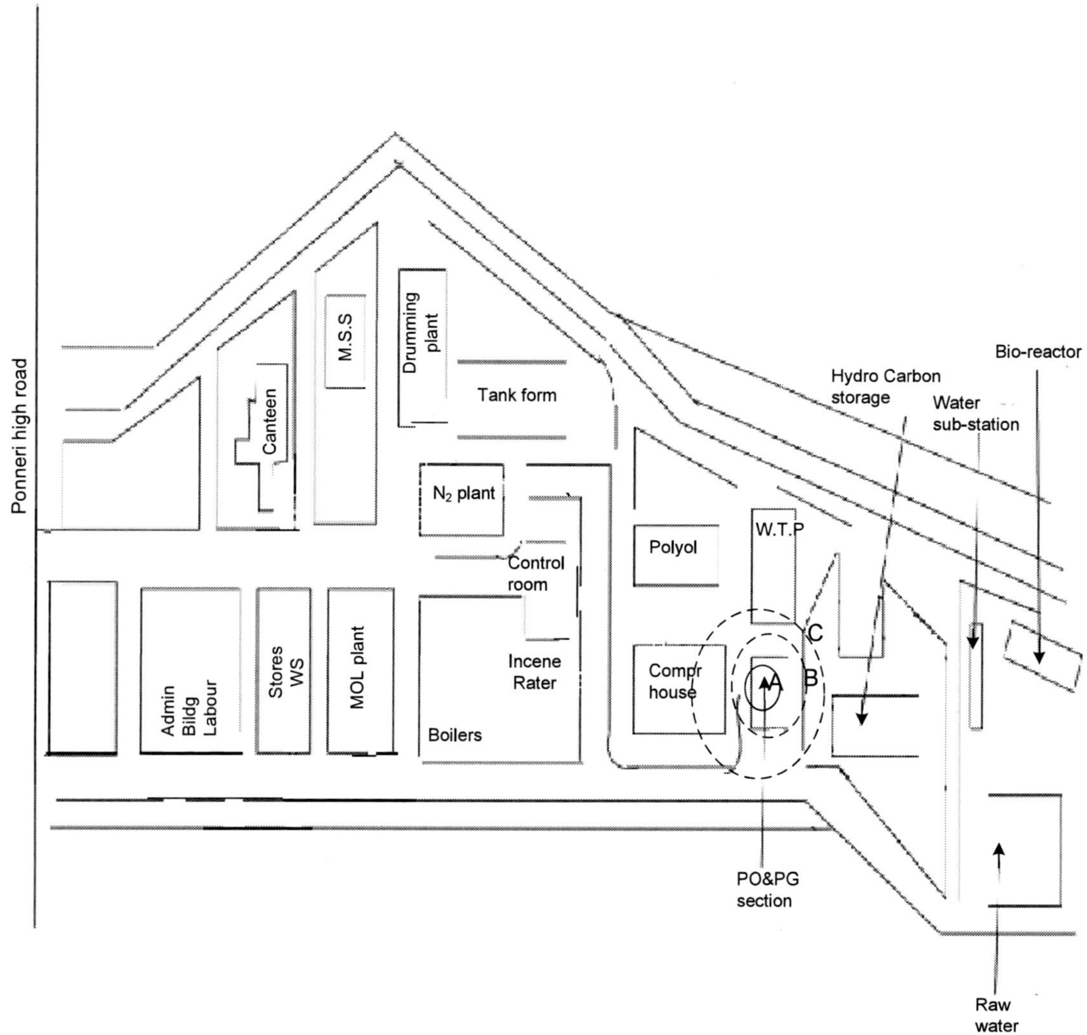

Figure 20.18. Risk contours for scenario 8 indicating the impact area due to severe risk (A), high risk (B), and moderate risk (C).

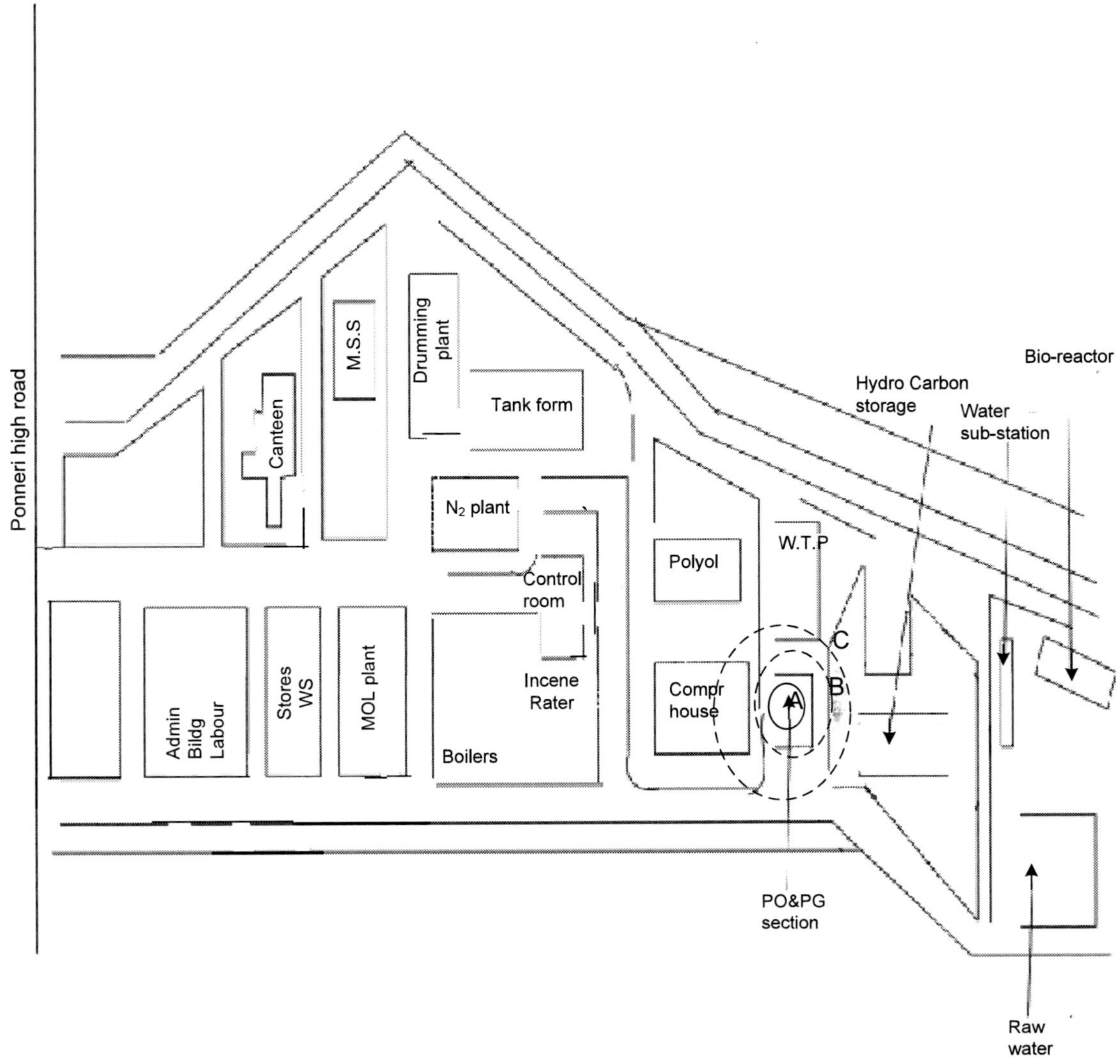

Figure 20.19. Risk contours for scenario 9 indicating the impact area due to severe risk (A), high risk (B), and moderate risk (C).

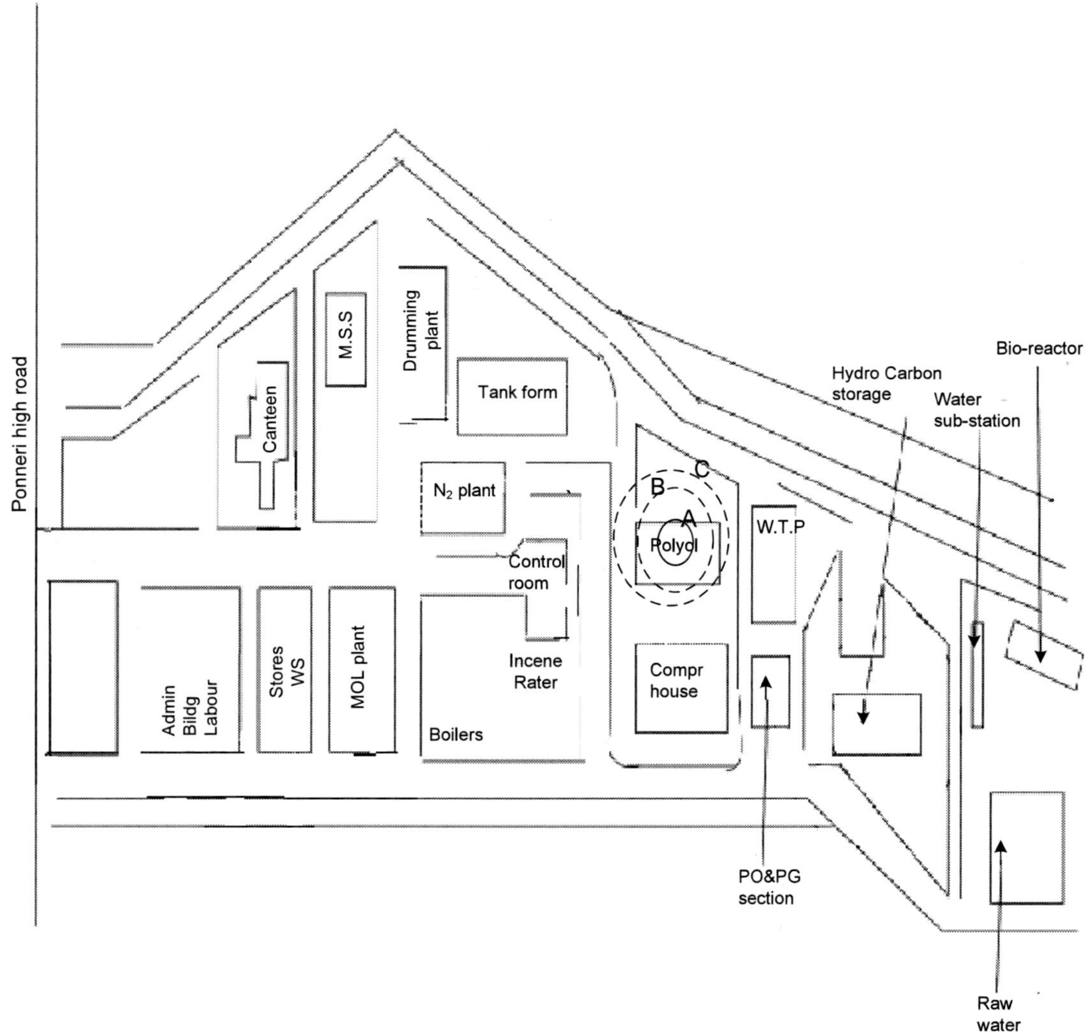

Figure 20.20. Risk contours for scenario 10 indicating the impact area due to severe risk (A), high risk (B), and moderate risk (C).

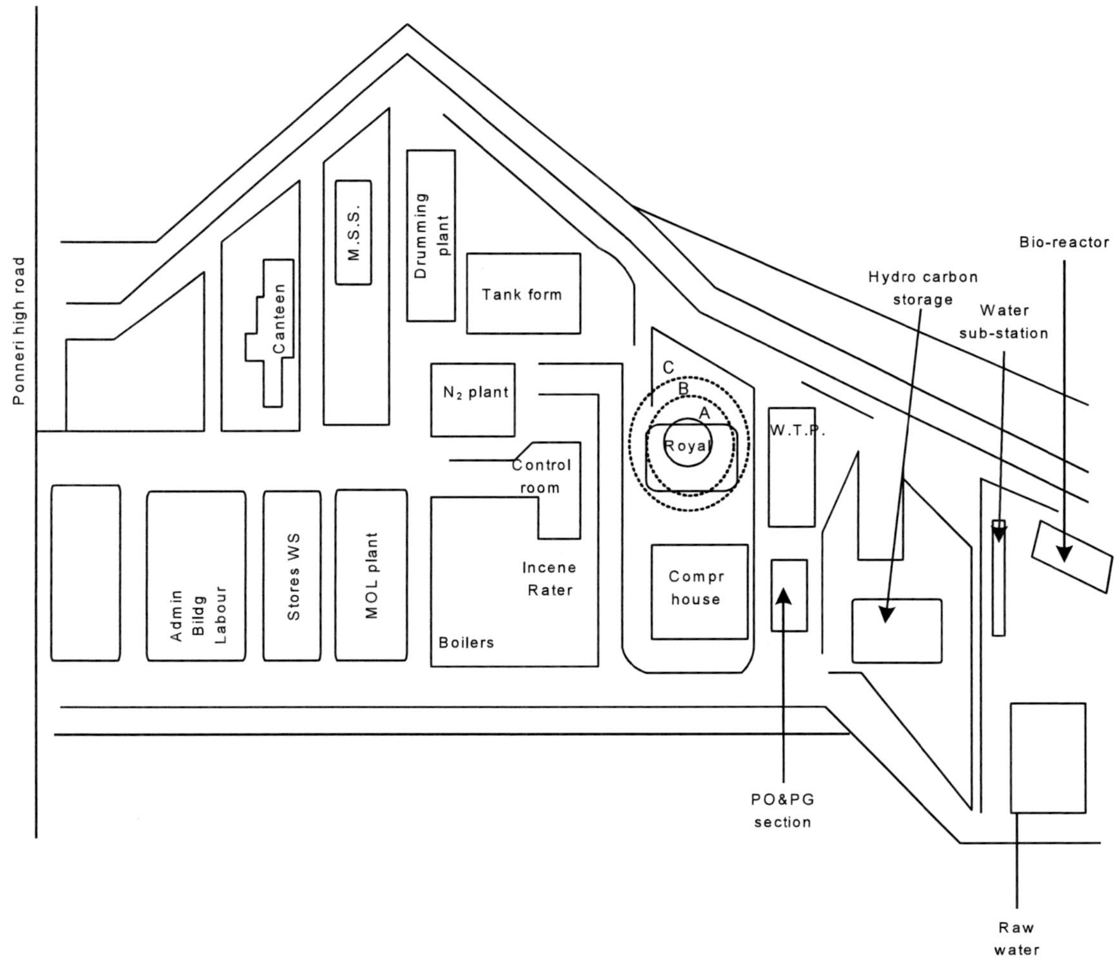

Figure 20.21. Risk contours for scenario 11 indicating the impact area due to severe risk (A), high risk (B), and moderate risk (C).

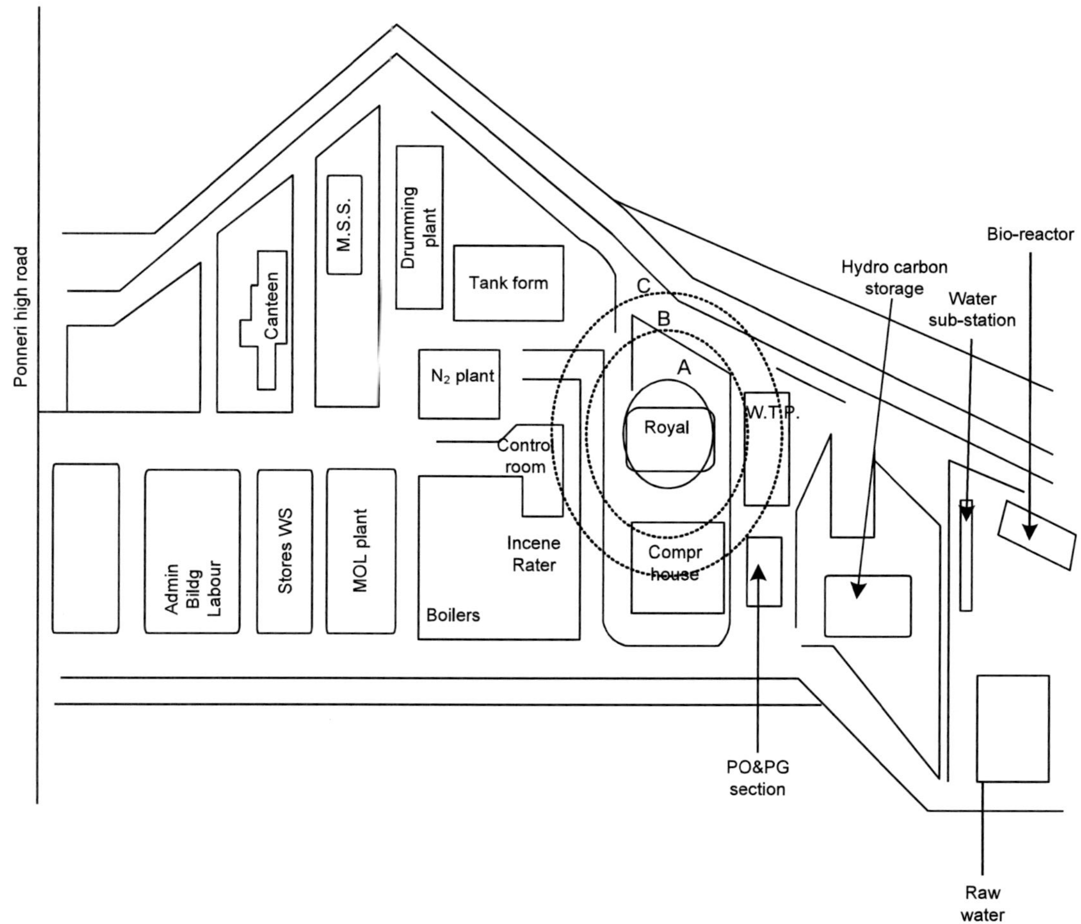

Figure 20.22. Risk contours for scenario 12 indicating the impact area due to severe risk (A), high risk (B), and moderate risk (C).

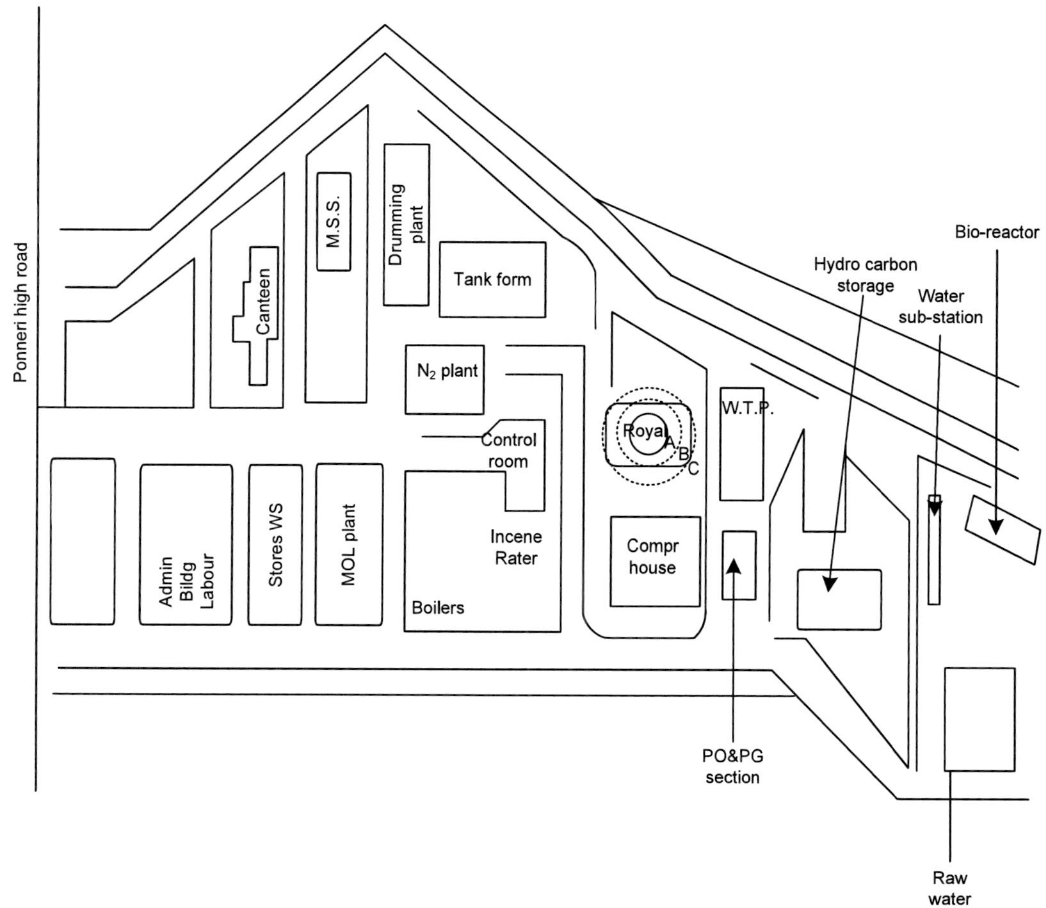

Figure 20.23. Risk contours for scenario 13 indicating the impact area due to severe risk (A), high risk (B), and moderate risk (C).

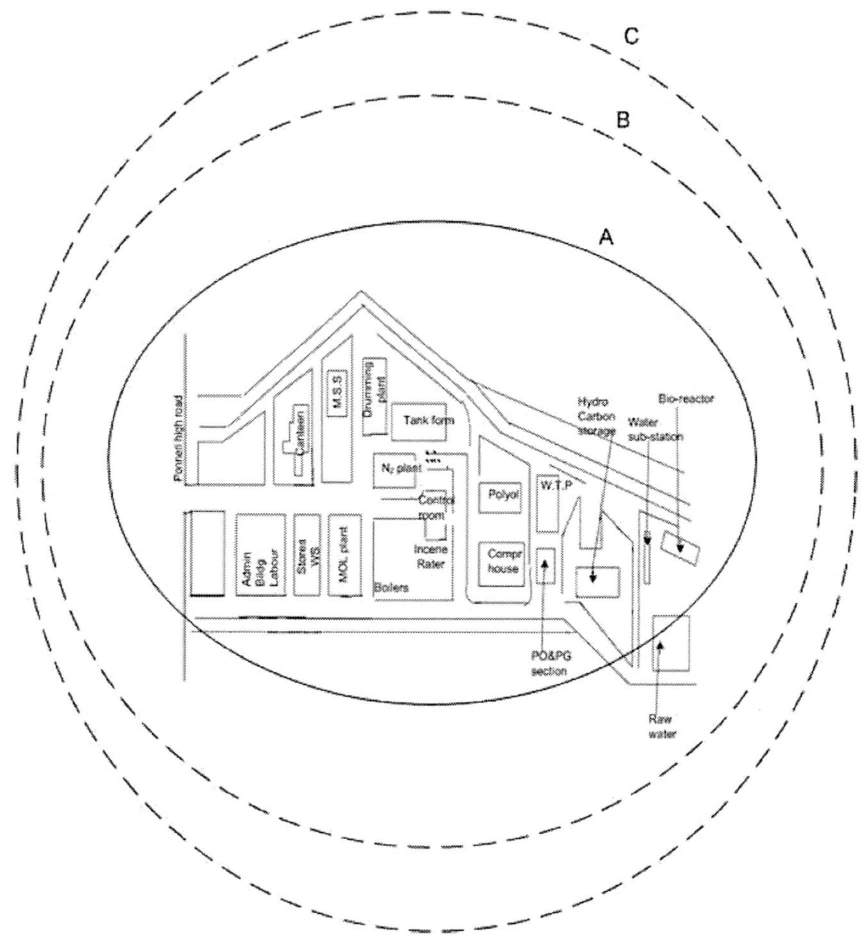

Figure 20.24. Risk contours for scenario 14 indicating the impact area due to severe risk (A), high risk (B), and moderate risk (C).

Chapter 21

REGIONAL RISK ASSESSMENT: STUDIES ON *DOMINO EFFECT* (CHAIN OF ACCIDENTS)

21.1. INTRODUCTION

Accidents in chemical process industries involving explosions and/or fires can damage nearby reactors, pipelines, or storage vessels. If the installations thus hit by the primary accident happen to contain hazardous chemicals, the damage my trigger fresh explosions/fires/toxic releases. This type of 'chain of accidents' may continue to propagate itself by taking into its fold other units within striking distance, which, on being hit, may become the cause of fresh accidents thereby perpetuating the chain.

In the domain of chemical process industries, occurrence of such a chain of accident is called 'domino effect'. The name is derived from the game of domino toppling in which large numbers of small flat blocks, often made of wood, are so arranged in patterns that toppling of one block over the next one starts a chain in which each toppling block makes the next one to topple and so on. Upto several thousand dominos kept down the line can be toppled by the initiating event of the toppling of the first domino.

A look at the history of accidents involving hazardous chemicals reveals that chain of accidents have been rather common. For example at Vishakhapatnam, India (Khan & Abbasi 1999, 2000), a leak in one of the eight Horton spheres at a refinery storing crude led to a fire/explosion which damaged a nearby sphere causing it to explode. The 'chain of accidents' spread to other storage vessels and soon there was fire and damage all around. Sixty people lost their lives and assets worth ~ Rs 60 million were damaged. In 1989, a vapour cloud explosion caused by a leak in the polyethylene plant of a Philips installation near Houston, Texas, USA, led to two other major explosions in isobutane storage vessel and polyethylene reactor. These in turn set off a series of seven more explosions. Similar instances of domino effect have been recorded at Mexico City, Mexico 1984; Pasadena, Texas 1989; and elsewhere (Lees 1996, Khan & Abbasi 1998b). Whereas in the earlier times, the industries handling hazardous chemicals were fewer in number and used to be located far away from populous neighbourhoods, in recent years the number of such industries has risen sharply – as a direct consequence of ever increasing variety of products and processes. Due to pressure on land, the distances between such industries and human settlements have become lesser and lesser. For example in Manali Industrial Complex (MIC), several major petroleum and petrochemical industries are situated within a rectangular area of about 10 km^2.

We may briefly recapitulate that MIC is about 3.5 km West of the Bay of Bengal, and is surrounded by the villages Thiruvottiyur and Ennore in the East, Ammulavoil and Vaikkadu in the North, Mathur and Madhavaram in the West, and Chinnasekkadu, Periyasekkadu, and Selavayal in the South (Figure 21.1). The terrain is flat, at mean sea level of 4.5 m. The mean roughness factor of the area is 0.3. Besides MIC, the area within 10km radius of the centre of MIC has populous neighbourhoods, agricultural lands, and barren lands. All these factors combine to make the study area susceptible to domino effect.

21.2. Events that Can Initiate Domino Effect

Domino effect may be initiated by one or more of these causative events:

a) Fire: pool fire, flash fire, fire ball, and jet fire
b) Explosions: confined vapour cloud explosions (CVCE), boiling liquid expanding vapour explosion (BLEVE), vented explosion, vapour cloud explosion, and dust explosion
c) Toxic release: light-as-air-gases; lighter-than-air gases, and heavier-than-air gases; theoretically toxic release can cause domino effect even though instances of it having happened, have not yet been reported.

In order to find out the probability of occurrence of domino effect, assess its damage potential and determine the sequence, in which chain of reactions may occur, deterministic models have to be used in conjunction with probabilistic analysis. The former type of models are useful in quantifying physical and chemical processes such as size of a leak, chemical release mode and rate, dispersion (including cloud formation), type of explosion and its intensity, heat/missile/shock wave loads generated, etc. The latter type, which are based on careful recapitulation and analysis of past accidents, and site characteristics, are needed to work out the probabilities of equipment failure, direction of missiles, direction of fire ball/jet fire, pattern of toxic plume dispersion etc (Latha et al., 1992; Raghvan and Swaminathan, 1996; Khan & Abbasi, 1998a).

While choosing the models, and conducting analysis based on them, one has to be conscious of the choice of models.

Choice of Models

The choice of models, and their use, must be done with care, keeping in view the following factors:

a) the computed magnitudes of incident heat and shock-wave effects would have a measure of uncertainty;

b) the stress/strain patterns in vessels affected by fire or fire-explosion combinations would be complex and difficult to quantify with precision;
c) external and transient factors such as wind direction can play a major role;
d) simultaneous occurrence of heat and mass transfer, made complex by transient and rapidly changing nature and magnitudes of the initiating events such as fire due to leakage in a storage vessel, can contribute to the uncertainty.

21.3. PROCEDURE FOR DEA (DOMINO EFFECT ANALYSIS)

The first step of DEA requires detailed and accurate (as much as possible) forecast of the impacts of the initiating event at the various points where other hazardous units are located. For this, latest models for assessing the impacts (missiles, heat load, shock waves etc) generated at the source of the secondary event by the primary event, directional probabilities (in case of missiles and fire jets), damage radii, scenarios of vessel failure etc should be utilised (Raghvan and Swaminathan, 1996; Khan and Abbasi, 1998a). The sequence of steps for conducting a typical domino effect analysis is presented in Figure 21.2. We have developed DOMIFFECT (DOMIno eFFECT) which is a computer-automated tool incorporating the DEA methodology (Khan & Abbasi♦ 1998a). DOMIFFECT is a menu-driven, interactive software, and it is capable of the following operations:

- assessing of likely impacts of accidents: fires, explosions, toxic release, and combination of these;
- study of interaction among different accidental events (generation of domino or cascading accident scenarios);
- estimation of domino effect probability;
- estimation of domino effect consequences.

The software incorporates state-of-the-art models for studying a) the hazards of fire, explosion, toxic release, or combination of these present in a chemical process industry; b) the damage potential of likely accidents, assessed on the basis of credible scenarios the tool develops; c) the likelihood of second accident being triggered by the first, d) the scenarios of the second accident, their damage potential, and the probability of their causing a third accident (steps similar to a-c above), and so on. A major feature of DOMIFFECT is the module automatic analysis. This module uses a knowledge-base built into DOMIFFECT, estimates the detailed consequences, and checks the probability of occurrence of secondary or higher order accidents at different locations.

DOMIFFECT further works out the probability of occurrence of domino effect on the basis of the location, contents, orientation, and material characteristics of the likely target units. The analysis employs, *inter alia*, vessel failure and probit models. If the probability of occurrence of a `hit' is sufficiently high then an accident scenario for the secondary accident is developed, followed by an estimation of its damage potential. This procedure is repeated till the full chain of likely accidents is covered.

♦ Khan, F.I., and Abbasi, S.A., (1998a). Models for domino effect analysis in chemical process industries, Process Safety Progress,- AIChE, USA, 17 (2), 107-113. Reported in Appendix

DEA is performed at two levels. In the first level, a screening of all the units of a chemical process industry is done in order to identify the units, which may come under the spell of domino effect. For this purpose threshold values of different damaging effects reported in standard professional literature are used. For example, an over pressure in excess of 0.689 bar can cause probable total destruction of buildings, and shall move and badly damage heavy machine tools (7000 lb; Davies 1993). Likewise, it has been reported that a heat load of certain intensity is sufficient to induce failure of certain kinds of vessels, and a missile (sharp edged) having a certain velocity has sufficient potential to penetrate a certain type of target unit provided that it collides with the unit. If the estimated values of these parameters at the location of the target unit are higher than the threshold values, then a detailed analysis, or the second level study is performed.

At the second level, further analysis is conducted to verify the existence of domino effect, using the damage potential of the primary event and the characteristics of the secondary unit. The following characteristics are considered:

- the shape and the constituent material of the unit,
- the chemicals involved and the operating conditions under which they are used,
- quantities and physical properties of the chemicals involved,
- location of the unit in terms of distance from other units, and
- meteorology, especially the prevailing wind direction.

21.4. Domino Effects Likely to Be Caused by the Mic Industries

The following credible accident scenarios were developed using DOMIFFECT, and the criteria described elsewhere (Khan and Abbasi, 2000).

Madras Refineries Limited (MRL)

Stabiliser unit
BLEVE followed by fireball

Compressor discharge unit
Fireball/jet fire

Depropanizer column
BLEVE followed by fireball

Reflux drum
CVCE followed by fireball

LPG storage
CVCE followed by fireball

Propylene storage
CVCE followed by fireball

Propane storage
BLEVE followed by Fireball

Madras Fertiliser Limited

Ammonia synthesis unit
BLEVE followed by fireball and dispersion of toxic gas

Secondary reformer unit
CVCE followed by fireball

Urea reactor
CVCE followed by fireball and dispersion of toxic gases

Primary absorber
BLEVE followed by toxic release

Pre-neutraliser
BLEVE followed by toxic release

Drier unit
BLEVE followed by fireball

SPIC Heavy Chemicals Division

Hydrogen treatment unit
BLEVE followed by fire ball

Hydrogen storage unit
CVCE followed by fireball

Ammonia storage unit
BLEVE followed by toxic release and dispersion

Manali Petrochemicals Ltd

Propylene oxide storage unit:
CVCE followed by fireball

Propylene storage unit
Instantaneous release causing BLEVE followed by fireball

Mono propylene glycol storage unit
Instantaneous release leading to pool fire

Fuel oil storage unit
Pool fire

Tamilnadu Petroproducts Limited

Propylene storage unit
CVCE followed by fireball

Allyal chloride storage unit
Pool fire

Fuel oil storage unit
Pool fire

Indian Organic Chemicals Ltd. (IOCL)

Polymerization autoclave
Pool fire
Monoethylene glycol storage
Pool fire

SPIC Organic Limited (SOL)

Propylene storage unit
BLEVE followed by fireball.
Propylene oxide storage unit
CVCE followed by fireball.
Ethylene oxide storage unit
BLEVE followed by fireball.

21.5. Results and Discussion

Domino Effect MRL

The detailed results of DEA (primary accident as well as likely chain of accidents) for a probable accident in stabiliser unit (accident scenario BLEVE followed by fireball) is presented in Table 21.1. It is evident that the intensity of shock waves, missiles, and heat load high enough over an area of ~200 m to initiate secondary and tertiary accidents in the units placed within this area. As the vacuum distillation column is placed 65 m from the stabilizer unit, it would be impacted by the primary accident; the most credible accident, which may result, is BLEVE followed by fireball. The intensity of shock wave and heat load generated due to this accident would envelop an area of over 100m radius. The impact of secondary accident would trigger a tertiary accident in the de-butanizer unit placed 45 m from the vacuum distillation column, generating a fireball (Table 21.2). The likely chain of events for this is presented in Figure 21.3.

A sudden release of butane from depropanizer column as BLEVE followed by fireball may cause development of severe shock waves, and heat load (Table 21.3). However, the severity of shock waves this scenario is comparatively lesser than accident scenario in stabilizer and storage units. The contours for 50% damage due to shock wave would be operative over an area of ~150 m radius, whereas heat radiation effects would be persistent over an area of ~100 m radius. Compared to previous scenario, this scenario has lesser' probability of causing domino effect. The likely chain of accident is presented in Figure 21.3. Vapour recover unit, crude distillation unit and vacuum distillation unit are likely to fail in secondary accidents, but there is low probability for tertiary accident.

The accident scenario for reflux drum is developed as CVCE followed by fireball. As the unit deals with high capacity of butane gases under high pressure, there is a higher probability of occurrence of CVCE. Unexploded hydrocarbons may burn as a fireball. The result for this scenario is depicted in Table 21.4. An intense over pressure load along with a high shock wave velocity would be observed over wide area (~ 450 m). Moreover a lethal heat load due to fire and explosion energy would be observed over a radial distance of more than ~300 m. The probable chain of accidents likely due to this accident is presented in Figure 21.3. It is evident that unifier hydrogen unit, and hydrodesulfurisation unit are likely to cause secondary accidents, while MEA absorption unit, fluidised catalytic cracking unit, and amine treating unit may fail in tertiary accidents.

Table 21.5 presents the results (output of DOMIFFECT) for a probable accident in LPG storage vessel (*confined vapour cloud explosion followed by fireball*). The shock wave generated due to CVCE can cause injury as well as second order accidents by seriously damaging other vessels (Table 21.5) over an area of more than ~400 m. The vapour cloud generated by CVCE on ignition may cause a fireball and hence severe heat radiation effect. Similarly the missiles generated by CVCE may hit nearby targets and may lead to secondary explosions or release of chemicals. The heat radiation effect with 50% probability of lethality would be observed over an area of ~400 m radius. The probability of domino effect occurrence due to this accident is so high that it is almost a certainty. The likely chain of accidents initiated by this accident is presented in Figure 21.3. It is clear that propylene storage units and most of the units of refinery I, and refinery II are under serious threats of secondary accident, while units of MPL, IAL, MFL and other industries are likely to fail tertiary accident.

As per accident scenario in propylene storage vessel, CVCE followed by fireball would give rise to shock waves, missile effects, and heat radiation effects (Table 21.6). The intensity of shock waves, high enough to cause complete damage to the objects coming in their way, would be persistent in an area of more than 300m radius. The intensity of heat radiation would be lethal over an area of ~400 m radius. Thus, the area of 300 m radius is under severe threat due to heat load as well as shock wave load. An accident in this unit too may cause domino effect. The probability of occurrence of such an effect due to this accident is estimated as unity. As may be seen from the likely chain of events depicted in Figure 21.3. LPG storage units, crude distillation unit, and units of Lube plant are very likely to fail as secondary accident while Vis breaking and thermal cracker, crude distillation of refinery II, and units of MFL are likely to meet with tertiary accidents.

The accident scenario for propane storage is generated as BLEVE followed by fireball. The summarised output DOMIFFECT for this scenario is presented in Table 21.7. The intensity of explosion would be lower compared to the previous cases. The intense shock wave capable of causing secondary accidents would affect an area of ~150 m radius, whereas the lethal heat load (50% probability of causing damage) would be observed over an area of ~ 100 m. Eventhough this unit too is likely to cause secondary and high order accidents, the probability is comparatively lower than in previous accident scenarios. The likely sequence of events triggered by this accident scenario is presented in Figure 21.3.

Domino Effect at MFL

The domino effect analysis (primary accident as well as likely chain of accidents) for a probable accident in ammonia synthesis unit (accident scenario BLEVE followed by fireball and dispersion of toxic gas ammonia) is presented in Table 21.8. It is evident that the intensity of shock waves, missiles, and overpressure is high enough over an area of ~450 m to initiate secondary and tertiary accidents in the units placed within this area. This cover most of the units of ammonia and urea plants and also the units of MPL and other neighbouring industries. As the primary absorber is placed 55 m from the ammonia synthesis unit, it would be impacted by the primary accident; the most credible accident, which may result, is BLEVE followed by fireball. The intensity of shock wave and heat load generated due to this accident would envelop an area of over ~200 m radius. The impact of secondary accident would trigger a tertiary accident in the ammonia stripper unit (primary and secondary) placed 90 m from the absorber, generating a fire ball (Tables 21.8 and 21.9). The likely chain of accident for this scenario is presented in 21.4.

A sudden release of hydrogen from secondary reformer, generating CVCE followed by fireball, may cause development of severe shock waves, and heat load (Table 21.10). The severity of shock waves for this scenario is similar to that has been observed for accident scenario in ammonia synthesis unit. The contours for 50% damage due to shock wave span an area of ~400 m radius, whereas heat radiation effects would devastate an area of ~200 m radius. This scenario too has higher probability of causing domino effect. The likely sequence of accidents is presented in Figure 21.4. Primary reformer, shift convertor and CO_2

absorber units are likely to fail in secondary accidents, whereas neighbouring industries such as MPL, units of refinery, and units of Madras Petro-Products Limited (MPPL) are likely to initiate tertiary and higher order accidents.

The absorber unit is the most important component of the urea plant. It is also the unit most susceptible to accidents, as it handles toxic chemicals at extreme conditions of temperature and pressure. The accident scenario for absorber unit is developed as BLEVE followed by dispersion of toxic gas. As the unit deals with large quantities of ammonia under high pressure, there is a high probability of occurrence of BLEVE. The released chemical would disperse in the atmosphere, building toxic load. The DEA results for this scenario are depicted in Table 21.11. An intense over-pressure load along with a high shock wave velocity would be observed over an area of ~150 m radius. The lethal toxic load would be operative over a radial distance of more than ~2500 m. The probable chain of accidents due to this accident in presented in Figure 21.4. It is evident that both primary and secondary reformer units, are likely to cause secondary accidents. The compression unit, the ammonia synthesis unit, and the surge gas condenser may fail in tertiary accidents.

Table 21.12 presents the DOMIFFECT output for a probable accident in urea reactor (*confined vapour cloud explosion followed by fireball and dispersion ox toxic gases*). The shock wave generated due to CVCE can cause injury as well as second order accidents by seriously damaging other units (Table 21.12). These impacts would span an area of more than 350 m radius. The vapour cloud generated by CVCE on ignition may cause a fireball and hence severe heat radiation effect. Similarly the missiles generated by CVCE may hit nearby targets and may lead to secondary explosions or release of chemicals. The heat radiation effect with 50% probability of lethality would be observed over an area of ~350 m radius. The probability of domino effect occurrence due to this accident is so high that it is almost a certainty. The chain of accidents likely to be initiated by this accident is presented in Figure 21.4. It is clear that storage unit of MPL and most of the units of ammonia and urea synthesis plant are under serious threats of secondary accident, while units of MPL, SPIC-HCD and MRL are likely to fail as tertiary accident.

As per accident scenario in pre-neutraliser unit, BLEVE followed by dispersion of toxic gases would give rise to shock waves, missile effects, and toxic load (Table 21.13). The intensity of shock waves, high enough to demolish all objects coming in their way, would be persistent in an area of more than 200 m radius. The intensity of toxic emissions would be lethal over an area of ~800 m radius. The probability of occurrence of domino effect due to this accident is estimated as unity. As may be seen from the likely chain of events depicted in Figure 21.4, drying unit, absorber unit, and units of urea plant are very likely to fail as secondary accident while ammonia plant and, units of MPL are likely to meet with tertiary accidents.

The accident scenario for drier unit is envisaged as BLEVE followed by fire ball. The summarised output of DOMIFFECT for this scenario is presented in Table 21.14. Intensity of explosion would be lower compared to the previous cases. The intense shock waves, capable of causing secondary accidents, would effect an area of ~100 m radius, whereas the lethal heat load (50% probability of causing damage) would be observed over an area of ~ 100 m. Eventhough this unit, too, is likely to cause secondary and high order accidents, the probability is comparatively lower than in previous accident scenarios. The sequence of events likely to be triggered in this accident scenario is presented in Figure 21.4.

Domino Effect at SPIC-HCD

The accident in hydrogen treatment unit (BLEVE with fire ball) would generate shock wave and heat load high enough to cause 50% damage over an area of more than ~100 m radius (Table 21.15). As a result, units placed close to the autoclave are likely to get affected and lead to secondary accidents as illustrated in Figure 21.5. The electrolyser, compressors, liquefaction unit and storage units are likely to cause higher order accidents but the probability of such accidents is low.

The results of domino effect analysis for a likely accident in hydrogen storage vessel (CVCE followed by fire ball) is presented in Table 21.16. It is evident that CVCE would generate great over pressure and shock wave over an area of more than 500 m radius, simultaneously, a fire ball would give rise to severe heat load which would cause damage (at 50% probability) over an area of more than ~400 m radius. The units placed within striking distance of these events would all fail, leading to fires and/or explosions. The escaped chemicals on ignition would further cause fire/explosion depending upon the type of chemical and mode of release. The probability of domino effect due to this accident scenario is estimated as one. The likely sequence of accidents is presented in Figure 21.5. It may be seen that electrolysis section, chlorine storage, and units of MPL are most likely to fail as secondary accident while several units of TPL and MFL may fail in tertiary accident.

Ammonia is a highly toxic and non-flammable chemical stored under pressurised conditions at SPIC-HCD. The results of DEA study for a probable accident scenario in the chlorine unit - BLEVE followed by toxic release and dispersion - are presented in Tables 21.17 and 21.18. It is evident signify that an instantaneous release of the gas (as BLEVE) develops intense shock waves, impact of which can be seen over wide areas (more than ~250 m). The damage impact of these shock waves accompanied by missile effect would initiate secondary accident in the units placed near to it. The impact of lethal toxic load would be operative over an area of more than ~1500 m.

Domino Effect at MPL

Propylene oxide (PO) is one of the reactive and highly flammable chemicals used at MPL it being the main raw material, is stored in large quantity. CVCE followed by fireball is the most credible accident scenario envisaged in the PO storage unit. The damage characteristics of this scenario have been studied in detail for the possibility of initiating domino effect. The output of the study is presented in Table 21.19. It is seen that CVCE shall develop powerful shock-waves and the fireball shall cause intense heat load. These impacts would reach an area of more than ~350 m radius; damaging other units in the area and sparking a cascade of accidents. The probability of this domino effect is close to unity (Tables 21.19 and 21.20) The sequence of accidents likely to occur due to this accident is in Figure 21.6. Propylene storage, ethylene oxide storage, propylene oxide reactor shall fail in secondary accidents and the polyol reactor, PG reactor and units of neighbouring hydrogen and chlorine plants of SPIC-HCD shall meet with tertiary accidents (Table 21.19).

The domino effect studies for an accident in propylene storage vessel are presented in Table 21.21. The likely accident (BLEVE followed by fireball) would cause development of highly intense shock waves and lethal heat load. Individual and/or combined impact of these effects would lead to failure of other units (secondary accident). The results of DOMIFFECT reveal that units placed 55-75 m away from propylene storage unit are vulnerable to secondary accident. The probabilities of occurrence of secondary accidents due to heat load, shock waves, and missiles, are all very high. The units involving storage of propylene oxide, propylene glycol fractionation, and PO purification shall meet with second order accidents. The impact of shock waves and heat load thus generated shall cause tertiary accident in the units of MFL, units of SPIC-HCD, and storage units of Madras Petrochemical Limited (MPPL). The adverse impacts of these accidents may lead to higher order accidents (Figure 21.6).

Accident in monopropylene glycol (MPG) storage unit of MPL – pool fire – would generate very high heat load, sufficient to cause 50% damage over an area of more than 250 m radius (Table 21.22). There may be boiling of chemicals (benzene, propane, ethylene, etc.) and consequent release in instantaneous or continuous fashion. The escaping chemicals on meeting with an ignition source would cause fire/explosion depending upon the type of chemical and the mode of release. Most of the units of MPL, and purification and electrolysis units of SPIC-HCD may be damaged perpetuating the chain of accidents effecting units of TPL, storage units of SPIC-HCD and eplichlorohydrin plant.

The results of DEA study for an accident in fuel oil storage unit of Manali Petrochemicals Limited is presented in Table 21.23. The results reveal that burning of released chemical (fuel oil) as pool fire would cause building up of intense heat load over an area ~250 m radius. The probability of occurrence of secondary and higher accidents due to this event in high enough to be taken as a certainty. Most of the units of MPPL, TPL and epichlorohydrin would be damaged in the higher order accidents.

Domino Effect at TPL

As with MPL, propylene oxide (PO) is the main raw material at TPL, and is stored in large quantities. The most credible accident scenario envisaged in this unit is *CVCE followed by fire ball*. The damage characteristics of this scenario have been studied in detail for the possibility of initiating domino effect. The output of the study is presented in Table 21.24. CVCE shall develop intense over pressure (shock wave) whereas fireball shall develop highly intense heat load. The impacts of these would penetrate over an area of more than ~300 m radius. The other units placed closer to the PO unit shall be involved in chain of accidents, of which probability is close to unity. The sequence of accidents likely to occur due to this accident, depicted in Figure 21.7, reveals that other storage units of TPL, epichlorophydrin plant, and units of SPIC-HCD shall encounter secondary accidents while chlorine storage of SPIC-HCD, polyol reactor, and other units of neighbouring MPL are likely meet with tertiary accident.

The accident scenario of pool fire in allyl chloride storage unit forecasts burning of allyl chloride high heat load over an area of more than ~100 m radius (Table 21.25). The probability of occurrence of secondary and higher order accident due to this initiating event is high but comparatively lower than the domino effect probabilities in other units of MPL and TPL.. The units that are likely to meet with secondary accidents are propylene oxide storage, reactors, fuel oil storage, and storage units of epichlorohyrin plant. There is low possibility of tertiary or higher order accidents.

The results of DEA study for an accident in fuel oil storage unit of TPL is presented in Tables 21.26 and 21.27. The results reveal that burning of fuel oil as pool fire would cause building up of intense heat load. Heat flux sufficient to cause damage or boiling of liquid would be observed over an area of more than ~200 m radius. Secondary and higher accidents due to pool fire in fuel oil storage unit are almost certain to occur (Table 21.27). The units likely to face secondary accidents are allyl chloride storage unit, reaction unit, chlorine storage, and also units of epichlorohyrin plant. The fractionation and other process units of MPL, SPIC-HCD, MPL and neighbouring industry IAL are likely to involve in tertiary and higher order accidents.

Domino Effect at IOCL

The results of domino effect analysis for an accident in monoethylene glycol (MEG) storage unit (accident scenario pool fire) is presented in Tables 21.28 and 21.29. It is evident that the pool fire would generate excessively high heat load which would cause damage (at 50% probability) over an area of more than ~250 m radius. Heat load generated due to pool fire would cause boiling of the chemicals stored near this unit (benzene, propane, ethylene, etc) and thereby cause instantaneous or continuous release of these chemicals. The escaped chemicals on ignition would further cause fire/explosion depending upon the type of chemical and mode of release. The probability of domino effect due to this accident scenario is estimated as one. The likely sequence of accidents is presented in Figure 21.8. It is observed from the figure that furnace oil storage, ammonia storage, and methanol storage units are most likely to fail as secondary accident while units such as polyol reactor, LSHS storage, and polymerisation autoclave may fail as tertiary accident.

The accident in polymerisation autoclave (pool fire) would generate heat load high enough to cause 50% damage over an area of more than ~150 m radius (Table 21.30). Therefore, units placed closer to the autoclave are likely to get affected and cause secondary accidents as illustrated in Figure 21.8. The units such as esterification reactor, polymer granulater, and extrude are likely to cause secondary accident but there is low probability of tertiary accidents.

Domino Effect at SOL

The domino effect studies for an accident in propylene storage vessel are presented in Tables 21.31 and 21.32. The likely accident (CVCE followed by fireball) would cause development of highly intense shock waves and lethal heat load. Individual and/or combined impact of these effects would lead to failure of other units (secondary accident). The results of DOMIFFECT reveal that units placed 55-75 m away from propylene storage unit are vulnerable to secondary accident. The probability of occurrence of secondary accident due to heat load is estimated to be high, while this value is even greater for shock waves and missile effects. The units involving storage of propylene, ethylene oxide, and propylene oxide are most likely to fail in second order accident. The impact of shock waves and heat load thus generated shall cause tertiary accident in the ethylene oxide reactor, Polyol reactor, chlorine storage vessel, and distillation column. The adverse impacts of these accidents may lead to high order accidents (Figure 21.9).

Propylene oxide (PO) is one of the reactive and highly flammable chemicals used in the UB Petrochemicals (UBP) plant. PO, being the main raw material, is stored in large quantity. CVCE followed by fire is the most credible accident scenario envisaged in the PO storage unit. The damage characteristics of this scenario has been studied in detail for the possibility of initiating domino effect. The output of the study is presented in Table 21.33. It is evident from the results that CVCE shall develop intense over pressure (shock wave) whereas fireball shall develop highly intense heat load. These impacts would be operative over an area of more than ~200 m. The other units placed closer to the PO unit are very likely to cause chain of accidents (probability is close to unity). The sequence of accidents likely to occur due to this accident is depicted in Figure 21.9. It is clear from the figure that propylene storage, ethylene oxide storage, and chlorine storage units are most likely to fail in secondary accidents while polyol reactor, propylene oxide reactor, and poly propylene glycol reactor are likely to fail in tertiary accident.

Like the PO unit, ethylene oxide storage unit may also cause severe damage (Table 21.34). The probability of occurrence of domino effect due to an accident in this unit is estimated as unity. The logical sequence of events likely to occur is depicted in Figure 21.9.

21.6. Risk Estimation

To visualise the severity and impact of the primary events (initiating events) Individual risk factors have been plotted over the MIC layout. Due to the limitation of space we are presenting discussions on the worst scenario for each of the industries.

The detailed calculation for accident scenario in LPG storage vessel of MRL reveals that severe risk contour (risk factor $>1*10^{-3}$) due to combined effects of explosion and fire would be observed over an area of ~500 m (Figure 21.10). This encompasses other vulnerable units of MRL (other LPG bullets, Refinery I and II, etc.) as well as other industries such as MPL, MFL, TPL, and BL. The likely secondary accident that may occur due to this event is presented in Figure 21.10. It is evident from the figure that this secondary accident would endanger other units and thus may initiate tertiary or higher order accidents. The individual risk contour would extend over the entire industrial complex (Figure 21.10).

The detailed calculation for accident scenario in ammonia synthesis unit of MFL reveals that severe risk contour (risk factor $>1*10^{-4}$) due to combined effects of explosion and fire would encompass an area of ~400 m (Figure 21.11). This would impact several vulnerable units of MFL (urea plant, NPK plant, storage farm) as well as other industries such as MPL, MRL, and TPL. The likely secondary accident that may occur due to this event is presented in Figure 21.11. It would endanger other units and thus may initiate tertiary or higher order accidents. The individual risk contour would extend over the entire industrial complex (Figure 21.11).

SPIC-HCD is located in the midst of several other industries and any mishap in SPIC-HDD would have far reaching consequences. This is evident from the damage calculations for an accident scenario in hydrogen storage vessel. The risk contours due to combined effect of heat and shock wave generated due to CVCE and fire-ball cover an area of ~500 m (Figure 21.12). This is a large area and encompasses a number of industries such as TPL, MPL, MRL, and MFL (Figure 21.12). The secondary accidents in these units would initiate tertiary and higher order accidents. The eventual area-of-impact would cover the entire Manali industrial complex.

Like SPIC-HCD, MPL too situated in the midst of industries. However, the extent of risk posed by this industry is lesser than SPIC-HCD. The risk contour for a probable accident in propylene oxide storage vessel covers an area of ~200 m radius (Figure 21.13). This encompasses the entire plant as well as process area of other industries (MPCL, SPIC, and MFL). The overall risk contour following domino effect envelops an area of more than 300 m radius (Figure 21.13) extending over most of the complex.

Assessment of the likely consequences of accident in propylene storage vessel of TPL (accident scenario CVCE followed by fire ball) reveals that an area of ~250 m radius is under severe risk due to combined effects of fire and explosion (Figure 21.14). This would encompasses many units of the parent industry and also the units of SPIC, and MRL. The likely secondary accident accidents in these units are depicted in Figure 21.14. It is evident from the figure that secondary accidents shall envelops other hazardous units and thus may initiate tertiary accidents. The overall risk contour considering domino effect is presented in Figure 21.14, indicating that most of the area of MIC would be under severe threat if this accident occurs.

IOCL is located in the southern corner of MIC, and has lesser congestion of industries around it in comparison to other industries in MIC. As mentioned earlier, it is a synthetic fibre industry and does not involve major hazards. However, the accident scenario in the monethylene glycol unit envisages development of heat load over an area of ~150 m. The impact of this accident shall reach upto ~80 m from the accident epicentre, effecting process units of the IOCL and the neighbouring industry CETEX. The likely secondary accidents in these units are depicted in Figure 21.15. The risk contours plotted, considering domino effect, are presented in Figure 21.15. It is clear from the figure that the severe risk contour for this unit is limited to a lower area than other industries.

SOL petrochemical is surrounded with BL and MRL storage tanks. The presence of MRL storage tank makes the location vulnerable. The adverse impacts of an accident in propylene storage vessel would reach an area of ~200 m. This envelops the other storage vessels of the parent industries and also the storage vessel of the MRL. The likely chain of secondary accidents due to this scenario is depicted in Figure 21.16. It is clear that the consequence of secondary accident in MRL storage tanks would be very dangerous; the harmful impact may be felt over entire MIC.

21.7. SUMMARY AND CONCLUSIONS

1) The paper highlights the need to make domino effect studies an integral parts of all risk assessment exercise. In such studies the probability of occurrence of domino effect should be assessed the impacts if such chain of accident occurs, should be forecast. The authors are prompted by the history of several past accidents including

the most recent one -- and one of the biggest ever -- which occurred at the HPCL's refinery at Vishakhapatnam, India, on 14 September, 1997.

2) Five types of primary events many initiate domino effect: fire, overpressure/shock waves, missiles, toxic load, and a combination of one or more of these.
3) Studies conducted by the authors, and others, indicate that it is not necessary that the unit, which has the maximum damage potential (if it meets with an accident) would also have the maximum probability of causing domino effect. It is because the probability of occurrence of domino effect depends not only upon the damage potential of the primary accident but also on a number of others factors such as characteristics of the secondary unit, location of the unit, and site characteristics.
4) Among the different industries MFL, SPIC-HCD, MPL, and TPL, and are the ones most likely to initiate domino effect. The consequences of the domino effect envisaged in the units of these industries would be far reaching, the severe risk contours for most of these accident scenarios are encompassing the complete MIC area.
5) A further detailed analysis reveals that accidents in the ammonia synthesis, secondary reformer, and urea reactor of MFL may cause domino effect. Similarly, storage of, propylene oxide, ethylene oxide and mono propylene glycol of MPL, hydrogen storage of SPIC-HCD, and Propylene oxide and fuel oil storage of TPL is also likely to cause domino effect.

REFERENCES

Abbasi, S.A., and Khan F.I. (1999). *EIA and Sustainability Studies in Manali Industrial Complex: Executive Summary of Final Report*, Centre for Pollution Control & Energy Technology, Pondicherry University, Pondicherry; 267 pages.

Abbasi, S.A., and Khan F.I. (2000): Computer-aided Environmental Management. *Discovery Publishing House, New Delhi, (in press).*

Khan, F.I., and Abbasi, S.A., (2000). Credible accident scenario generation, *Journal of Loss Prevention in Process Industries*, U K, (in press).

Khan, F.I., and Abbasi, S.A., (1999a). Major accidents in process industries and analysis of their causes and consequences, *Journal of Loss Prevention in Process Industries*, U K, 12, 361-378.

Khan, F.I., and Abbasi, S.A., (1999b).The worst chemical industry accident of 1990s – What happened and what might have been : A quantitative study, *Process Safety Progress*, 18, 135-145.

Khan, F.I., and Abbasi, S.A., (1998a). Models for domino effect analysis in chemical process industries, *Process Safety Progress*,- AIChE, USA, 17 (2), 107-113.

Khan, F.I., and Abbasi, S.A., (1998b) DOMIFFECT (DOMino eFFECT) : a new software for domino effect analysis in chemical process industries, *Environment Modelling and Software*, The Netherlands, 13, 163-177,.

Latha, P, Gautam, G, and Raghavan, K V, (1992). Strategies for quantification of thermally initiated cascade effects, *J Loss Prevention Process Industries*, 5(1), 15-21.

Lees, F P, (1996). Loss prevention in process industries, (2nd edition), *Butterworths*, 1-3, London.

Raghavan, K V, and Swaminathan, G, (1996). Hazard assessment and disaster mitigation in petrochemical industries, *Oxford Publishing Company*, Chennai.

Table 21.1. The output of DOMIFFECT for the primary accident in stabilizer unit of Refinery-I of MRL and the subsequent domino effects

DOMIFFECT
(c) F.I. Khan & S.A. Abbasi, Pondicherry-605 014, India

Parameters	Values
PRIMARY EVENT	
Unit: Stabilizer unit	
Scenario: BLEVE followed by Fire ball	
Explosion : BLEVE	
Total energy released	(kJ) : 2.89E+06
Peak over pressure	(kPa) : 1244.3
Variation of over pressure in air	(kPa/s): 515.04
Shock wave velocity	(m/s) : 987.6
Duration of shock wave	(ms) : 14.68
Missile characteristics (general)	
Initial velocity of fragment	(m/s) : 196.90
Kinetic energy of fragment	(kJ) : 105452.68
Penetration ability at study point	
Concrete structure	(m) : 0.06
Brick structure	(m) : 0.11
Steel structure	(m) : 0.01
Fire Ball	
Radius of the fire ball	(m) : 85.7
Duration of the fire ball	(s) : 34.5
Energy released by fire ball	(kJ) : 1.24E+07
Radiation heat flux	(kJ/sq.m) : 8787.2
Domino effect –I: Secondary accident	
Unit: Vacuum distillation column	
Distance of the vulnerable unit from the primary event	(m) : 65.0
Domino effect probabilities:	
due to heat load	(%) : 100
due to explosion is	(%) : 85
due to missile	(%) : 40
Domino effect impacts:	
Scenario: BLEVE followed by fireball	

Table 21.1. (Continued)

Parameters	Values
Explosion : BLEVE	
Total energy released	(kJ) : 1.55E+06
Peak over pressure	(kPa) : 702.0
Variation of over pressure in air	(kPa/s) : 350.0
Shock wave velocity	(m/s) : 685.5
Duration of shock wave	(s) : 14.68
Missile characteristics	
Initial velocity of a typical fragment	(m/s) : 95.90
Kinetic energy a typical of fragment	(kJ) : 41241.68
Final velocity of a typical fragment	(m/s) : 65.90
Penetration ability at the location of secondary accident	
Concrete structure	(m) : 0.01
Brick structure	(m) : 0.05
Steel structure:	(m) : 0.00
Fire Ball	
Radius of the fire ball	(m) : 75.5
Duration of the fire ball	(s) : 25.5
Energy released by fire ball	(kJ) : 4.75E+06
Radiation heat intensity	(kJ/sq.m) : 6305.45
Domino effect-II: Tertiary accident	
Unit: De-butaniser unit	
Distance of the vulnerable unit from the primary event	(m) : 45.0
Domino effect probabilities:	
due to heat load	(%) : 85
due to explosion	(%) : 100
due to missile	(%) : 65
Domino effect Impacts	
Scenario : instantaneous release followed by fire ball	
Fire : Fire ball	
Radius of the fire ball	(m) : 175.3
Duration of the fire ball	(s) : 542.7
Energy released by fire ball	(kJ) : 2.02E+09
Radiation heat intensity	(kJ/sq.m) : 19985.7

Table 21.2. Impacts of initiating events and the probabilities of causing domino effect in stabiliser unit of MRL

Parameters	Scenario and their likely impacts
	BLEVE followed by fire ball in stabilizer unit
Explosion:	BLEVE
Total energy released, kJ	2.89E+06
Peak overpressure developed, kPa	1244.3
Variation of overpressure in air, kPa/s	515.0
Shock wave velocity, m/s	887.6
Duration of shock wave, ms	14.6
Missile Characteristics:	
Initial velocity of fragment, m/s	196.9
Kinetic energy of fragment, kJ	105452.6

Table 21.2 (Continued)

Penetration ability at study point	
Concrete structure, m	0.06
Brick structure, m	0.11
Steel structure, m	0.01
Parameters	Scenario and their likely impacts
Fire	
Radius of fireball, m	85.7
Duration of fireball, s	34.5
Energy released by fire ball, kJ	1.24E+07
Radius of pool fire, m	Not applicable
Burning area, sq.m	Not applicable
Burning rate, kg/s	Not applicable
Radiation heat flux, kJ/sq.m	8787.2
Domino Checking	
Location of the unit from primary event, m	75.0
Domino effect due to heat load	
Total heat received, kJ	1.11E+06
Heat intensity, kJ/sq.m	3102.5
Probability of domino effect due to fire	1.0
Domino effect due to over pressure:	
Explosion energy, kJ	1.24E+04
Peak overpressure, kPa	73.6
Probability of domino effect due to overpressure	0.6
Domino effect due to missile	
Explosion energy, kJ	1.01E+03
Missile velocity, m/s	8.3E+01
Probability of domino effect due to missile after meeting the target	0.24
Toxic release and dispersion	
Instantaneous/continuous	
Puff/Plume Characteristics:	
Puff concentration at centre of cloud, kg/cu.m	Not applicable
Concentration at cloud edge, kg/cu.m	
Downwind distance, m	
Distance where maximum concentration occurs, m	

Table 21.3. Impacts of initiating events and the probabilities of causing domino effect in depropanizer unit of MRL

Parameters	**Scenario and their likely impacts**
	BLEVE with fireball in depropanizer unit
Explosion:	BLEVE
Total energy released, kJ	9.45E+05
Peak overpressure developed, kPa	695.3
Variation of overpressure in air, kPa/s	344.04
Shock wave velocity, m/s	682.6
Duration of shock wave, ms	14.6
Missile Characteristics:	
Initial velocity of fragment, m/s	95.90
Kinetic energy of fragment, kJ	41241.6

Table 21.3. (Continued)

Penetration ability at study point Concrete structure, m Brick structure, m Steel structure, m	 0.01 0.05 0.00
Fire Radius of fireball, m Duration of fireball, s Energy released by fire ball, kJ Radius of pool fire, m Burning area, sq.m Burning rate, kg/s Radiation heat flux, kJ/sq.m	 67.30 21.45 4.50E+06 Not applicable Not applicable Not applicable 6285.4
Parameters	**Scenario and their likely impacts**
Domino Checking Location of the unit from primary event, m	 75.0
Domino effect due to heat load Total heat received, kJ Heat intensity, kJ/sq.m Probability of domino effect due to fire	 2.21E+05 609.5 0.75
Domino effect due to over pressure: Explosion energy, kJ Peak overpressure, kPa Probability of domino effect due to overpressure	 1.25E+03 53.55 0.4
Domino effect due to missile Explosion energy, kJ Missile velocity, m/s Probability of domino effect due to missile after meeting the target	 8.54E+02 19.3 0.0
Toxic release and dispersion Instantaneous/continuous Puff/Plume Characteristics: Puff concentration at centre of cloud, kg/cu.m Concentration at cloud edge, kg/cu.m Downwind distance, m Distance where maximum concentration occurs, m	Not applicable

Table 21.4. Impacts of initiating events and the probabilities of causing domino effect in reflux drum of MRL

Parameters	Scenario and their likely impacts
	CVCE followed by fireball in reflux drum
Explosion: Total energy released, KJ Peak overpressure developed, kPa Variation of overpressure in air, kPa/s Shock wave velocity, m/s Duration of shock wave, ms	CVCE 1.25E+07 1673.4 642.4 985.3 21.7

Table 21.4 (Continued)

Missile Characteristics:	
Initial velocity of fragment, m/s	428.81
Kinetic energy of fragment, kJ	1.66E+06
Penetration ability at study point	
Concrete structure, m	0.30
Brick structure, m	0.70
Steel structure, m	0.02
Fire	
Radius of fireball, m	215.5
Duration of fireball, s	475.5
Energy released by fire ball, kJ	7.06E+09
Radius of pool fire, m	Not applicable
Burning area, sq.m	Not applicable
Burning rate, kg/s	Not applicable
Radiation heat flux, kJ/sq.m	93841.5
Parameters	Scenario and their likely impacts
Domino Checking	
Location of the unit from primary event, m	75.0
Domino effect due to heat load	
Total heat received, kJ	6.59E+07
Heat intensity, kJ/sq.m	8118.2
Probability of domino effect due to fire	1.0
Domino effect due to over pressure:	
Explosion energy, kJ	5.25E+05
Peak overpressure, kPa	344.8
Probability of domino effect due to overpressure	1.0
Domino effect due to missile	
Explosion energy, kJ	3.54E+05
Missile velocity, m/s	119.3
Probability of domino effect due to missile after meeting the target	1.0
Toxic release and dispersion Instantaneous/continuous Puff/Plume Characteristics: Puff concentration at centre of cloud, kg/cu.m Concentration at cloud edge, kg/cu.m Downwind distance, m Distance where maximum concentration occurs, m	Not applicable

Table 21.5. Impacts of initiating events and the probabilities of causing domino effect in LPG storage vessel of MRL

Parameters	Scenarios and their likely impacts
	CVCE followed by fireball in LPG storage vessel

Table 21.5 (Continued)

Explosion: Total energy released, kJ Peak overpressure developed, kPa Variation of overpressure in air, kPa/s Shock wave velocity, m/s Duration of shock wave, ms	CVCE 2.45E+07 1785.5 685.8 1021.3 21.7
Missile Characteristics: Initial velocity of fragment, m/s Kinetic energy of fragment, kJ	 475.5 1.97E+06
Penetration ability at study point Concrete structure, m Brick structure, m Steel structure, m	 0.35 0.75 0.03
Fire Radius of fireball, m Duration of fireball, s Energy released by fire ball, kJ Radius of pool fire, m Burning area, sq.m Burning rate, kg/s Radiation heat flux, kJ/sq.m	 225.7 484.2 7.15E+08 Not applicable Not applicable Not applicable 93925.5
Domino Checking Location of the unit from primary event, m	 75.0
Parameters	Scenarios and their likely impacts
Domino effect due to heat load Total heat received, kJ Heat intensity, kJ/sq.m Probability of domino effect due to fire	 6.85E+07 8245.2 1.0
Domino effect due to over pressure: Explosion energy, kJ Peak overpressure, kPa Probability of domino effect due to overpressure	 5.31E+05 351.8 1.0
Domino effect due to missile Explosion energy, kJ Missile velocity, m/s Probability of domino effect due to missile after meeting the target	 3.65E+05 128.5 1.0
Toxic release and dispersion Instantaneous/continuous Puff/Plume Characteristics: Puff concentration at centre of cloud, kg/cu.m Concentration at cloud edge, kg/cu.m Downwind distance, m Distance where maximum concentration occurs, m	Not applicable

Table 21.6. Impacts of initiating events and the probabilities of causing domino effect in propylene storage vessel of MRL

Parameters	Scenario and their likely impacts

Table 21.6 (Continued)

	CVCE followed by fireball in propylene storage vessel
Explosion: Total energy released, KJ Peak overpressure developed, kPa Variation of overpressure in air, kPa/s Shock wave velocity, m/s Duration of shock wave, ms	CVCE 1.11E+07 1556.7 575.7 865.5 14.5
Missile Characteristics: Initial velocity of fragment, m/s Kinetic energy of fragment, kJ	 405.4 1.51E+06
Penetration ability at study point Concrete structure, m Brick structure, m Steel structure, m	 0.30 0.70 0.02
Fire Radius of fireball, m Duration of fireball, s Energy released by fire ball, kJ Radius of pool fire, m Burning area, sq.m Burning rate, kg/s Radiation heat flux, kJ/sq.m	 185.21 347 6.56E+09 Not applicable Not applicable Not applicable 82325.4
Domino Checking Location of the unit from primary event, m	 75.0
Parameters	Scenario and their likely impacts
Domino effect due to heat load Total heat received, kJ Heat intensity, kJ/sq.m Probability of domino effect due to fire	 5.45E+07 6222.2 1.0
Domino effect due to over pressure: Explosion energy, kJ Peak overpressure, kPa Probability of domino effect due to overpressure	 4.31E+05 296.5 1.0
Domino effect due to missile Explosion energy, kJ Missile velocity, m/s Probability of domino effect due to missile after meeting the target	 3.04E+05 85.25 1.0
Toxic release and dispersion Instantaneous/continuous Puff/Plume Characteristics: Puff concentration at centre of cloud, kg/cu.m Concentration at cloud edge, kg/cu.m Downwind distance, m Distance where maximum concentration occurs, m	Not applicable

Table 21.7. Impacts of initiating events and the probabilities of causing domino effect in propane storage vessel of MRL

Parameters	**Scenarios and their likely impacts**
	BLEVE followed by fireball in propane storage vessel
Explosion: Total energy released, KJ Peak overpressure developed, kPa Variation of overpressure in air, kPa/s Shock wave velocity, m/s Duration of shock wave, ms	BLEVE 7.55E+05 587.3 301.04 631.6 14.5
Missile Characteristics: Initial velocity of fragment, m/s Kinetic energy of fragment, kJ	 82.2 20511.5
Penetration ability at study point Concrete structure, m Brick structure, m Steel structure, m	 0.01 0.03 0.00
Fire Radius of fireball, m Duration of fireball, s Energy released by fire ball, kJ Radius of pool fire, m Burning area, sq.m Burning rate, kg/s Radiation heat flux, kJ/sq.m	 82.5 34 1.21E+07 Not applicable Not applicable Not applicable 8775.2
Domino Checking Location of the unit from primary event, m	 75.0
Parameters	**Scenarios and their likely impacts**
Domino effect due to heat load Total heat received, kJ Heat intensity, kJ/sq.m Probability of domino effect due to fire	 1.06E+06 3025.4 1.0
Domino effect due to over pressure: Explosion energy, kJ Peak overpressure, kPa Probability of domino effect due to overpressure	 1.15E+03 51.4 0.4
Domino effect due to missile Explosion energy, kJ Missile velocity, m/s Probability of domino effect due to missile after meeting the target	 7.8E+03 1.5E+01 0.14
Toxic release and dispersion Instantaneous/continuous Puff/Plume Characteristics: Puff concentration at centre of cloud, kg/cu.m Concentration at cloud edge, kg/cu.m Downwind distance, m Distance where maximum concentration occurs, m	Not applicable

Table 21.8. The output of DOMIFFECT for the primary accident in ammonia synthesis unit of MFL and subsequent domino effects

DOMIFFECT
(c) F.I. Khan & S.A. Abbasi, Pondicherry-605 014, India

Parameters	Values
Primary Event Unit: Ammonia Synthesis unit Scenario: BLEVE followed by fireball and dispersion of toxic gas	
Explosion: BLEVE	
Total energy released	(kJ): 4.38E+09
Peak over pressure	(kPa): 20454.8
Variation of over pressure in air	(kPa/s): 10475.7
Shock wave velocity	(m/s): 3963.8
Duration of shock wave	(ms): 41.4
Missile characteristics	
Initial velocity of fragment	(m/s): 7253.2
Kinetic energy of fragment	(kJ): 1.31E+08
Penetration ability at study point Concrete structure	(m) : 0.50
Brick structure	(m) : 0.68
Steel structure	(m) : 0.10
Fire Ball	
Radius of the fire ball	(m) : 185.4
Duration of the fire ball	(s) : 87.0
Energy released by fire ball	(kJ) : 6.59E+09
Radiation heat flux	(kJ/sq.m) : 65413.5
Toxic release and dispersion	
Box instantaneous model: Elevated source	
Concentration at distance of 200 (m),	(kg/cu.m) : 3.452E-04
Parameters	Values
Heavy gas puff Characteristics Ground level concentration of puff	(kg/cu.m) : 3.452E-04
Ground level concentration on puff axis	(kg/cu.m) : 3.452E-03
Cloud radius	(m) : 1.167E+02
Maximum distance traveled by the cloud	(m) : 7.874E+02
Maximum ground level concentration	(kg/cu.m) : 1.151E-01
DOMINO EFFECT –I : Secondary accident Unit: Primary reformer unit	
Distance of the vulnerable unit from the primary event	(m) : 55.0
Domino effect probabilities:	
due to heat load	(%) : 100
due to explosion	(%) : 100
due to missile	(%) : 45
Domino effect impacts Scenario: BLEVE followed by fire ball	
Explosion : BLEVE Total energy released	(kJ) : 12890.0
Peak over pressure	(kPa) : 65.8
Variation of over pressure in air	(kPa/s) : 0.67
Shock velocity of air	(m/s) : 389.0
Duration of shock wave	(ms) : 300

Table 21.8 (Continued)

No missile effect	
Fire: Fire ball Radius of the fire ball	(m) : 88.1
Duration of the fire ball	(s) : 36.4
Energy released by fire ball	(kJ) : 1.2E+06
Radiation heat flux	(kJ/sq.m) : 745.6
DOMINO EFFECT –II: Secondary accident Unit: Catacarb reboiler unit	
Distance of the vulnerable unit from the primary event	(m) : 70.0
Domino effect probabilities:	
due to heat load	(%) : 60
due to explosion	(%) : 45
due to missile	(%) : 15
Domino effect impacts Scenario: continuous release followed by pool fire	
Fire: Pool fire Radius of the pool fire	(m) : 47.0
Burning area	(sq.m) : 91781.8
Burning rate	(kg/min) : 574.8
Heat flux	(kJ/sq.m) : 332
DOMINO EFFECT –III: Tertiary effect Unit: Ammonia stripper unit	
Distance of the vulnerable unit from the secondary event	(m): 90.0
Domino effect probabilities:	
due to heat load	(%) : 25
due to explosion	(%) : 30
due to missile	(%) : 11
Domino effect impacts Scenario: BLEVE followed by toxic load build-up	
Explosion: BLEVE Energy released during explosion	(kJ) : 2.4E+06
Peak over pressure	(kPa) : 1151.4
Variation of over pressure in air	(kPa/s) : 481.4
Shock velocity of air	(m/s) : 795.6
Duration of shock wave	(ms) : 41.2
Missile characteristics:	
Initial velocity of a typical fragment	(m/s) : 235.6
Kinetic energy associated with a typical fragment	(kJ): 1.5E+05
Parameters	Values
Penetration ability at 50 m from the location of tertiary accident: Concrete structure	(m) : 0.06
Brick structure	(m) : 0.11
Steel structure	(m) : 0.01
Toxic release and dispersion	
Box instantaneous model: Elevated source Concentration at distance of 200 (m),	(kg/cu.m) : 2.154E-04
Heavy gas puff Characteristics Ground level concentration of puff	(kg/cu.m) : 2.001E-04
Ground level concentration on puff axis	(kg/cu.m) : 2.001E-03
Cloud radius	(m) : 1.897E+02
Maximum distance traveled by the cloud	(m) : 6.504E+02
Maximum ground level concentration	(kg/cu.m) : 1.562E-01

Table 21.9. Impacts of initiating events and the probabilities of causing domino effect in ammonia synthesis unit of MFL

Parameters	**Scenario and their likely impacts**
	BLEVE with fire ball and dispersion of toxic gas in ammonia synthesis unit
Explosion:	BLEVE
Total energy released, KJ	4.38E+09
Peak overpressure developed, kPa	20454.8
Variation of overpressure in air, kPa/s	10475.7
Shock wave velocity, m/s	3963.8
Duration of shock wave, ms	41.4
Missile Characteristics:	
Initial velocity of fragment, m/s	7253.2
Kinetic energy of fragment, kJ	1.31E+8
Penetration ability at study point	
Concrete structure, m	0.50
Brick structure, m	0.68
Steel structure, m	0.10
Fire	FIREBALL
Radius of fireball, m	185.4
Duration of fireball, s	87
Energy released by fire ball, kJ	6.59E+09
Radius of pool fire, m	not applicable
Burning area, sq.m	not applicable
Burning rate, kg/s	not applicable
Radiation heat flux, kJ/sq.m	65413.5
Domino Checking	
Location of the unit from primary event, m	75.0
Domino effect due to heat load	
Total heat received, kJ	1.30E+08
Heat intensity, kJ/sq.m	19306.0
Probability of domino effect due to fire	1.0
Domino effect due to over pressure:	
Explosion energy, kJ	2.75E+07
Peak overpressure, kPa	9876.4
Probability of domino effect due to overpressure	1.0
Parameters	Scenario and their likely impacts
Domino effect due to missile	
Explosion energy, kJ	3.86E+07
Missile velocity, m/s	1444.5
Probability of domino effect due to missile after meeting the target	1.0
Toxic release and dispersion	
Instantaneous/continuous	Instantaneous
Puff/Plume Characteristics:	Puff characteristics
Puff concentration at centre of cloud, kg/cu.m	3.452E-03
Concentration at cloud edge, kg/cu.m	3.452E-04
Maximum ground level concentration, kg/cu.m	1.15E-01
Distance where maximum concentration occurs, m	787.4

Table 21.10. Impacts of initiating events and the probabilities of causing domino effect in secondary reformer unit of MFL

Parameters	Scenarios and their likely impacts
	CVCE followed by fire ball in secondary reformer
Explosion:	CVCE
Total energy released, KJ	4.05E+08
Peak overpressure developed, kPa	11354.8
Variation of overpressure in air, kPa/s	8675.5
Shock wave velocity, m/s	1673.5
Duration of shock wave, ms	20.7
Missile Characteristics:	
Initial velocity of fragment, m/s	1553.2
Kinetic energy of fragment, kJ	2.45E+07
Penetration ability at study point	
Concrete structure, m	0.35
Brick structure, m	0.52
Steel structure, m	0.08
Fire	FIREBALL
Radius of fireball, m	75.5
Duration of fireball, s	29.4
Energy released by fire ball, kJ	1.54E+08
Radius of pool fire, m	not applicable
Burning area, sq.m	not applicable
Burning rate, kg/s	not applicable
Radiation heat flux, kJ/sq.m	6070.7
Domino Checking	
Location of the unit from primary event, m	75.0
Domino effect due to heat load	
Total heat received, kJ	7.45E+07
Heat intensity, kJ/sq.m	2105.4
Probability of domino effect due to fire	1.0
Domino effect due to over pressure:	
Explosion energy, kJ	2.11E+06
Peak overpressure, kPa	7431.4
Probability of domino effect due to overpressure	1.0
Domino effect due to missile	
Explosion energy, kJ	2.65E+06
Missile velocity, m/s	878.4
Probability of domino effect due to missile after meeting the target	1.0
Parameters	Scenarios and their likely impacts
Toxic release and dispersion	
Instantaneous/continuous	Not applicable
Puff/Plume Characteristics:	
Puff concentration at centre of cloud, kg/cu.m	
Concentration at cloud edge, kg/cu.m	
Maximum ground level concentration, kg/cu.m	
Distance where maximum concentration occurs, m	

Table 21.11. Impacts of initiating events and the probabilities of causing domino effect in absorber unit of MFL

Parameters	Scenario and their likely impacts
	BLEVE followed by toxic release in absorber unit
Explosion:	BLEVE
Total energy released, KJ	2.55E+06
Peak overpressure developed, kPa	587.3
Variation of overpressure in air, kPa/s	301.04
Shock wave velocity, m/s	631.6
Duration of shock wave, ms	14.5
Missile Characteristics:	
Initial velocity of fragment, m/s	182.2
Kinetic energy of fragment, kJ	1.12E+05
Penetration ability at study point	
Concrete structure, m	0.01
Brick structure, m	0.00
Steel structure, m	0.00
Fire	Not applicable
Radius of fireball, m	Not applicable
Duration of fireball, s	Not applicable
Energy released by fire ball, kJ	Not applicable
Radius of pool fire, m	Not applicable
Burning area, sq.m	Not applicable
Burning rate, kg/s	Not applicable
Radiation heat flux, kJ/sq.m	Not applicable
Domino Checking	
Location of the unit from primary event, m	75.0
Domino effect due to heat load	
Total heat received, kJ	Not applicable
Heat intensity, kJ/sq.m	Not applicable
Probability of domino effect due to fire	Not applicable
Domino effect due to over pressure:	
Explosion energy, kJ	4.43E+04
Peak overpressure, kPa	178.4
Probability of domino effect due to overpressure	0.6
Domino effect due to missile	
Explosion energy, kJ	5.41E+04
Missile velocity, m/s	356.4
Probability of domino effect due to missile after meeting the target	0.14
Parameters	Scenario and their likely impacts
Toxic release and dispersion	
Instantaneous/continuous	Instantaneous
Puff/Plume Characteristics:	Puff characteristics
Puff concentration at centre of cloud, kg/cu.m	1.26E-02
Concentration at cloud edge, kg/cu.m	1.42E-03
Maximum ground level concentration, kg/cu.m	2.45E-01
Distance where maximum concentration occurs, m	1050.4

Table 21.12. Impacts of initiating events and the probabilities of causing domino effect in urea reactor of MFL

Parameters	Scenarios and their likely impacts
	CVCE followed by fire ball in urea reactor
Explosion:	CVCE
Total energy released, KJ	1.11E+07
Peak overpressure developed, kPa	1356.7
Variation of overpressure in air, kPa/s	575.7
Shock wave velocity, m/s	865.5
Duration of shock wave, ms	14.5
Missile Characteristics:	
Initial velocity of fragment, m/s	1405.4
Kinetic energy of fragment, kJ	1.51E+07
Penetration ability at study point	
Concrete structure, m	0.30
Brick structure, m	0.41
Steel structure, m	0.07
Fire	
Radius of fireball, m	191.6
Duration of fireball, s	154
Energy released by fire ball, kJ	2.71E+10
Radius of pool fire, m	Not applicable
Burning area, sq.m	Not applicable
Burning rate, kg/s	Not applicable
Radiation heat flux, kJ/sq.m	82325.4
Domino Checking	
Location of the unit from primary event, m	75.0
Domino effect due to heat load	
Total heat received, kJ	5.45E+07
Heat intensity, kJ/sq.m	11215.4
Probability of domino effect due to fire	1.0
Domino effect due to over pressure:	
Explosion energy, kJ	3.32E+05
Peak overpressure, kPa	677.4
Probability of domino effect due to overpressure	1.0
Domino effect due to missile	
Explosion energy, kJ	4.78E+05
Missile velocity, m/s	515.4
Probability of domino effect due to missile after meeting the target	1.0
Toxic release and dispersion Instantaneous/continuous Puff/Plume Characteristics: Puff concentration at centre of cloud, kg/cu.m Concentration at cloud edge, kg/cu.m Maximum ground level concentration, kg/cu.m Distance where maximum concentration occurs, m	Not applicable

Table 21.13. Impacts of initiating events and the probabilities of causing domino effect in pre-neutralizer unit of MFL

Parameters	Scenarios and their likely impacts
	BLEVE followed by toxic release in pre-neutralizer of NPK plant
Explosion:	BLEVE
Total energy released, KJ	2.35E+06
Peak overpressure developed, kPa	517.3
Variation of overpressure in air, kPa/s	271.4
Shock wave velocity, m/s	611.6
Duration of shock wave, ms	14.5
Missile Characteristics:	
Initial velocity of fragment, m/s	155.1
Kinetic energy of fragment, kJ	1.05E+05
Penetration ability at study point	
Concrete structure, m	0.01
Brick structure, m	0.00
Steel structure, m	0.00
Fire	
Radius of fireball, m	Not applicable
Duration of fireball, s	Not applicable
Energy released by fire ball, kJ	Not applicable
Radius of pool fire, m	Not applicable
Burning area, sq.m	Not applicable
Burning rate, kg/s	Not applicable
Radiation heat flux, kJ/sq.m	Not applicable
Domino Checking	
Location of the unit from primary event, m	75.0
Domino effect due to heat load	Not applicable
Total heat received, kJ	Not applicable
Heat intensity, kJ/sq.m	Not applicable
Probability of domino effect due to fire	Not applicable
Domino effect due to over pressure:	
Explosion energy, kJ	8.73E+04
Peak overpressure, kPa	195.5
Probability of domino effect due to overpressure	0.75
Domino effect due to missile	
Explosion energy, kJ	9.11E+04
Missile velocity, m/s	371.4
Probability of domino effect due to missile after meeting the target	0.23
Toxic release and dispersion	
Instantaneous/continuous	not applicable
Puff/Plume Characteristics:	
Puff concentration at centre of cloud, kg/cu.m	1.90E-04
Concentration at cloud edge, kg/cu.m	2.13E-05
Maximum ground level concentration, kg/cu.m	1.54E-02
Distance where maximum concentration occurs, m	120.0

Table 21.14. Impacts of initiating events and the probabilities of causing domino effect in drier unit of MFL

Parameters	Scenarios and their likely impacts
	BLEVE followed by fireball in drier unit
Explosion:	BLEVE
Total energy released, KJ	1.45E+05
Peak overpressure developed, kPa	355.2
Variation of overpressure in air, kPa/s	144.0
Shock wave velocity, m/s	382.6
Duration of shock wave, ms	9.5
Missile Characteristics:	
Initial velocity of fragment, m/s	65.9
Kinetic energy of fragment, kJ	4.1E+03
Penetration ability at study point	
Concrete structure, m	0.00
Brick structure, m	0.00
Steel structure, m	0.00
Fire	
Radius of fireball, m	67.30
Duration of fireball, s	21.45
Energy released by fire ball, kJ	4.50E+06
Radius of pool fire, m	Not applicable
Burning area, sq.m	Not applicable
Burning rate, kg/s	Not applicable 4285.4
Radiation heat flux, kJ/sq.m	
Domino Checking	
Location of the unit from primary event, m	75.0
Domino effect due to heat load	
Total heat received, kJ	8.45E+04
Heat intensity, kJ/sq.m	609.5
Probability of domino effect due to fire	0.75
Domino effect due to over pressure:	
Explosion energy, kJ	1.31E+03
Peak overpressure, kPa	085.4
Probability of domino effect due to overpressure	0.10
Domino effect due to missile	
Explosion energy, kJ	1.53E+03
Missile velocity, m/s	275
Probability of domino effect due to missile after meeting the target	0.0
Toxic release and dispersion	
Instantaneous/continuous	not applicable
Puff/Plume Characteristics:	
Puff concentration at centre of cloud, kg/cu.m	
Concentration at cloud edge, kg/cu.m	
Maximum ground level concentration, kg/cu.m	
Distance where maximum concentration occurs, m	

Table 21.15. Impacts of initiating events and the probabilities of causing domino effect in hydrogen treatment unit of SPIC-HCD

Parameters	Scenarios and their likely impacts
	Fireball in hydrogen treatment unit
Explosion:	BLEVE
Total energy released, KJ	1.2E+06
Peak overpressure developed, kPa	262.0
Variation of overpressure in air, kPa/s	169.0
Shock wave velocity, m/s	342.2
Duration of shock wave, ms	21.0
Missile Characteristics:	
Initial velocity of fragment, m/s	111.5
Kinetic energy of fragment, kJ	2.11E+05
Penetration ability at study point	
Concrete structure, m	0.00
Brick structure, m	0.01
Steel structure, m	0.00
Fire	
Radius of fireball, m	75.6
Duration of fireball, s	31.5
Energy released by fire ball, kJ	6.52E+07
Radius of pool fire, m	not applicable
Burning area, sq.m	not applicable
Burning rate, kg/s	not applicable
Radiation heat flux, kJ/sq.m	5441.6
Domino Checking	
Location of the unit from primary event, m	75.0
Domino effect due to heat load	
Total heat received, kJ	1.14×10^6
Heat intensity, kJ/sq.m	1141.3
Probability of domino effect due to fire	1.0
Domino effect due to over pressure:	
Explosion energy, kJ	7.15E+04
Peak overpressure, kPa	164.3
Probability of domino effect due to overpressure	0.64
Domino effect due to missile	
Explosion energy, kJ	8.45E+04
Missile velocity, m/s	304.4
Probability of domino effect due to missile after meeting the target	0.17
Toxic release and dispersion Instantaneous/continuous Puff/Plume Characteristics: Puff concentration at centre of cloud, kg/cu.m Concentration at cloud edge, kg/cu.m Maximum ground level concentration, kg/cu.m Distance where maximum concentration occurs, m	Not applicable

Table 21.16. Impacts of initiating events and the probabilities of causing domino effect in hydrogen storage vessel unit of SPIC-HCD

Parameters	Scenarios and their likely impacts
	CVCE followed by fireball in hydrogen storage vessel
Explosion:	CVCE
Total energy released, KJ	5.45E+09
Peak overpressure developed, kPa	22550.1
Variation of overpressure in air, kPa/s	15475.5
Shock wave velocity, m/s	4255.8
Duration of shock wave, ms	106
Missile Characteristics:	
Initial velocity of fragment, m/s	8415.3
Kinetic energy of fragment, kJ	2.46E+08
Penetration ability at study point	
Concrete structure, m	0.55
Brick structure, m	0.78
Steel structure, m	0.17
Fire	FIREBALL
Radius of fireball, m	215.4
Duration of fireball, s	128.6
Energy released by fire ball, kJ	9.54E+09
Radius of pool fire, m	not applicable
Burning area, sq.m	not applicable
Burning rate, kg/s	not applicable
Radiation heat flux, kJ/sq.m	755150.3
Domino Checking	
Location of the unit from primary event, m	75.0
Domino effect due to heat load	
Total heat received, kJ	1.30E+08
Heat intensity, kJ/sq.m	345541.2
Probability of domino effect due to fire	1.0
Domino effect due to over pressure:	
Explosion energy, kJ	3.15E+07
Peak overpressure, kPa	2105.3
Probability of domino effect due to overpressure	1.0
Domino effect due to missile	
Explosion energy, kJ	5.45E+08
Missile velocity, m/s	7.14E+03
Probability of domino effect due to missile after meeting the target	1.0
Toxic release and dispersion	
Instantaneous/continuous	
Puff/Plume Characteristics:	
Puff concentration at centre of cloud, kg/cu.m	Not applicable
Concentration at cloud edge, kg/cu.m	
Maximum ground level concentration, kg/cu.m	
Distance where maximum concentration occurs, m	

Table 21.17. Impacts of initiating events and the probabilities of causing domino effect in ammonia storage unit of SPIC-HCD

Parameters	Scenarios and their likely impacts
	BLEVE followed by toxic release in ammonia storage vessel
Explosion:	BLEVE
Total energy released, KJ	8.33E+07
Peak overpressure developed, kPa	8764.3
Variation of overpressure in air, kPa/s	4010.5
Shock wave velocity, m/s	1205.2
Duration of shock wave, ms	41.5
Missile Characteristics:	
Initial velocity of fragment, m/s	1065.4
Kinetic energy of fragment, kJ	1.9E+07
Penetration ability at study point	
Concrete structure, m	0.31
Brick structure, m	0.41
Steel structure, m	0.07
Fire	Not applicable
Radius of fireball, m	Not applicable
Duration of fireball, s	Not applicable
Energy released by fire ball, kJ	Not applicable
Radius of pool fire, m	Not applicable
Burning area, sq.m	Not applicable
Burning rate, kg/s	Not applicable
Radiation heat flux, kJ/sq.m	Not applicable
Domino Checking	
Location of the unit from primary event, m	75.0
Domino effect due to heat load	
Total heat received, kJ	Not applicable
Heat intensity, kJ/sq.m	Not applicable
Probability of domino effect due to fire	Not applicable
Domino effect due to over pressure:	
Explosion energy, kJ	5.15E+05
Peak overpressure, kPa	710.4
Probability of domino effect due to overpressure	1.0
Domino effect due to missile	
Explosion energy, kJ	6.35E+05
Missile velocity, m/s	542.4
Probability of domino effect due to missile after meeting the target	1.0
Toxic release and dispersion	
Instantaneous/continuous	Instantaneous
Puff/Plume Characteristics:	Puff characteristics
Puff concentration at centre of cloud, kg/cu.m	5.145E-02
Concentration at cloud edge, kg/cu.m	5.145E-03
Maximum ground level concentration, kg/cu.m	1.54E-01
Distance where maximum concentration occurs, m	875.4

Table 21.18. The detailed results of DOMIFFECT for accident in ammonia storage unit of SPIC-HCD and subsequent domino effects in other units

DOMIFFECT
(c) F.I. Khan & S.A. Abbasi, Pondicherry-605 014, India

Parameters	Values
Ammonia storage vessel	
BLEVE followed by toxic release and dispersion	
Explosion: BLEVE	
Total energy released	(kJ) : 8.33E+07
Peak over pressure	(kPa) : 8764.3
Variation of over pressure in air	(kPa/s) : 4010.5
Shock wave velocity	(m/s) : 1205.2
Duration of shock wave	(ms) : 18.5
ity of fragment	(m/s) : 1065.4
Kinetic energy of fragment	(kJ) : 1.91E+07
ability at study point	
Concrete structure	(m) : 0.31
Brick structure	(m) : 0.41
Steel structure	(m) : 0.07
(m),	(kg/cu.m) : 5.225E-03
Ground level concentration of puff	(kg/cu.m) : 5.145E-03
Ground level concentration on puff axis	(kg/cu.m) : 5.145E-02
Cloud radius	(m) : 1.52E+02
Maximum distance traveled by the cloud	(m) : 8.752E+02
Maximum ground level concentration	(kg/cu.m) : 1.544E-01
DOMINO EFFECT –I: Secondary accident	
Unit: Compressor unit	
Distance of the vulnerable unit from primary unit	(m) : 55.0
Domino effect probabilities:	
due to heat load	(%) : 100
due to explosion	(%) : 100
due to missile	(%) : 100
Domino effect impacts	
Scenario: Instantaneous release and dispersion of chlorine	
Box instantaneous model: Elevated source	
Concentration at distance of 200 (m),	(kg/cu.m) : 4.151E-04
Heavy gas puff Characteristics	
Ground level concentration of puff	(kg/cu.m) : 4.011E-04
Ground level concentration on puff axis	(kg/cu.m) : 4.011E-03
Cloud radius	(m) : 2.145E+02
Maximum distance traveled by the cloud	(m) : 8.012E+02
Maximum ground level concentration	(kg/cu.m) : 3.751E-01
DOMINO EFFECT –II: Secondary accident	
Unit: Hydrogen storage unit	
Distance of the vulnerable unit from primary unit	(m) : 65.0
Domino effect probabilities:	
due to heat load	(%) : ---

Table 21.18 (Continued)

due to explosion	(%) : 100
due to missile	(%) : 100
Domino effect impacts	
Scenario: BLEVE followed by fireball	
Explosion : BLEVE	
Parameters	Values
Total energy released	(kJ) : 3.33e+07
Peak over pressure	(kPa) : 904.7
Variation of over pressure in air	(kPa/s) : 1010.5
Shock velocity of air	(m/s) : 1705.2
Duration of shock wave	(ms) : 21.5
Missile characteristics	
Initial velocity of fragment	(m/s) : 665.4
Kinetic energy of fragment	(kJ) : 5.9e+06
Penetration ability at 50 m from the location of secondary accident	
Concrete structure	(m) : 0.67
Brick structure	(m) : 0.80
Steel structure	(m) : 0.15
Fire : Fire ball	
Radius of the fire ball	(m) : 97.3
Duration of the fire ball	(s) : 39.7
Energy released by fire ball	(kJ) : 9.06E+07
Radiation heat flux	(kJ/sq.m) : 11384.5
DOMINO EFFECT –III: Tertiary effect	
Unit: Chlorine storage unit	
Distance of the vulnerable unit from secondary unit	(m) : 90.0
Domino effect probabilities:	
due to heat load	(%) : ---
due to explosion	(%) : 100
due to missile	(%) : 100
Domino effect impacts	
Scenario: BLEVE followed by toxic load build-up	
Explosion: BLEVE	
Energy released during explosion	(kJ) : 1.4E+06
Peak over pressure	(kPa) : 515.4
Variation of over pressure in air	(kPa/s) : 281.4
Shock velocity of air	(m/s) : 495.6
Duration of shock wave	(ms) : 24.2
Missile characteristics:	
No missile effect	
Toxic release and dispersion	
Box instantaneous model: Elevated source	
Concentration at distance of 200 (m),	(kg/cu.m) : 7.145E-04
Heavy gas puff Characteristics	
Ground level concentration of puff	(kg/cu.m) : 7.071E-04
Ground level concentration on puff axis	(kg/cu.m) : 7.071E-03
Cloud radius	(m) : 1.214E+03
Maximum distance traveled by the cloud	(m) : 6.454E+03
Maximum ground level concentration	(kg/cu.m) : 7.345E-01

Table 21.19. The detailed results of DOMIFFECT for accident in Propylene oxide storage unit of MPL and subsequent domino effects in other units

DOMIFFECT
(c) F I Khan & S A Abbasi, Pondicherry-605 014, India

Parameters	Values
PRIMARY EVENT	
Unit: Propylene oxide storage vessel Scenario: CVCE followed by fireball	
Explosion: CVCE	
Total energy released	(kJ) : 5.45e+07
Peak over pressure	(kPa) : 17743.4
Variation of over pressure in air	(kPa/s) : 9424.3
Shock wave velocity	(m/s) : 3155.5
Duration of shock wave	(ms) : 34.5
Missile characteristics	
Initial velocity of fragment	(m/s) : 1894.2
Kinetic energy of fragment	(kJ) : 4.54e+07
Penetration ability at study point	
Concrete structure	(m) : 0.35
Brick structure	(m) : 0.52
Steel structure	(m) : 0.08
Fire Ball	
Radius of the fire ball	(m) : 155.5
Duration of the fire ball	(s) : 50.0
Energy released by fire ball	(kJ) : 1.06e+09
Radiation heat flux	(kJ/sq.m) : 112841.5
DOMINO EFFECT –I: Secondary accident Unit: Ethylene oxide reactor	
Distance of the vulnerable unit from primary unit	(m) : 55.0
Domino effect probabilities:	
due to heat load	(%) : 100
due to explosion	(%) : 100
due to missile	(%) : 100
Domino effect impacts Scenario: BLEVE followed by fire ball	
Explosion: BLEVE	
Total energy released	(kJ) : 3.33e+07
Peak over pressure	(kPa) : 904.7
Variation of over pressure in air	(kPa/s) : 1010.5
Shock velocity of air	(m/s) : 1705.2
Duration of shock wave	(ms) : 21.5
Missile characteristics Initial velocity of fragment	(m/s) : 665.4
Kinetic energy of fragment	(kJ) : 5.9e+06
Penetration ability at 50 m from the location of secondary accident Concrete structure	(m) : 0.67
Brick structure	(m) : 0.80
Steel structure	(m) : 0.15
Fire: Fire ball Radius of the fire ball	(m) : 97.3
Duration of the fire ball	(s) : 39.7

Table 21.19. (Continued)

Energy released by fire ball	(kJ) : 9.06E+07
Radiation heat flux	(kJ/sq.m): 11384.5
DOMINO EFFECT –III: Tertiary effect Unit: Chlorine storage unit	
Distance of the vulnerable unit from secondary unit	(m): 90.0
Parameters	Values
Domino effect probabilities:	
due to heat load	(%): 100
due to explosion	(%): 100
due to missile	(%): 65
Domino effect impacts Scenario: BLEVE followed by toxic load build-up	
Explosion : BLEVE Energy released during explosion	**(kJ)**: 2.4E+06
Peak over pressure	(kPa): 1151.4
Variation of over pressure in air	(kPa/s): 481.4
Shock velocity of air	(m/s): 795.6
Duration of shock wave	(ms): 41.2
Missile characteristics: Initial velocity of fragment	(m/s): 235.6
Kinetic energy associated with fragment	(kJ): 1.5E+05
Penetration at 50 meter from the location of tertiary accident Concrete structure	(m) : 0.06
Brick structure	(m) : 0.11
Steel structure	(m) : 0.01
Toxic release and dispersion Box instantaneous model : Elevated source	
Concentration at distance of 200 (m),	(kg/cu.m) : 4.754E-03
Heavy gas puff Characteristics Ground level concentration of puff	(kg/cu.m) : 4.311E-03
Ground level concentration on puff axis	(kg/cu.m) : 4.311E-02
Cloud radius	(m) : 7.315E+01
Maximum distance travelled by the cloud	(m) : 8.215E+01
Maximum ground level concentration	(kg/cu.m) : 9.882E-01

Table 21.20. Impacts of initiating events and the probabilities of causing domino effect in propylene oxide storage unit of MPL

Parameters	Scenarios and their likely impacts
	CVCE with fire ball in propylene oxide storage vessel
Explosion:	CVCE
Total energy released, KJ	5.45E+07
Peak overpressure developed, kPa	17443.4
Variation of overpressure in air, kPa/s	9424.3
Shock wave velocity, m/s	3155.5
Duration of shock wave, ms	34.5
Missile Characteristics:	
Initial velocity of fragment, m/s	1894.2
Kinetic energy of fragment, kJ	4.54E+07

Table 21.20 (Continued)

Penetration ability at study point Concrete structure, m Brick structure, m Steel structure, m	 0.35 0.52 0.08
Fire Radius of fireball, m Duration of fireball, s Energy released by fire ball, kJ Radius of pool fire, m Burning area, sq.m Burning rate, kg/s Radiation heat flux, kJ/sq.m	 155.5 50.0 1.06E+09 Not applicable Not applicable Not applicable 112841.5
Parameters	Scenarios and their likely impacts
Domino Checking Location of the unit from primary event, m	 75.0
Domino effect due to heat load Total heat received, kJ Heat intensity, kJ/sq.m Probability of domino effect due to fire	 1.07E+05 1105 1.0
Domino effect due to over pressure: Explosion energy, kJ Peak overpressure, kPa Probability of domino effect due to overpressure	 1.45E+06 797.3 1.0
Domino effect due to missile Explosion energy, kJ Missile velocity, m/s Probability of domino effect due to missile after meeting the target	 1.97E+06 831.2 1.0
Toxic release and dispersion Instantaneous/continuous Puff/Plume Characteristics: Puff concentration at centre of cloud, kg/cu.m Concentration at cloud edge, kg/cu.m Maximum ground level concentration, kg/cu.m Distance where maximum concentration occurs, m	Not applicable

Table 21.21. Impacts of initiating events and the probabilities of causing domino effect in propylene storage unit of MPL

Parameters	Scenarios and their likely impacts
	BLEVE with fire ball in propylene storage vessel
Explosion: Total energy released, KJ Peak overpressure developed, kPa Variation of overpressure in air, kPa/s Shock wave velocity, m/s Duration of shock wave, ms	BLEVE 5.45E+07 17443.4 9424.3 3155.5 34.5
Missile Characteristics: Initial velocity of fragment, m/s Kinetic energy of fragment, kJ	 1894.2 4.54E+07

Table 21.21. (Continued)

Penetration ability at study point Concrete structure, m Brick structure, m Steel structure, m	 0.35 0.52 0.08
Fire Radius of fireball, m Duration of fireball, s Energy released by fire ball, kJ Radius of pool fire, m Burning area, sq.m Burning rate, kg/s Radiation heat flux, kJ/sq.m	 155.5 50 1.06E+09 Not applicable Not applicable Not applicable 112841.5
Domino Checking Location of the unit from primary event, m	 75.0
Domino effect due to heat load Total heat received, kJ Heat intensity, kJ/sq.m Probability of domino effect due to fire	 7.69E+07 8914.5 1.0
Parameters	Scenarios and their likely impacts
Domino effect due to over pressure: Explosion energy, kJ Peak overpressure, kPa Probability of domino effect due to overpressure	 3.47E+06 7615.4 1.0
Domino effect due to missile Explosion energy, kJ Missile velocity, m/s Probability of domino effect due to missile after meeting the target	 3.57E+06 901.5 1.0
Toxic release and dispersion Instantaneous/continuous Puff/Plume Characteristics: Puff concentration at centre of cloud, kg/cu.m Concentration at cloud edge, kg/cu.m Maximum ground level concentration, kg/cu.m Distance where maximum concentration occurs, m	Not applicable

Table 21.22. Impacts of initiating events and the probabilities of causing domino effect in MPG storage unit of MPL

Parameters	Scenarios and their likely impacts
	Pool fire in mono propylene glycol storage vessel
Explosion: Total energy released, KJ Peak overpressure developed, kPa Variation of overpressure in air, kPa/s Shock wave velocity, m/s Duration of shock wave, ms	Not applicable

Table 21.22 (Continued)

Missile Characteristics: Initial velocity of fragment, m/s Kinetic energy of fragment, kJ	Not applicable
Penetration ability at study point Concrete structure, m Brick structure, m Steel structure, m	Not applicable
Fire Radius of fireball, m Duration of fireball, s Energy released by fire ball, kJ Radius of pool fire, m Burning area, sq.m Burning rate, kg/s Radaitaion heat flux, kJ/sq.m	POOLFIRE 5.0 not applicable not applicable not applicable 78.5 21.5 89659.8
Domino Checking Location of the unit from primary event, m	 75.0
Domino effect due to heat load Total heat received, kJ Heat intensity, kJ/sq.m Probability of domino effect due to fire	 1.83E+08 10404.5 1.0

Parameters	Scenarios and their likely impacts
Domino effect due to over pressure: Explosion energy, kJ Peak overpressure, kPa Probability of domino effect due to overpressure	Not applicable
Domino effect due to missile Explosion energy, kJ Missile velocity, m/s Probability of domino effect due to missile after meeting the target	not applicable
Toxic release and dispersion Instantaneous/continuous Puff/Plume Characteristics: Puff concentration at centre of cloud, kg/cu.m Concentration at cloud edge, kg/cu.m Maximum ground level concentration, kg/cu.m Distance where maximum concentration occurs, m	not applicable

Table 21.23. Impacts of initiating events and the probabilities of causing domino effect in fuel oil storage unit of MPL

Parameters	Scenarios and their likely impacts
	Pool fire in fuel oil storage vessel

Table 21.23. (Continued)

Explosion: Total energy released, KJ Peak overpressure developed, kPa Variation of overpressure in air, kPa/s Shock wave velocity, m/s Duration of shock wave, ms	Not applicable
Missile Characteristics: Initial velocity of fragment, m/s Kinetic energy of fragment, kJ	Not applicable
Penetration ability at study point Concrete structure, m Brick structure, m Steel structure, m	Not applicable
Fire Radius of fireball, m Duration of fireball, s Energy released by fire ball, kJ Radius of pool fire, m Burning area, sq.m Burning rate, kg/s Radaitaion heat flux, kJ/sq.m	POOLFIRE 5.0 not applicable not applicable not applicable 78.5 35.0 154446.5
Domino Checking Location of the unit from primary event, m	75.0
Domino effect due to heat load Total heat received, kJ Heat intensity, kJ/sq.m Probability of domino effect due to fire	 5.15e+08 21518.3 1.0
Domino effect due to over pressure: Explosion energy, kJ Peak overpressure, kPa Probability of domino effect due to overpressure	Not applicable
Parameters	Scenarios and their likely impacts
Domino effect due to missile Explosion energy, kJ Missile velocity, m/s Probability of domino effect due to missile after meeting the target	Not applicable
Toxic release and dispersion Instantaneous/continuous Puff/Plume Characteristics: Puff concentration at centre of cloud, kg/cu.m Concentration at cloud edge, kg/cu.m Maximum ground level concentration, kg/cu.m Distance where maximum concentration occurs, m	Not applicable

Table 21.24. Impacts of initiating events and the probabilities of causing domino effect in propylene storage unit of TPL

Parameters	Scenarios and their likely impacts
	CVCE followed by fireball in propylene storage vessel
Explosion:	CVCE
Total energy released, KJ	3.25E+07
Peak overpressure developed, kPa	1541.2
Variation of overpressure in air, kPa/s	801.4
Shock wave velocity, m/s	875.6
Duration of shock wave, ms	24.0
Missile Characteristics:	
Initial velocity of fragment, m/s	495.4
Kinetic energy of fragment, kJ	2.25E+06
Penetration ability at study point	
Concrete structure, m	0.28
Brick structure, m	0.40
Steel structure, m	0.07
Fire	
Radius of fireball, m	175.4
Duration of fireball, s	85
Energy released by fire ball, kJ	6.15E+09
Radius of pool fire, m	Not applicable
Burning area, sq.m	Not applicable
Burning rate, kg/s	Not applicable
Radiation heat flux, kJ/sq.m	64431.2
Domino Checking	
Location of the unit from primary event, m	75.0
Domino effect due to heat load	
Total heat received, kJ	1.15E+08
Heat intensity, kJ/sq.m	9105.4
Probability of domino effect due to fire	1.0
Domino effect due to over pressure:	
Explosion energy, kJ	3.25E+07
Peak overpressure, kPa	8674.4
Probability of domino effect due to overpressure	1.0
Domino effect due to missile	
Explosion energy, kJ	3.25E+07
Missile velocity, m/s	275.0
Probability of domino effect due to missile after meeting the target	1.0
Parameters	Scenarios and their likely impacts
Toxic release and dispersion Instantaneous/continuous Puff/Plume Characteristics: Puff concentration at centre of cloud, kg/cu.m Concentration at cloud edge, kg/cu.m Maximum ground level concentration, kg/cu.m Distance where maximum concentration occurs, m	Not applicable

Table 21.25. Impacts of initiating events and the probabilities of causing domino effect in allyl chloride storage unit of TPL

Parameters	Scenarios and their likely impacts
	Pool fire in allyl chloride storage vessel
Explosion: Total energy released, KJ Peak overpressure developed, kPa Variation of overpressure in air, kPa/s Shock wave velocity, m/s Duration of shock wave, ms	Not applicable
Missile Characteristics: Initial velocity of fragment, m/s Kinetic energy of fragment, kJ	Not applicable
Penetration ability at study point Concrete structure, m Brick structure, m Steel structure, m	Not applicable
Fire Radius of fireball, m Duration of fireball, s Energy released by fire ball, kJ Radius of pool fire, m Burning area, sq.m Burning rate, kg/s Radiation heat flux, kJ/sq.m	POOLFIRE 5.0 Not applicable Not applicable Not applicable 78.5 15.0 6896.1
Domino Checking Location of the unit from primary event, m	 75.0
Domino effect due to heat load Total heat received, kJ Heat intensity, kJ/sq.m Probability of domino effect due to fire	 7.45E+07 5004.8 1.0
Domino effect due to over pressure: Explosion energy, kJ Peak overpressure, kPa Probability of domino effect due to overpressure	Not applicable
Domino effect due to missile Explosion energy, kJ Missile velocity, m/s Probability of domino effect due to missile after meeting the target	Not applicable
Toxic release and dispersion Instantaneous/continuous Puff/Plume Characteristics: Puff concentration at centre of cloud, kg/cu.m Concentration at cloud edge, kg/cu.m Maximum ground level concentration, kg/cu.m Distance where maximum concentration occurs, m	Not applicable

Table 21.26. Impacts of initiating events and the probabilities of causing domino effect in fuel oil storage unit of TPL

Parameters	Scenarios and their likely impacts
	Pool fire in fuel oil storage vessel
Explosion: Total energy released, KJ Peak overpressure developed, kPa Variation of overpressure in air, kPa/s Shock wave velocity, m/s Duration of shock wave, ms	Not applicable
Missile Characteristics: Initial velocity of fragment, m/s Kinetic energy of fragment, kJ	Not applicable
Penetration ability at study point Concrete structure, m Brick structure, m Steel structure, m	Not applicable
Fire	POOLFIRE
Radius of fireball, m	5.0
Duration of fireball, s	not applicable
Energy released by fire ball, kJ	not applicable
Radius of pool fire, m	not applicable
Burning area, sq.m	78.5
Burning rate, kg/s	34.0
Radiation heat flux, kJ/sq.m	14418.9
Domino Checking Location of the unit from primary event, m	75.0
Domino effect due to heat load	
Total heat received, kJ	5.05E+08
Heat intensity, kJ/sq.m	20105.3
Probability of domino effect due to fire	1.0
Domino effect due to over pressure: Explosion energy, kJ Peak overpressure, kPa Probability of domino effect due to overpressure	Not applicable
Domino effect due to missile Explosion energy, kJ Missile velocity, m/s Probability of domino effect due to missile after meeting the target	Not applicable
Toxic release and dispersion Instantaneous/continuous Puff/Plume Characteristics: Puff concentration at centre of cloud, kg/cu.m Concentration at cloud edge, kg/cu.m Maximum ground level concentration, kg/cu.m Distance where maximum concentration occurs, m	Not applicable

Table 21.27. The output of DOMIFFECT for the primary accident in fuel oil storage unit of TPL and the subsequent domino effects

DOMIFFECT
(c) F.I. Khan & S.A. Abbasi, Pondicherry-605 014, India

Parameters	Values
PRIMARY EVENT	
Unit: Fuel oil storage unit	
Scenario: Pool fire	
Pool fire	
Radius of pool fire	(m) : 5.0
Burning area	(sq.m) : 78.5
Burning rate,	(kg/s) : 34.0
Radiation heat flux	(kj/sq.m) : 144181.9
DOMINO EFFECT –I: Secondary accident Unit: Allyl chloride storage	
Distance of the vulnerable unit from the primary event	(m) : 50.0
Domino effect probabilities:	
due to heat load	(%) : 100
due to explosion is	(%) : 0
due to missile	(%) : 0
Domino effect impacts: Scenario: Pool fire	
Fire : Pool fire	
Radius of pool fire	(m) : 7.5
Burning area	(sq.m) : 176.6
Burning rate,	(kg/s) : 1.75
Radiation heat flux	(kj/sq.m) : 234551.2
DOMINO EFFECT-II: Tertiary accident Unit: fractionation unit	
Distance of study from secondary accident unit	(m) : 95.0
Domino effect probabilities:	
due to heat load	(%) : 100
due to explosion	(%) : 0
due to missile	(%) : 0
Domino effect impacts: Scenario: BLEVE followed by fireball	
Explosion: BLEVE	
Total energy released	(kJ) : 1.55E+05
Peak over pressure	(kPa) : 702.0
Variation of over pressure in air	(kPa/s) : 350.0
Shock wave velocity	(m/s) : 685.5
Duration of shock wave	(ms) : 14.68
Missile characteristics Initial velocity of a typical fragment	(m/s) : 95.90
Kinetic energy a typical of fragment	(kJ) : 41241.68
Final velocity of a typical fragment	(m/s) : 65.90
Penetration ability at 50 m from the location of tertiary accident	
Concrete structure	**(m) : 0.01**
Brick structure	(m) : 0.05

Table 21.27 (Continued)

Steel structure	(m) : 0.00
Fire Ball	
Radius of the fire ball	(m) : 75.5
Duration of the fire ball	(s) : 25.5
Energy released by fire ball	(kJ) : 4.75E+06
Radiation heat intensity	(kJ/sq.m) : 6305.45

Table 21.28. The output of DOMIFFECT for the primary accident in MEG storage unit of IOCL and the subsequent domino effects

DOMIFFECT
(c) F.I. Khan & S.A. Abbasi, Pondicherry-605 014, India

Parameters	Values
PRIMARY EVENT	
Unit: MEG storage unit	
Scenario: Pool fire	
Pool fire	
Radius of pool fire	(m) : 5.0
Burning area	(sq.m) : 78.5
Burning rate,	(kg/s) : 2.69
Radiation heat flux	(kj/sq.m) : 114991.4
DOMINO EFFECT –I: Secondary accident Unit: Furnace oil	
Distance of the vulnerable unit from the primary event	(m) : 70.0
Domino effect probabilities:	
due to heat load	(%) : 100
due to explosion is	(%) : 0
due to missile	(%) : 0
Domino effect impacts: Scenario: Pool fire	
Fire : Pool fire	
Radius of pool fire	(m) : 7.5
Burning area	(sq.m) : 176.6
Burning rate,	(kg/s) : 1.75
Radiation heat flux	(kj/sq.m) : 234551.2
DOMINO EFFECT-II: Secondary accident Unit: Ammonia storage vessel	
Distance of study from secondary accident unit	(m) : 45.0
Domino effect probabilities:	
due to heat load	(%) : 100
due to explosion	(%) : 0
due to missile	(%) : 0
Domino effect Impacts Scenario: instantaneous release through vent valve	

Table 21.28 (Continued)

Box instantaneous model: Elevated source	
Concentration at distance of 200 (m),	(kg/cu.m) : 3.451E-03
Heavy gas puff Characteristics Ground level concentration of puff	(kg/cu.m) : 3.154E-03
Ground level concentration on puff axis	(kg/cu.m) : 3.154E-02
Cloud radius	(m) : 2.15E+02
Maximum distance travelled by the cloud	(m) : 8.75E+02
Maximum ground level concentration	(kg/cu.m) : 4.1

Table 21.29. Impacts of initiating events and the probabilities of causing domino effect in MEG storage unit of IOCL

Parameters	Scenarios and their likely impacts
	Pool fire in mono ethylene glycol storage unit
Explosion: Total energy released, KJ Peak overpressure developed, kPa Variation of overpressure in air, kPa/s Shock wave velocity, m/s Duration of shock wave, ms	Not applicable
Missile Characteristics: Initial velocity of fragment, m/s Kinetic energy of fragment, kJ Final velocity of fragment, m/s	Not applicable
Penetration ability at study point Concrete structure, m Brick structure, m Steel structure, m	Not applicable
Fire Radius of fireball, m Duration of fireball, s Energy released by fire ball, kJ Radius of pool fire, m Burning area, sq.m Burning rate, kg/s Radaitaion heat flux, kJ/sq.m	POOLFIRE 5.0 Not applicable Not applicable Not applicable 78.5 2.69 114991.4
Domino Checking Location of the unit from primary event, m	 750.0
Domino effect due to heat load Total heat received, kJ Heat intensity, kJ/sq.m Probability of domino effect due to fire	 3.25E+07 2745.4 1.0
Domino effect due to over pressure: Explosion energy, kJ Peak overpressure, kPa Probability of domino effect due to overpressure	Not applicable

Table 21.29 (Continued)

Domino effect due to missile Explosion energy, kJ Missile velocity, m/s Probability of domino effect due to missile after meeting the target	Not applicable
Toxic release and dispersion Instantaneous/continuous Puff/Plume Characteristics: Puff concentration at centre of cloud, kg/cu.m Concentration at cloud edge, kg/cu.m Downwind distance, m Distance where maximum concentration occurs, m	Not applicable

Table 21.30. Impacts of initiating events and the probabilities of causing domino effect in autoclave unit of IOCL

Parameters	**Scenarios and their likely impacts**
	Pool fire in autoclave unit
Explosion: Total energy released, KJ Peak overpressure developed, kPa Variation of overpressure in air, kPa/s Shock wave velocity, m/s Duration of shock wave, ms	Not applicable
Missile Characteristics: Initial velocity of fragment, m/s Kinetic energy of fragment, kJ Final velocity of fragment, m/s	Not applicable
Penetration ability at study point Concrete structure, m Brick structure, m Steel structure, m	Not applicable
Fire	Pool Fire
Radius of fireball, m	7.0
Duration of fireball, s	not applicable
Energy released by fire ball, kJ	not applicable
Radius of pool fire, m	not applicable
Burning area, sq.m	176.6
Burning rate, kg/s	4.01
Radiation heat flux, kJ/sq.m	32756.6
Domino Checking	
Location of the unit from primary event, m	75.0
Domino effect due to heat load	
Total heat received, kJ	7.15E+07
Heat intensity, kJ/sq.m	8518.3
Probability of domino effect due to fire	1.0

Table 21.30 (Continued)

Domino effect due to over pressure: Explosion energy, kJ Peak overpressure, kPa Probability of domino effect due to overpressure	Not applicable
Domino effect due to missile Explosion energy, kJ Missile velocity, m/s Probability of domino effect due to missile after meeting the target	Not applicable
Toxic release and dispersion Instantaneous/continuous Puff/Plume Characteristics: Puff concentration at centre of cloud, kg/cu.m Concentration at cloud edge, kg/cu.m Downwind distance, m Distance where maximum concentration occurs, m	Not applicable

Table 21.31. The output of DOMIFFECT for the primary accident in propylene storage unit of UB Petro-products limited and the subsequent domino effects

DOMIFFECT
(c) F.I. Khan & S.A. Abbasi, Pondicherry-605 014, India

Parameters	Values
Primary Event Unit: Propylene storage vessel Scenario: BLEVE followed by fireball	
Explosion: CVCE	
Total energy released	(kJ) : 1.32E+07
Peak over pressure	(kPa) : 1745.4
Variation of over pressure in air	(kPa/s) : 814.1
Shock wave velocity	(m/s) : 1245.2
Duration of shock wave	(ms) : 21.4
Missile characteristics	
Initial velocity of fragment	(m/s) : 445.3
Kinetic energy of fragment	(kJ) : 1.66E+06
Penetration ability at study point Concrete structure	(m) : 0.50
Brick structure	(m) : 0.68
Steel structure	(m) : 0.10
Fire Ball	
Radius of the fire ball	(m) : 97.3
Duration of the fire ball	(s) : 39.7
Energy released by fire ball	(kJ) : 9.06E+07
Radiation heat flux	(kJ/sq.m) : 11384.5
DOMINO EFFECT –I: Secondary accident Unit: Ethylene oxide storage vessel	
Distance of the vulnerable unit from the primary event	(m) : 55.0
Domino effect probabilities:	

Table 21.31 (Continued)

due to heat load	(%) : 100
due to explosion	(%) : 100
due to missile	(%) : 100
Domino effect impacts Scenario: BLEVE followed by fire ball	
Explosion: BLEVE Total energy released	(kJ) : 3.33E+07
Peak over pressure	(kPa) : 1904.7
Variation of over pressure in air	(kPa/s) : 1010.5
Shock velocity of air	(m/s) : 1705.2
Duration of shock wave	(ms) : 21.5
Missile characteristics	
Initial velocity of a typical fragment	(m/s) : 665.4
Kinetic energy of a typical fragment	(kJ) : 5.9E+06
Penetration ability at study point Concrete structure	(m) : 0.67
Brick structure	(m) : 0.80
Steel structure	(m) : 0.15
Fire: Fire ball Radius of the fire ball	(m) : 97.3
Duration of the fire ball	(s) : 39.7
Energy released by fire ball	(kJ) : 9.06E+07
Radiation heat flux	(kJ/sq.m) : 11384.0
DOMINO EFFECT –II : Secondary accident Unit: Chlorine transportation line	
Distance of study for secondary unit	(m) : 75.0
Domino effect probabilities:	
due to heat load	(%) : 100
Parameters	Values
due to explosion	(%) : 100
due to missile	(%) : 100
Domino effect impacts	
Scenario: instantaneous release of chlorine through vent valve due to over pressurisation	
Toxic release and dispersion	
Box instantaneous model : Elevated source	
Concentration at distance of 200 (m),	(kg/cu.m) : 1.225E-04
Heavy gas puff Characteristics Ground level concentration of puff	(kg/cu.m) : 1.114E-04
Ground level concentration on puff axis	(kg/cu.m) : 1.112E-03
Cloud radius	(m) : 1.52E+02
Maximum distance travelled by the cloud	(m) : 5.454E+02
Maximum ground level concentration	(kg/cu.m) : 5.4
DOMINO EFFECT-II: Tertiary accident Unit: POLYOL reactor	
Distance of vulnerable unit from the secondary accident unit (m): 55.0	
Domino effect probabilities:	
due to heat load	(%) : 75
due to explosion	(%) : 100
due to missile	(%) : 45

Table 21.31 (Continued)

Domino effect Impacts Scenario : instantaneous release followed by fire ball	
Fire : Fireball	
Radius of the fire ball	(m) : 115.3
Duration of the fire ball	(s) : 375.7
Energy released by fire ball	(kJ) : 1.55E+08
Radiation heat intensity	(kJ/sq.m) : 16651.7

Table 21.32. Impacts of initiating events and the probabilities of causing domino effect in propylene storage vessel of UBP

Parameters	Scenarios and their likely impacts
	CVCE followed by fireball in propylene storage vessel
Explosion: Total energy released, KJ Peak overpressure developed, kPa Variation of overpressure in air, kPa/s Shock wave velocity, m/s Duration of shock wave, ms	BLEVE 1.32E+07 1745.4 814.1 1245.2 21.4
Missile Characteristics: Initial velocity of fragment, m/s Kinetic energy of fragment, kJ	 445.3 1.66E+06
Penetration ability at study point Concrete structure, m Brick structure, m Steel structure, m	 0.50 0.68 0.10
Parameters	Scenarios and their likely impacts
Fire Radius of fireball, m Duration of fireball, s Energy released by fire ball, kJ Radius of pool fire, m Burning area, sq.m Burning rate, kg/s Radiation heat flux, kJ/sq.m	 97.3 39.7 9.06E+07 Not applicable Not applicable Not applicable 11384.5
Domino Checking Location of the unit from primary event, m	 75.0
Domino effect due to heat load Total heat received, kJ Heat intensity, kJ/sq.m Probability of domino effect due to fire	 3.10E+06 3840.5 1.0
Domino effect due to over pressure: Explosion energy, kJ Peak overpressure, kPa Probability of domino effect due to overpressure	 7.19E+05 384.5 1.0

Table 21.32 (Continued)

Domino effect due to missile Explosion energy, kJ Missile velocity, m/s Probability of domino effect due to missile after meeting the target	 6.19E+05 1.91E+02 1.0
Toxic release and dispersion Instantaneous/continuous Puff/Plume Characteristics: Puff concentration at centre of cloud, kg/cu.m Concentration at cloud edge, kg/cu.m Downwind distance, m Distance where maximum concentration occurs, m	Not applicable

Table 21.33. Impacts of initiating events and the probabilities of causing domino effect in propylene oxide storage vessel of UBP

Parameters	Scenarios and their likely impacts
	CVCE followed by fireball in propylene oxide storage vessel
Explosion: Total energy released, KJ Peak overpressure developed, kPa Variation of overpressure in air, kPa/s Shock wave velocity, m/s Duration of shock wave, ms	CVCE 5.84E+06 1295.3 540.2 1082.6 14.6
Missile Characteristics: Initial velocity of fragment, m/s Kinetic energy of fragment, kJ	 214.9 115458.6
Penetration ability at study point Concrete structure, m Brick structure, m Steel structure, m	 0.09 0.12 0.02
Parameters	Scenarios and their likely impacts
Fire Radius of fireball, m Duration of fireball, s Energy released by fire ball, kJ Radius of pool fire, m Burning area, sq.m Burning rate, kg/s Radiation heat flux, kJ/sq.m	 145.0 59.0 2.02E+09 Not applicable Not applicable Not applicable 77314.0
Domino Checking Location of the unit from primary event, m	 75.0
Domino effect due to heat load Total heat received, kJ Heat intensity, kJ/sq.m Probability of domino effect due to fire	 3.10E+06 3840.5 1.0

Table 21.33 (Continued)

Domino effect due to over pressure: Explosion energy, kJ Peak overpressure, kPa Probability of domino effect due to overpressure	 1.95E+05 184.5 1.0
Domino effect due to missile Explosion energy, kJ Missile velocity, m/s Probability of domino effect due to missile after meeting the target	 1.84E+03 95.0 1.0
Toxic release and dispersion Instantaneous/continuous Puff/Plume Characteristics: Puff concentration at centre of cloud, kg/cu.m Concentration at cloud edge, kg/cu.m Downwind distance, m Distance where maximum concentration occurs, m	Not applicable

Table 21.34. Impacts of initiating events and the probabilities of causing domino effect in ethylene oxide storage vessel of UBP

Parameters	Scenarios and their likely impacts
	BLEVE with fire ball in ethylene oxide storage vessel
Explosion: Total energy released, KJ Peak overpressure developed, kPa Variation of overpressure in air, kPa/s Shock wave velocity, m/s Duration of shock wave, ms	BLEVE 3.33E+07 1904.7 1010.5 1705.2 21.5
Missile Characteristics: Initial velocity of fragment, m/s Kinetic energy of fragment, kJ	 665.4 5.9E+06
Penetration ability at study point Concrete structure, m Brick structure, m Steel structure, m	 0.67 0.80 0.15
Parameters	Scenarios and their likely impacts
Fire Radius of fireball, m Duration of fireball, s Energy released by fire ball, kJ Radius of pool fire, m Burning area, sq.m Burning rate, kg/s Radaitaion heat flux, kJ/sq.m	 92.0 38.0 3.82E+08 Not applicable Not applicable Not applicable 14583.0
Domino Checking Location of the unit from primary event, m	 75.0

Table 21.34 (Continued)

Domino effect due to heat load Total heat received, kJ Heat intensity, kJ/sq.m Probability of domino effect due to fire	 3.10E+06 3840.5 1.0
Domino effect due to over pressure: Explosion energy, kJ Peak overpressure, kPa Probability of domino effect due to overpressure	 8.95E+05 441.5 1.0
Domino effect due to missile Explosion energy, kJ Missile velocity, m/s Probability of domino effect due to missile after meeting the target	 8.75E+05 225.4 1.0
Toxic release and dispersion Instantaneous/continuous Puff/Plume Characteristics: Puff concentration at centre of cloud, kg/cu.m Concentration at cloud edge, kg/cu.m Downwind distance, m Distance where maximum concentration occurs, m	Not applicable

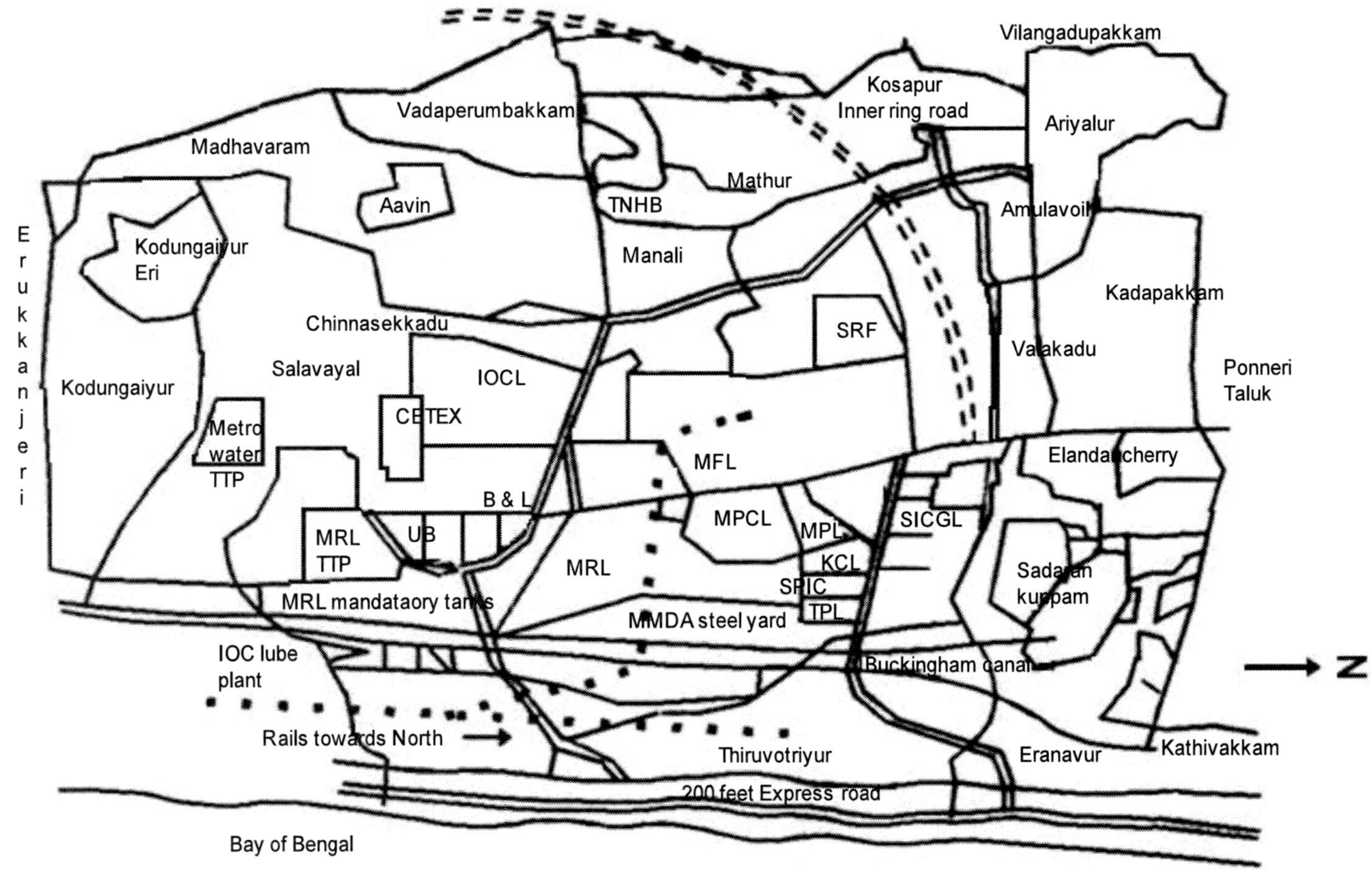

Figure 21.1. Location of various industries in the Manali Industrial Complex (MIC).

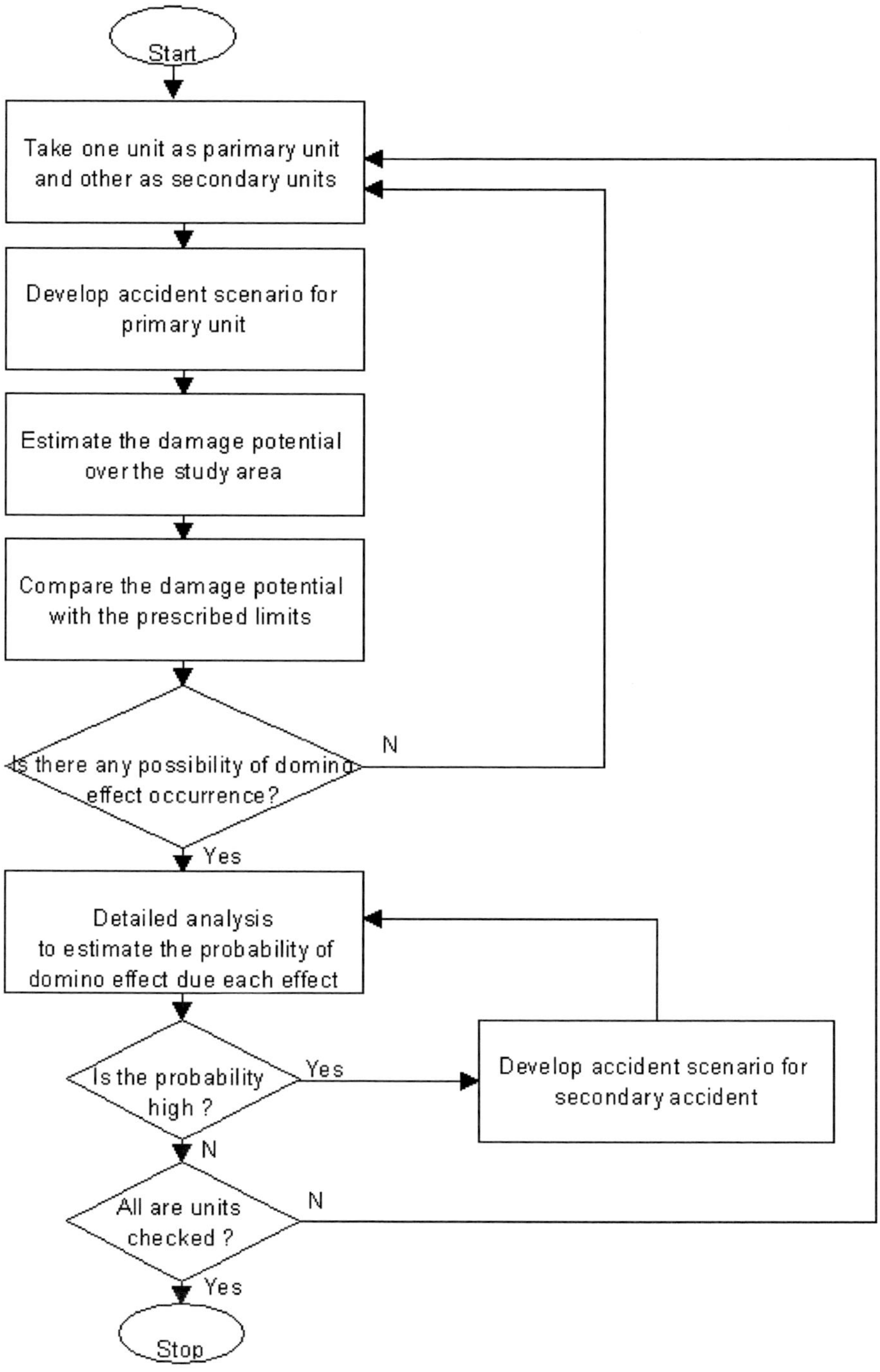

Figure 21.2. Domino effect analysis procedure.

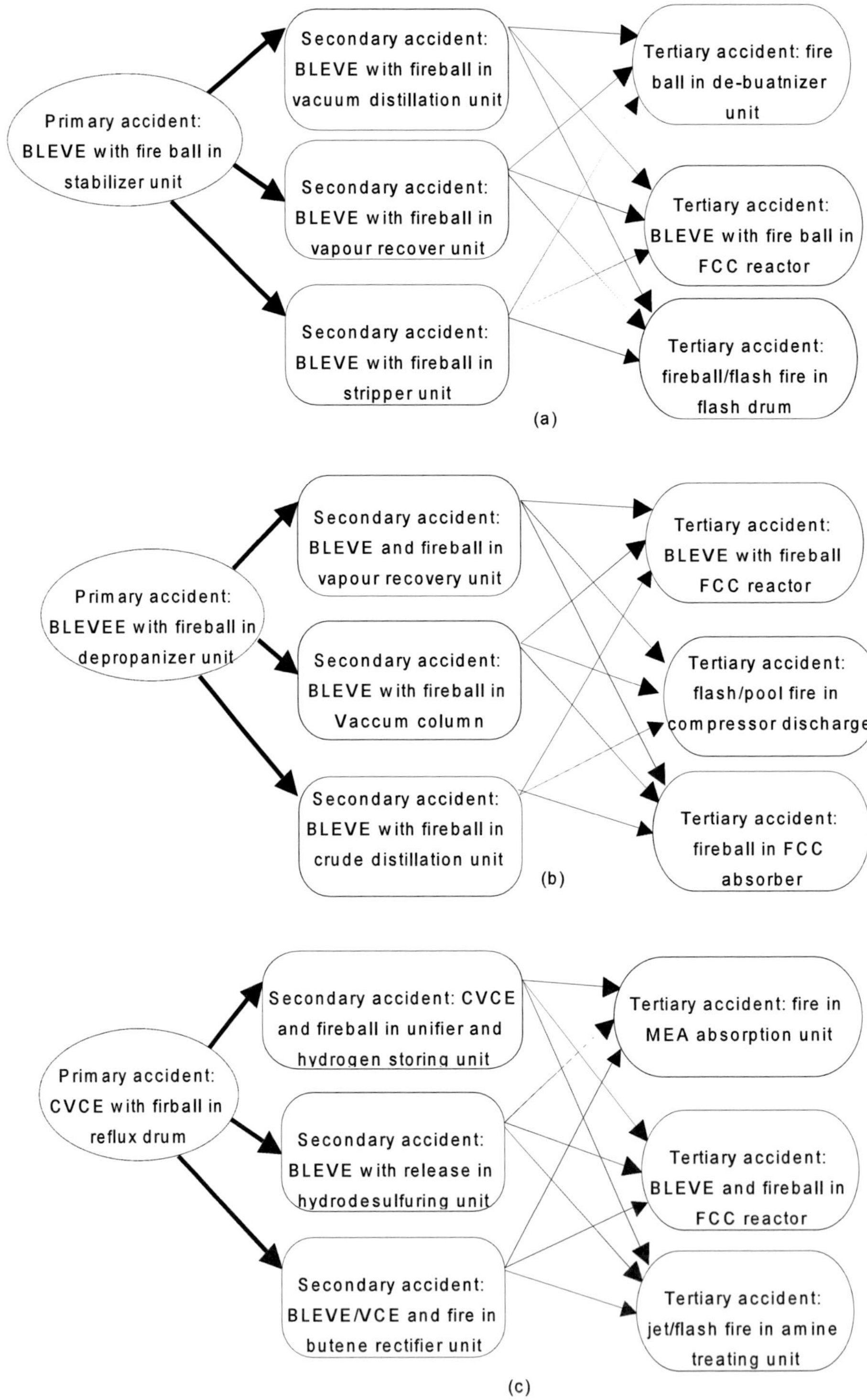

Figure 21.3. The likely sequence of accidents that may be triggered by an accident in (a) stabiliser unit, (b) debutaniser unit, (c) reflux drum, (d) LPG storage unit, (e) Propylene storage and (f) propane storage units.

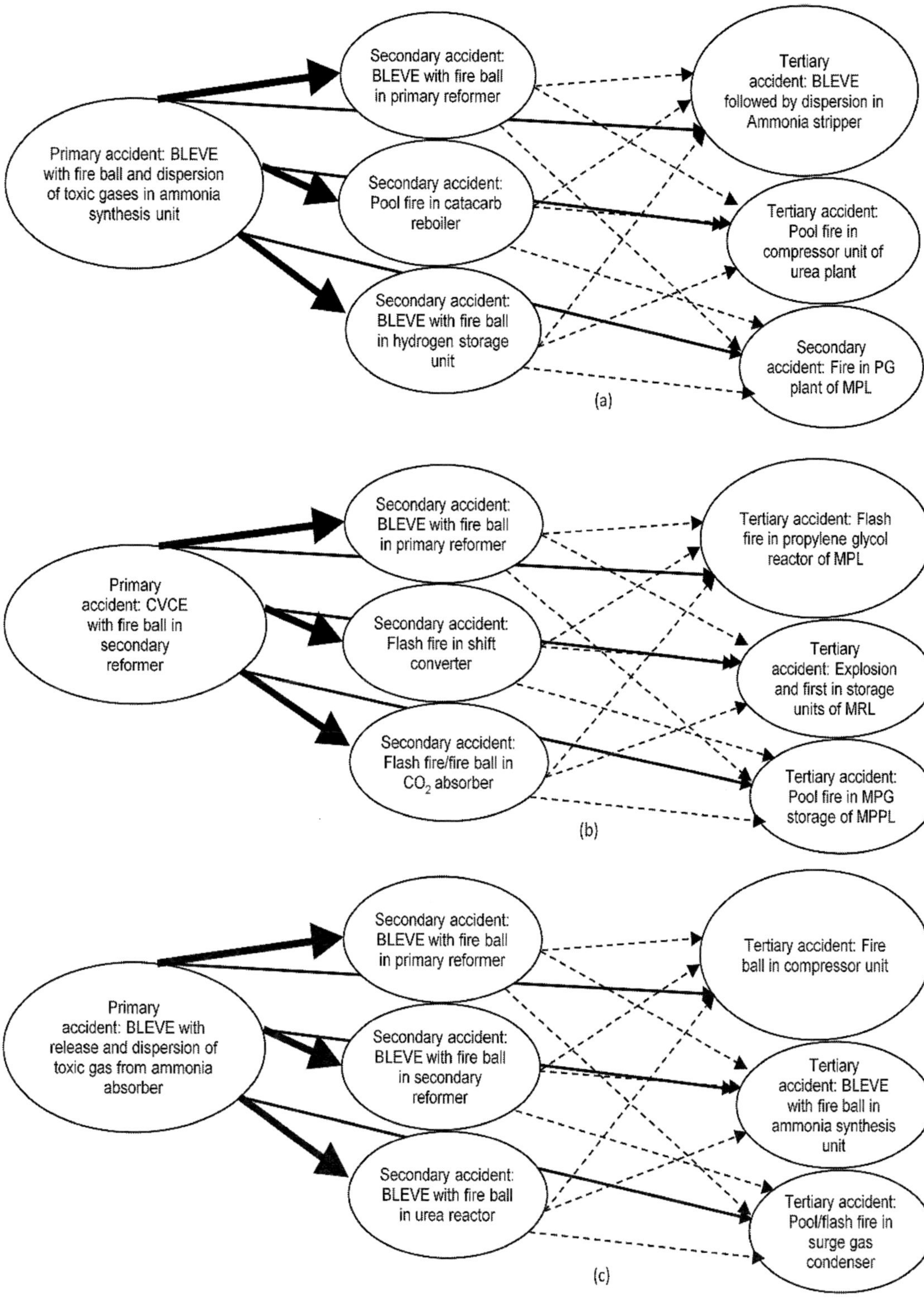

Figure 21.4. (Continued).

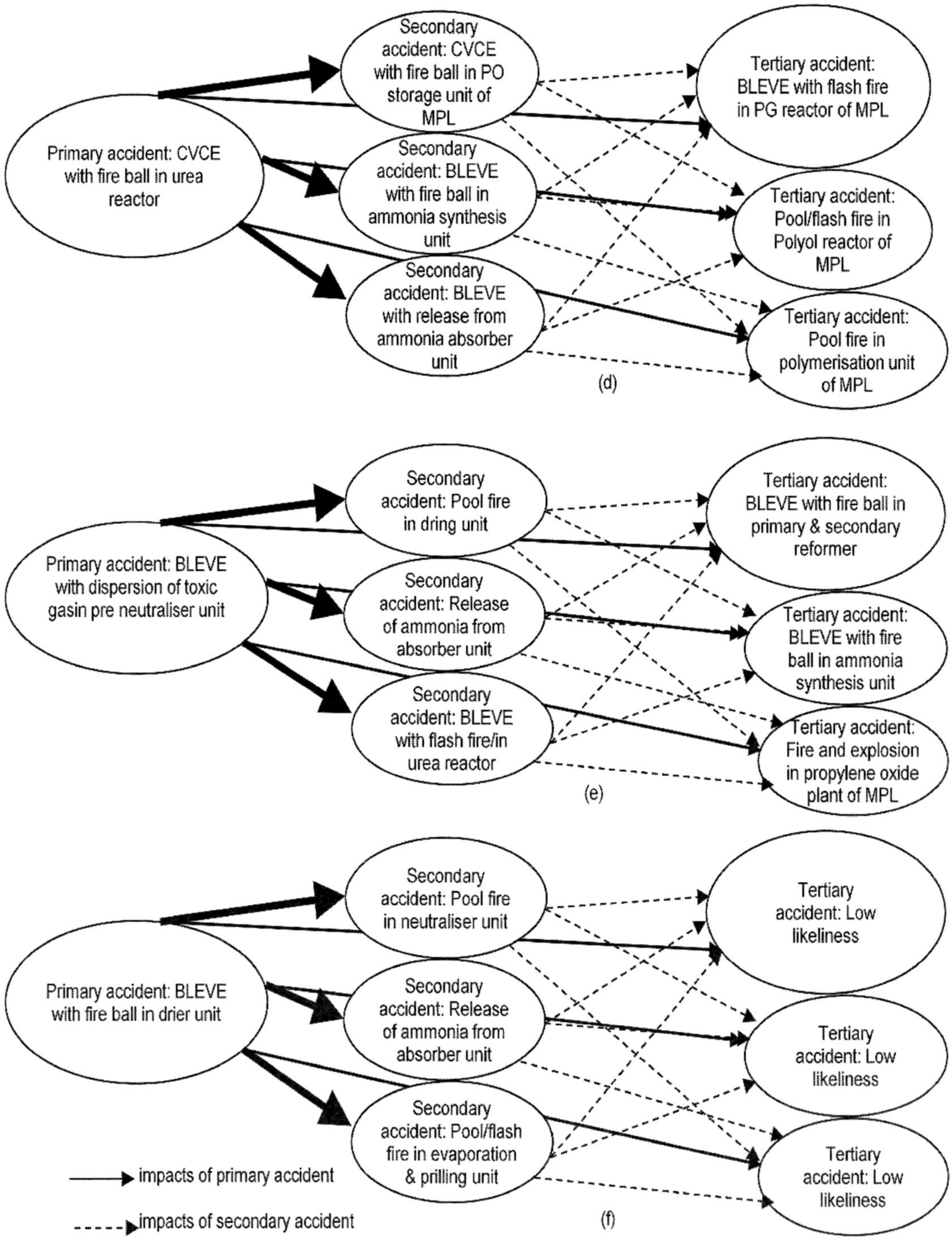

Figure 21.4. A typical sequence of accidents that may be triggered by an accident in (a) ammonia synthesis unit, (b) secondary reformer, (c) ammonia absorber, (d) urea reactor, (e) pre-neutraliser, and (f) drier units.

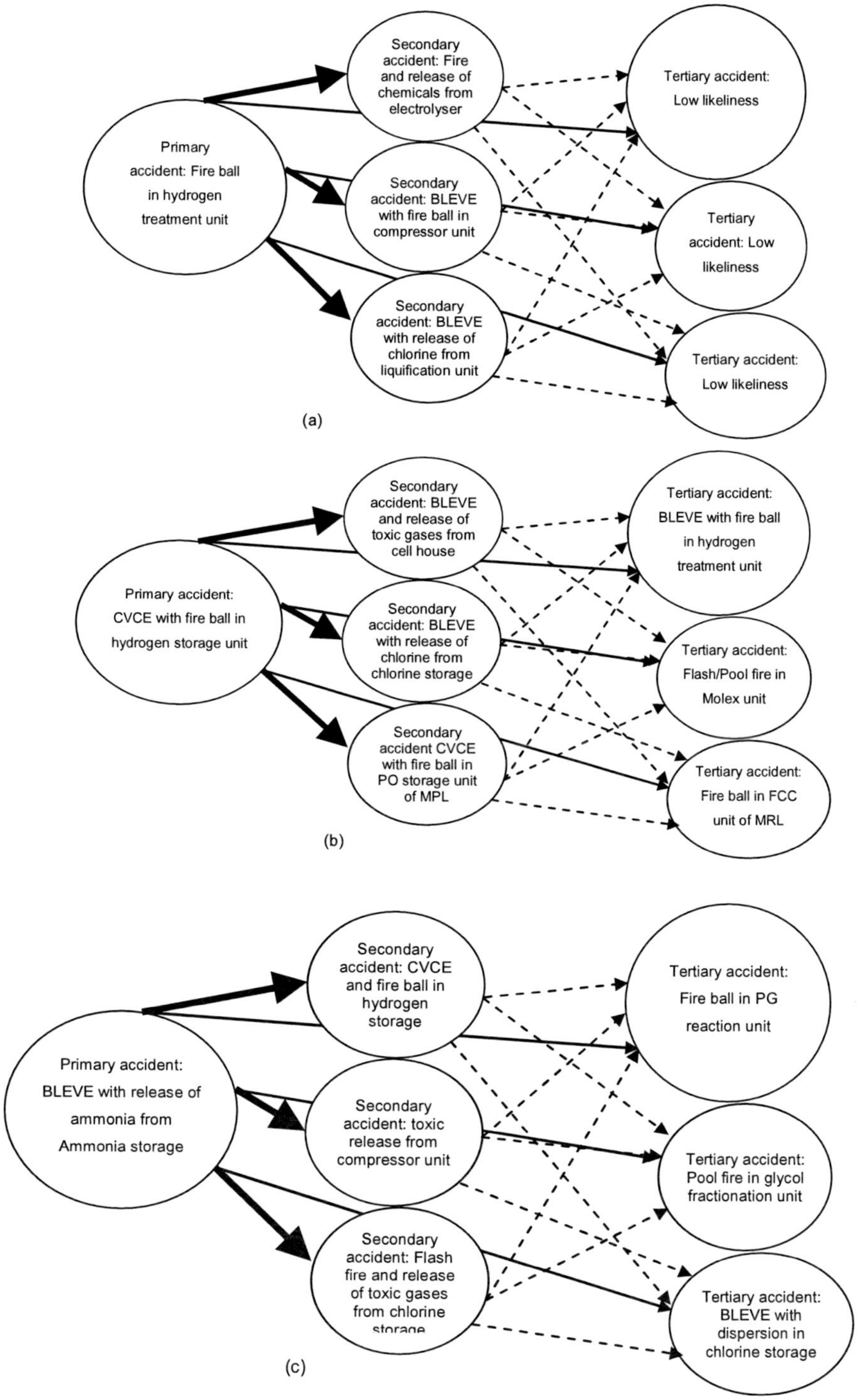

Figure 21.5. A typical sequence of accidents that may be triggered by an accident in (a) hydrogen treatment unit, (b) hydrogen storage, and (c) Ammonia storage.

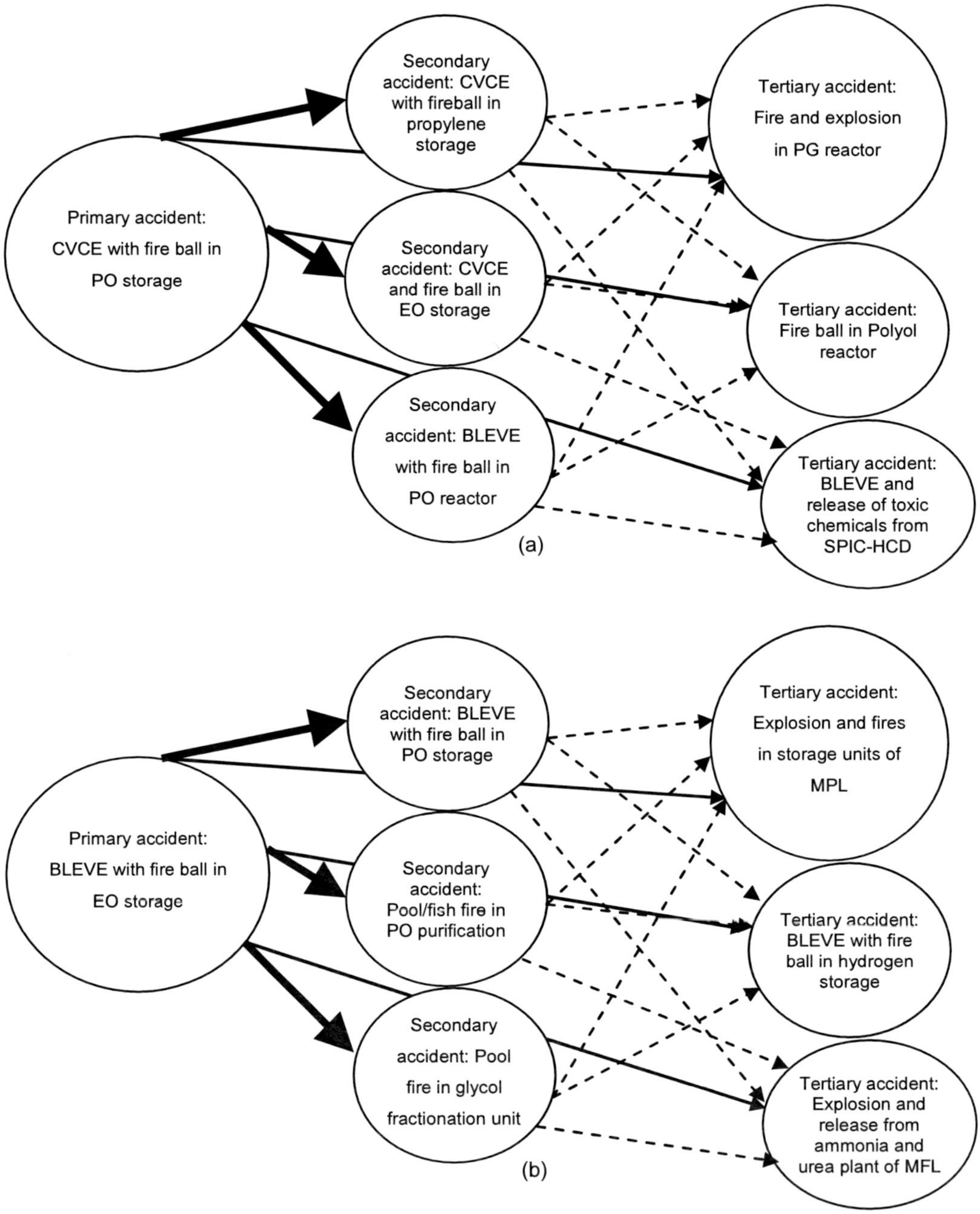

Figure 21.6. (Continued).

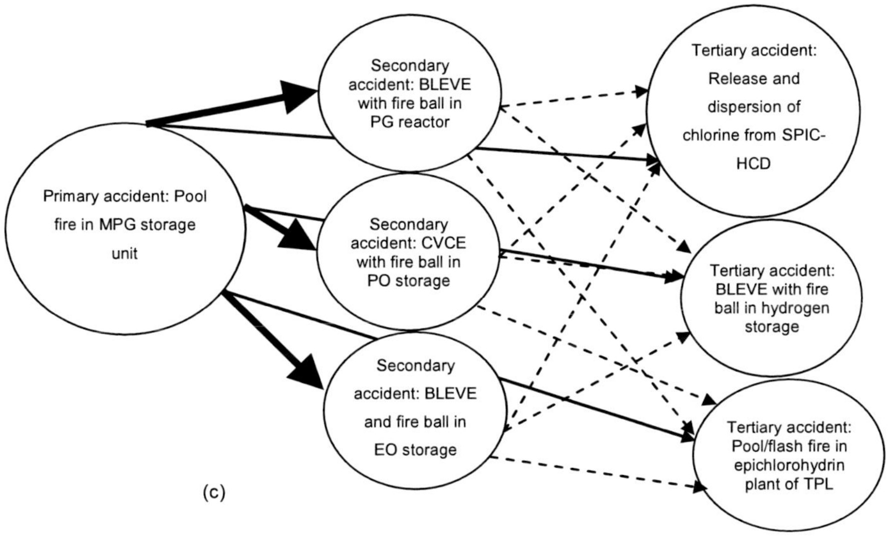

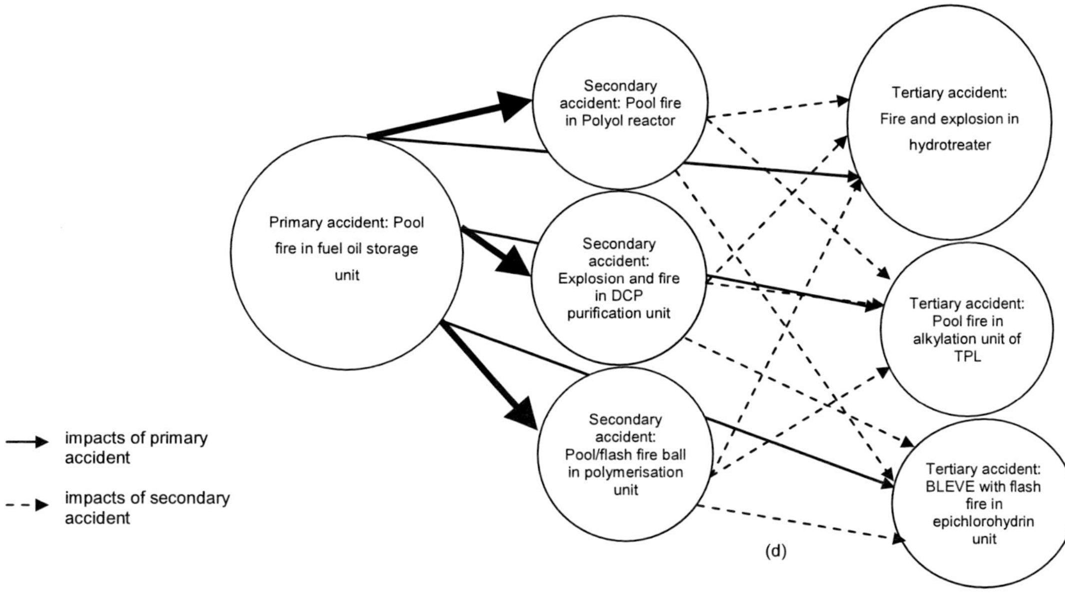

Figure 21.6. A typical sequence of accidents that may be triggered by an accident in (a) propylene oxide (PO) storage, (b) Ethylene oxide (EO) storage, and (c) Mono propylene glycol (MPG) storage, and (d) fuel oil storage.

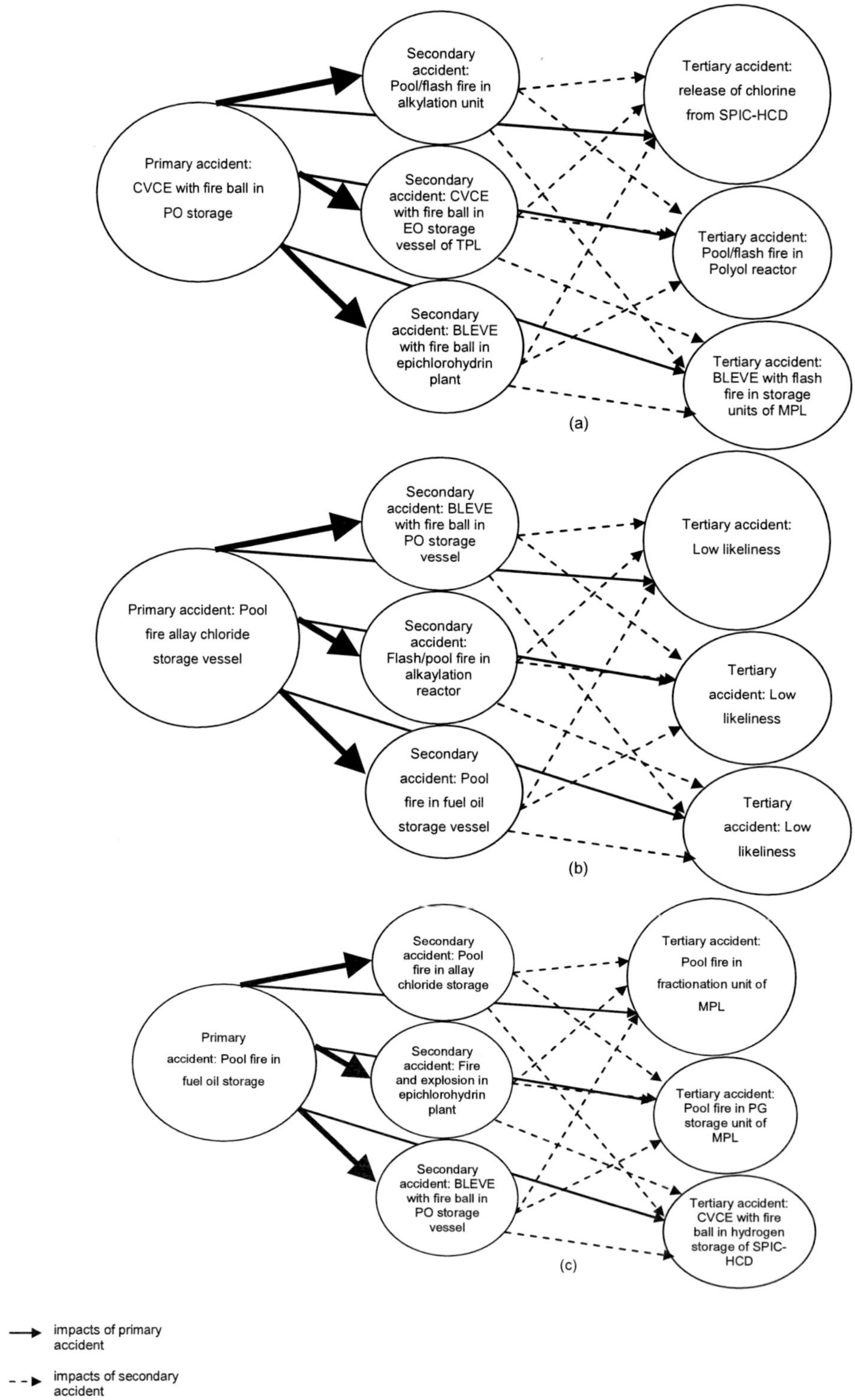

Figure 21.7. A typical sequence of accidents that may be triggered by an accident in (a) Propylene oxide storage, (b) Allyl chloride storage, and (c) fuel oil storage.

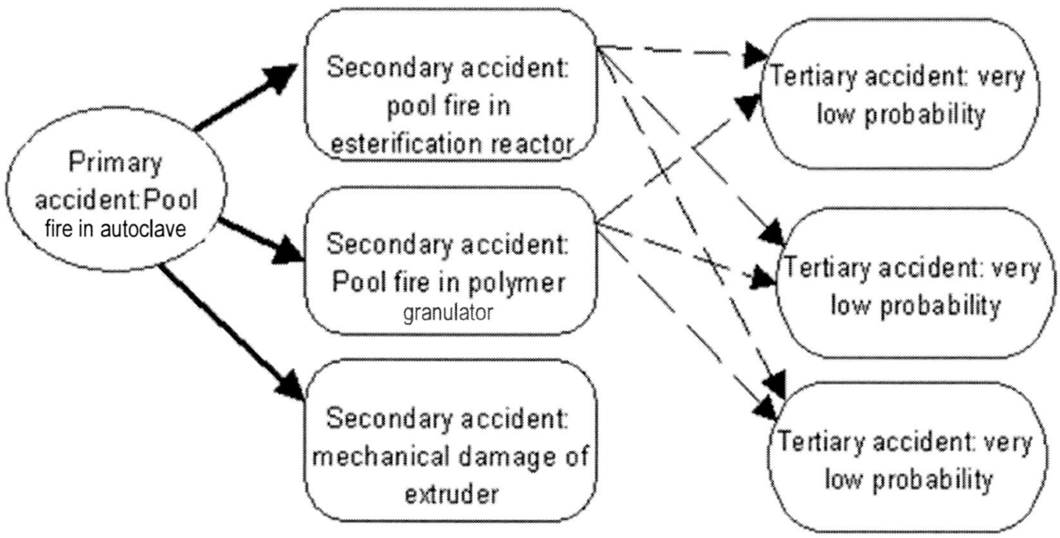

(a)

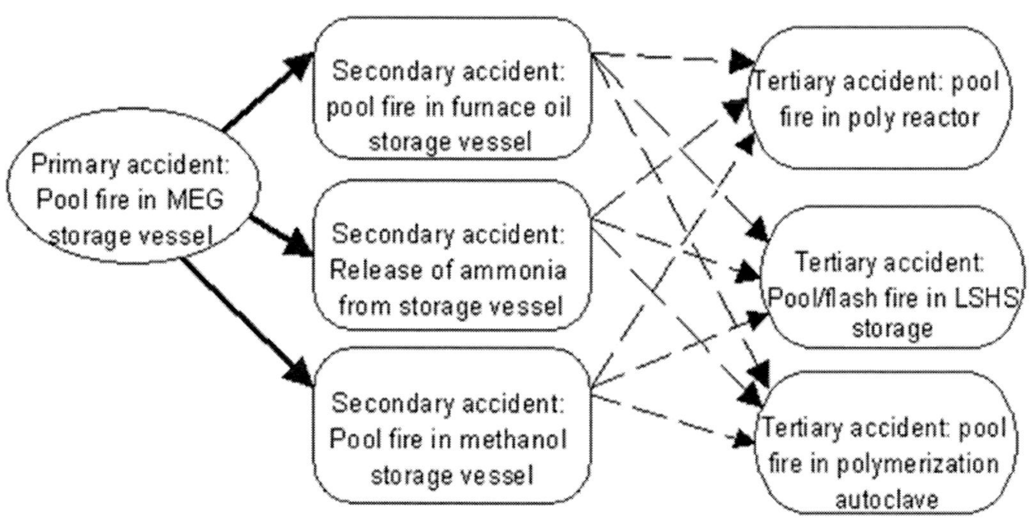

(b)

Figure 21.8. The likely sequence of accidents that may be triggered by an accident in (a) autoclave and (b) MEG storage.

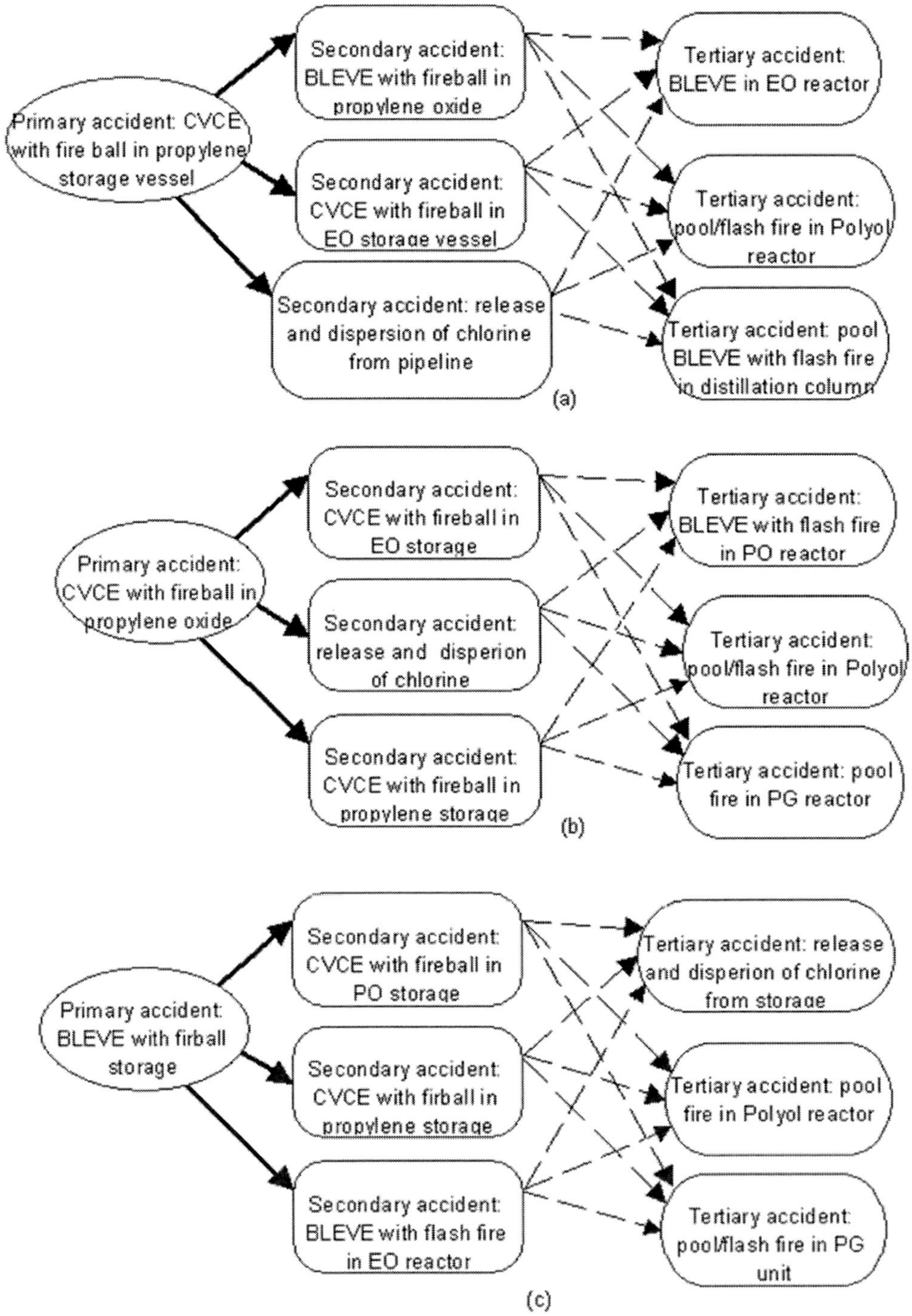

Figure 21.9. The likely sequence of accidents that may be triggered by an accident in (a) propylene storage, (b) propylene oxide storage and (c) ethylene storage units.

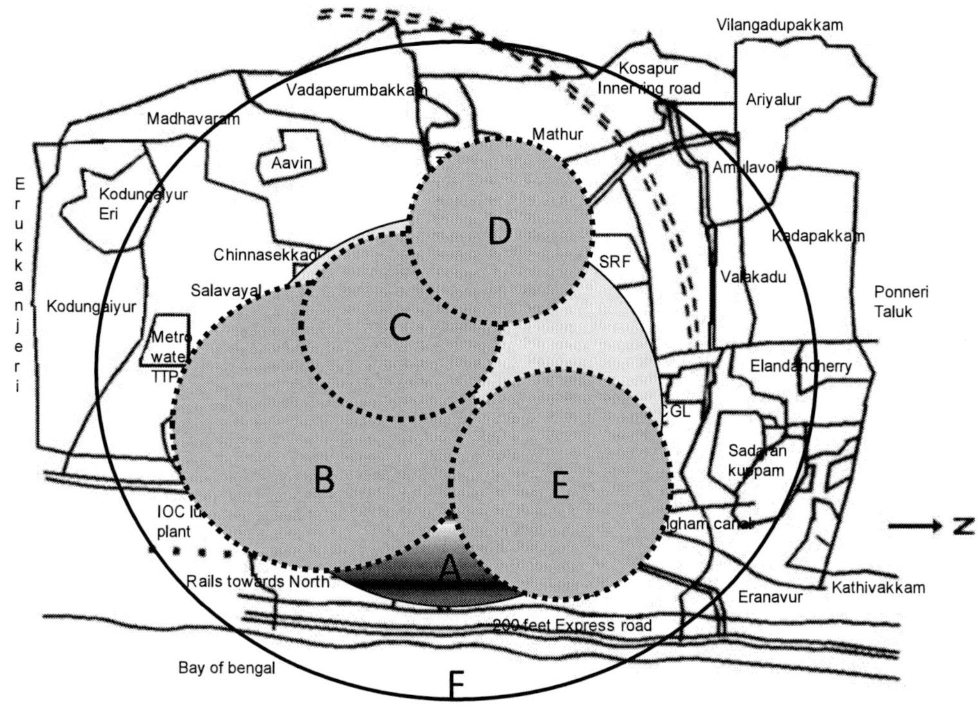

Figure 21.10. Risk contours and likely chain of accidents in MRL initiated by an accident in LPG storage unit; Curve A depicts risk contour for primary accident; curves B, C, D, and E represent risk contours for accidents triggered by A. The overall impact area is represented by F.

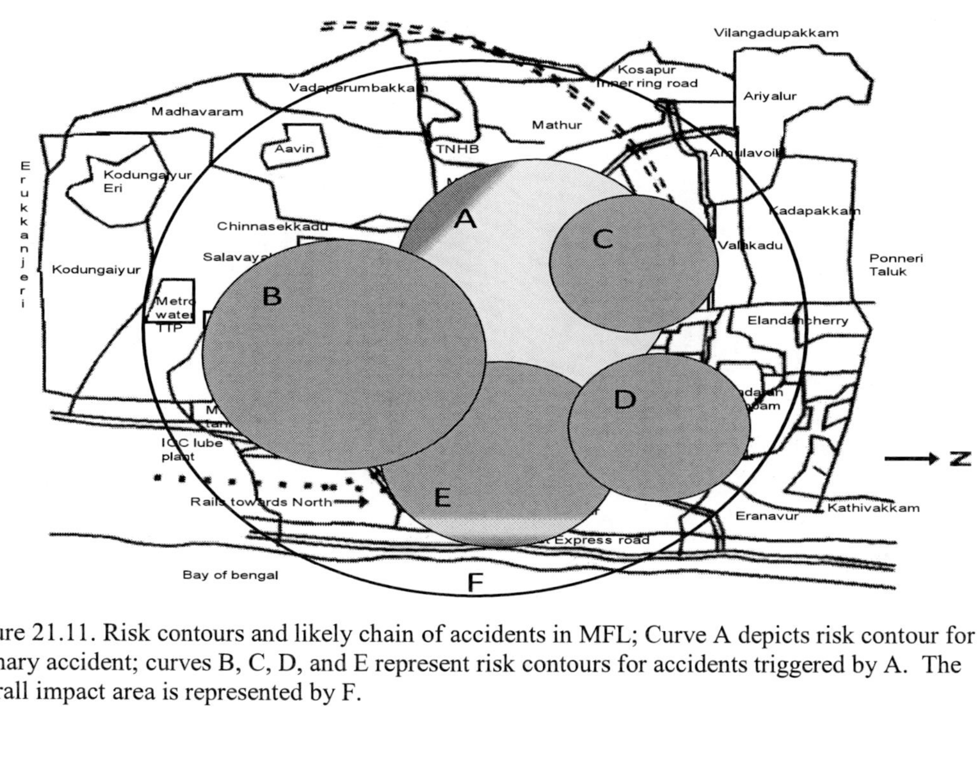

Figure 21.11. Risk contours and likely chain of accidents in MFL; Curve A depicts risk contour for primary accident; curves B, C, D, and E represent risk contours for accidents triggered by A. The overall impact area is represented by F.

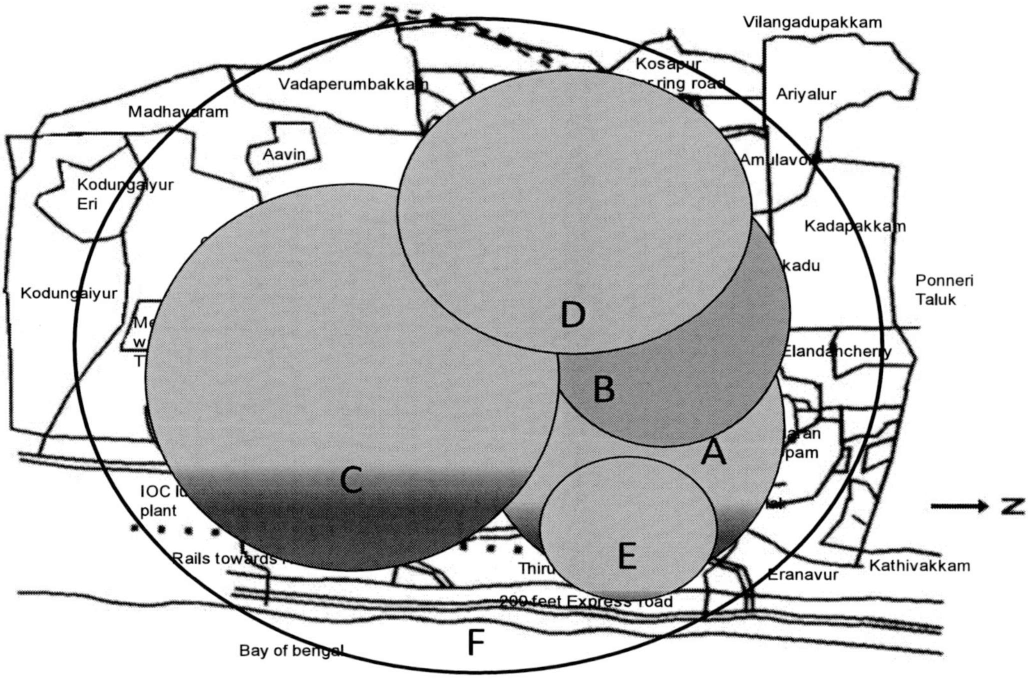

Figure 21.12. Risk contours and likely chain of accidents in SPIC-HCD; Curve A depicts risk contour for primary accidents, Curves B, C, D and E represent risk contours for accidents triggered by A. The overall impact area is represented by F.

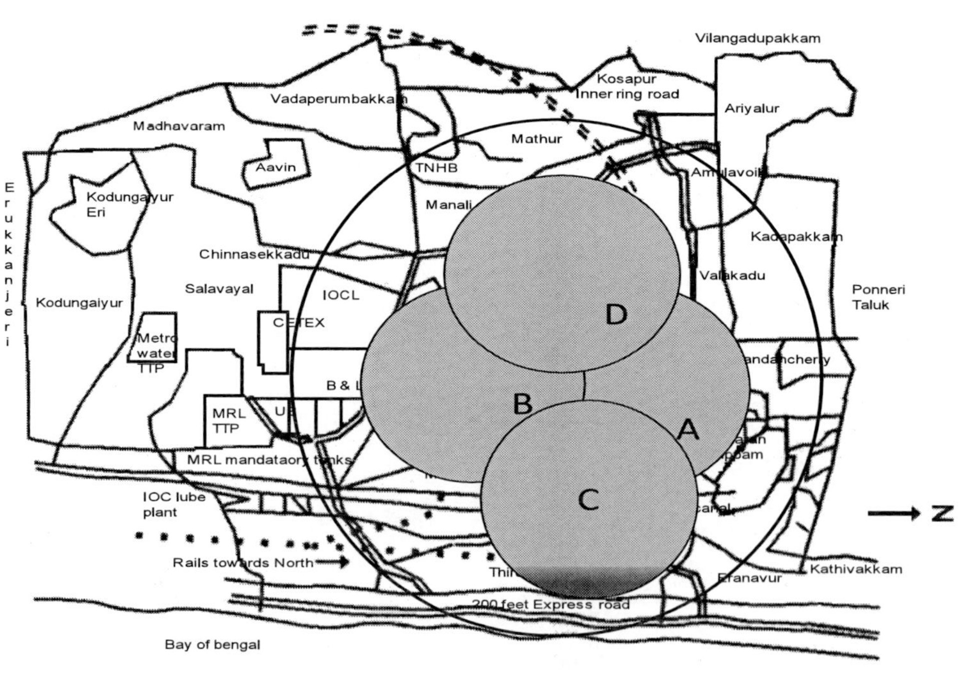

Figure 21.13. Risk contours and likely chain of accidents in MPL; Curve A depicts risk contour for primary accidents; Curves B, C and D represent risk contours triggered by A. The overall impact area is represented by E.

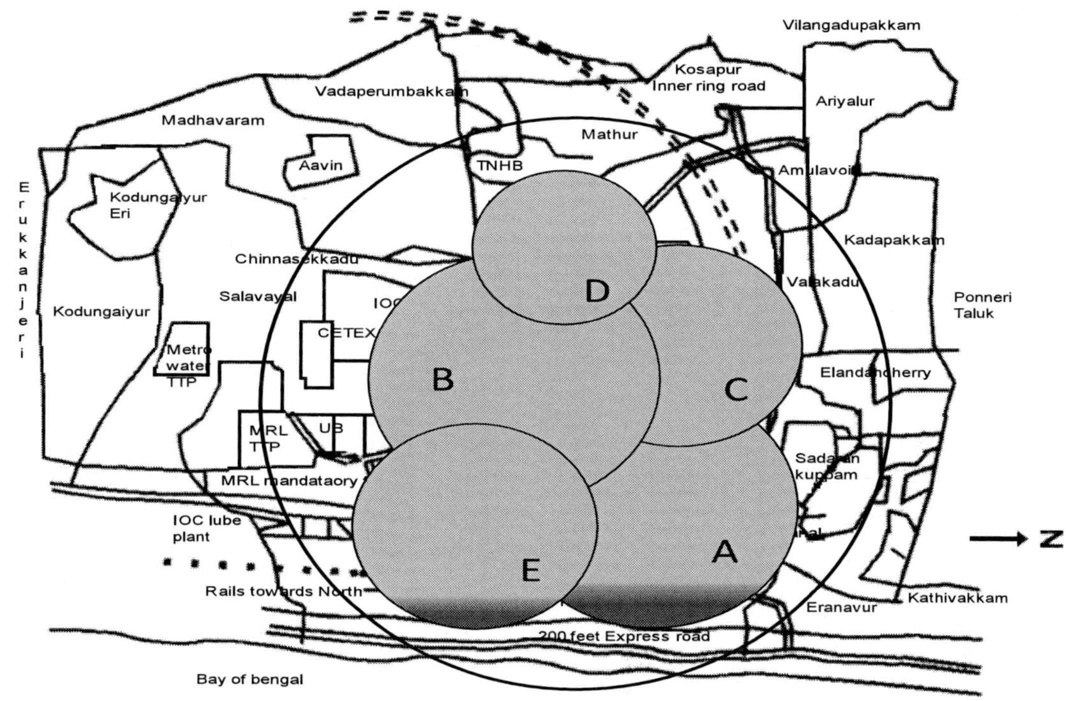

Figure 21.14. Risk contours and likely chain of accidents in TPL; Curve A depicts risk contour for primary accidents; Curves B, C, D and E represent risk contours triggered by A. The overall impact area is represented by F.

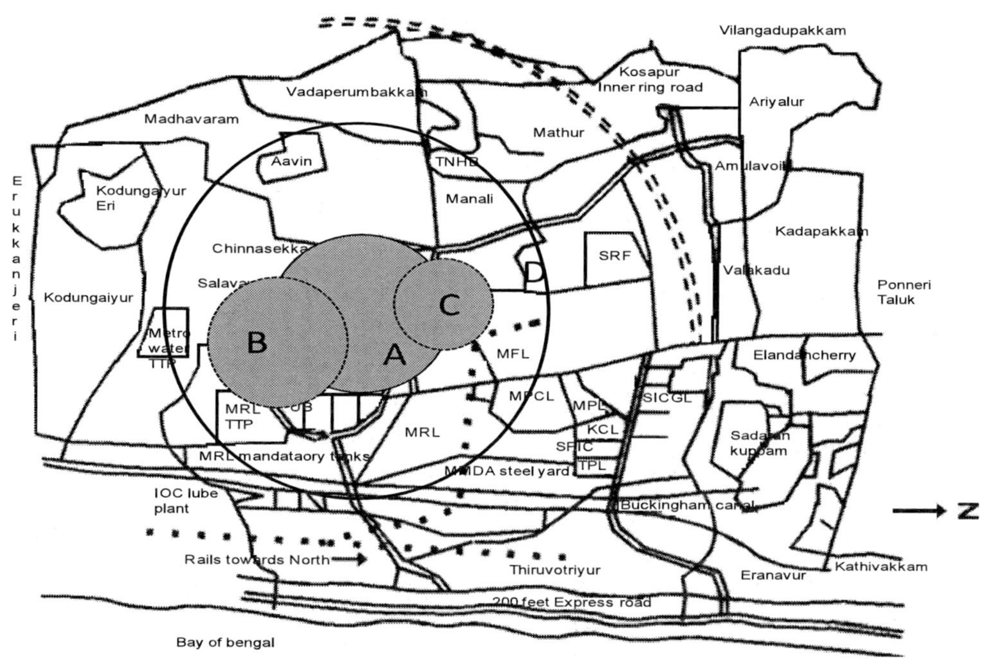

Figure 21.15. Risk contours and likely chain of accidents in IOCL; Curve A depicts risk contour for primary accidents; Curves B and C represent risk contours triggered by A. The overall impact area is represented by D.

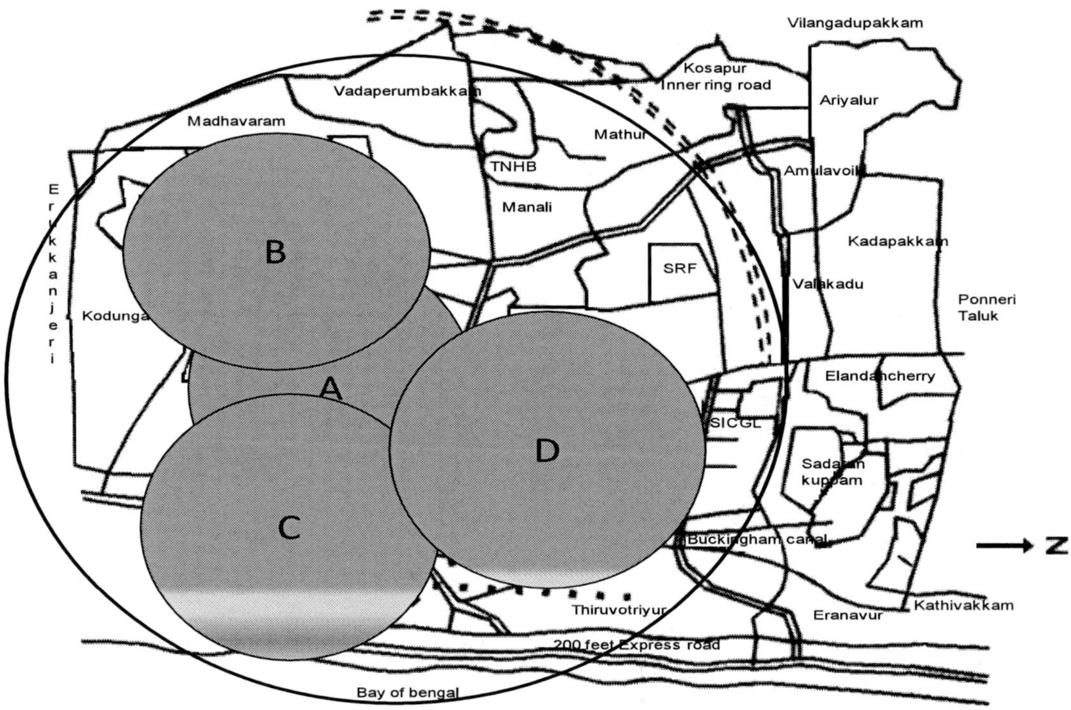

Figure 21.16. Risk contours and likely chain of accidents in UBP; Curve A depicts risk contour for primary accidents, Curves B,C, and D represent risk contours for accidents triggered by A. The overall impact area is represented by E.

INDEX

A

B

C

D

E

Q

R

S

T

U

V

W

Y